Intelligent Decision Support in Process Environments

NATO ASI Series

Advanced Science Institutes Series

A series presenting the results of activities sponsored by the NATO Science Committee, which aims at the dissemination of advanced scientific and technological knowledge, with a view to strengthening links between scientific communities.

The Series is published by an international board of publishers in conjunction with the NATO Scientific Affairs Division

A Life Sciences **B Physics**	Plenum Publishing Corporation London and New York
C Mathematical and **Physical Sciences**	D. Reidel Publishing Company Dordrecht, Boston and Lancaster
D Behavioural and **Social Sciences** **E Applied Sciences**	Martinus Nijhoff Publishers Boston, The Hague, Dordrecht and Lancaster
F Computer and **Systems Sciences** **G Ecological Sciences**	Springer-Verlag Berlin Heidelberg New York Tokyo

Series F: Computer and Systems Sciences Vol. 21

Intelligent Decision Support in Process Environments

Edited by

Erik Hollnagel
Computer Resources International A/S
Vesterbrogade 1 A, DK-1620 Copenhagen V/DENMARK

Giuseppe Mancini
Commission of European Communities
Joint Research Centre, Ispra Establishment
I-21020 ISPRA (Varese)/ITALY

David D. Woods
Westinghouse Research & Development Center
Beulah Road, Pittsburgh, PA 15235/USA

Springer-Verlag Berlin Heidelberg GmbH
Published in cooperation with NATO Scientific Affairs Division

Proceedings of the NATO Advanced Study Institute on Intelligent Decision Support in Process Environments held in San Miniato, Italy, September 16–27, 1985

ISBN 978-3-642-50331-3   ISBN 978-3-642-50329-0 (eBook)
DOI 10.1007/978-3-642-50329-0

Library of Congress Cataloging in Publication Data. NATO Advanced Study Institute on Intelligent Decision Support in Process Environments (1985 : San Miniato, Italy) Intelligent decision support in process environments. (NATO ASI series. Series F, Computer and system sciences ; vol. 21) "Proceedings of the NATO Advanced Study Institute on Intelligent Decision Support in Process Environments held in San Miniato, Italy, September 16–27, 1985"—T.p. verso. 1. Decision-making—Congresses. 2. Artificial intelligence—Congresses. I. Hollnagel, Erik, 1941-. II. Mancini, Giuseppe, 1940-. III. Woods, David D., 1952-. IV. North Atlantic Treaty Organization. Scientific Affairs Division. V. Title. VI. Series: NATO ASI series. Series F, Computer and system sciences ; no. 21. T57.95.N37 1985 658.4'03 86-6742

2145/3140-543210

NATO ADVANCED STUDY INSTITUTE

INTELLIGENT DECISION SUPPORT IN PROCESS ENVIRONMENTS

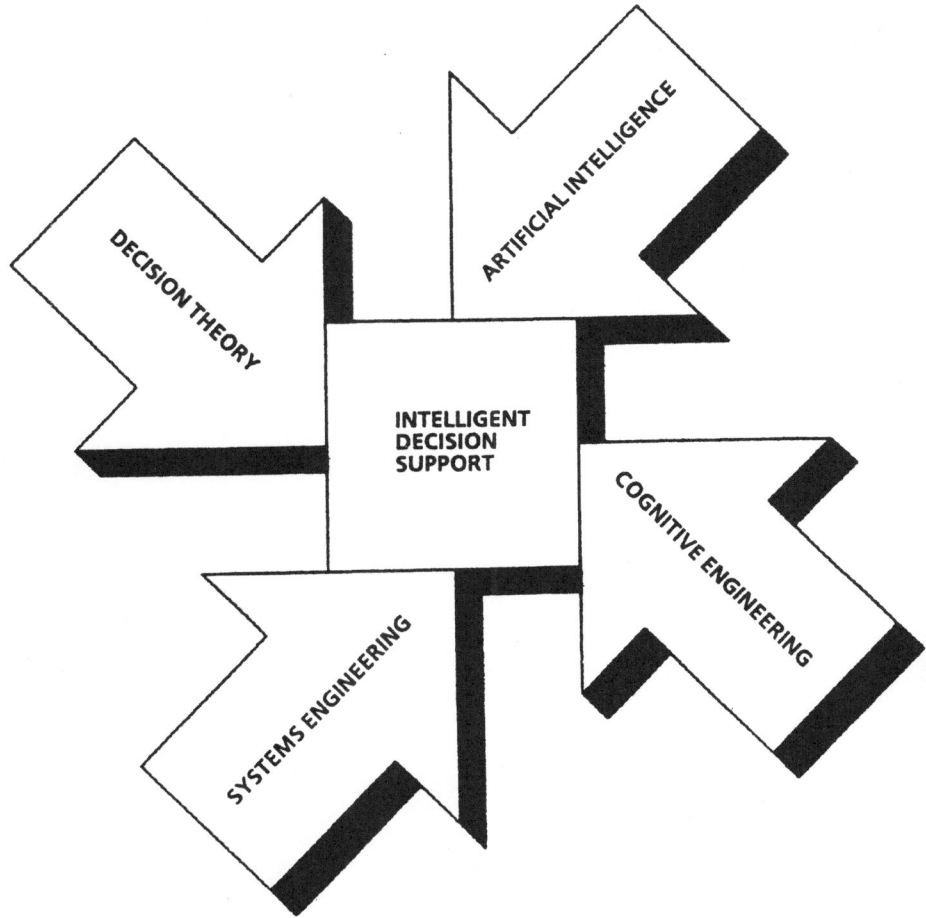

September 16 – 27, 1985, San Miniato, Italy

Sponsored by

DELPHI (I), IntelliCorp (USA)

Systems Designers (UK), TEMA (I)

NATO Scientific Affairs Division

Preface

This book contains the essence of the papers and the discussions from a two week long Advanced Study Institute (ASI) held in September 1985. The idea of having an ASI about the topic of Intelligent Decision Support (IDS) systems goes back to a NATO workshop on Human Error held at the Rockefeller Center in Bellagio, Italy, in September 1983. In a way it goes even further back since the Bellagio workshop continued a series of NATO ASIs and workshops on advanced human-machine systems that started a decade ago in 1976. The present ASI therefore represents neither a beginning nor an end, but is rather a snapshot of the frontiers of thinking about man-machine systems as they looked in 1985.

One problem that was important in 1983, and which remains so today, was to improve the understanding of human decision making and behaviour in particularly demanding applications, such as the control of complex industrial processes. One way of trying to solve this problem was to organise a high-level meeting where the relevant different scientific disciplines could converge and interact. Consequently three of the participants in the Bellagio workshop – Erik Hollnagel, Giuseppe Mancini, and David Woods, supported by a technical programme committee consisting of Don Norman, Jens Rasmussen, James Reason, and Giuseppe Volta – undertook to prepare and direct such a meeting. During the initial discussions the topic of the meeting was defined to be the application of intelligent decision aids in process environments, and this in turn became the title of the ASI. The *Scientific Affairs Division of NATO* awarded a grant for the organisation of an Advanced Study Institute. Because the topic of the ASI was considered to be of industrial as well as academic interest, it was included in the recently established NATO Double Jump programme, which specifically aims at promoting cooperation between universities and industries in NATO countries. This permitted the ASI directors to obtain additional funds from the following four sponsors, to supplement the financial support provided by NATO: *DELPHI* (Italy), *IntelliCorp* (USA), *Systems Designers* (U. K.), and *TEMA* (Italy). The ASI entitled 'Intelligent Decision Aids in Process Environments' took place at 'I Cappuccini' in San Miniato (Italy) from 16th to 27th of September 1985.

The ASI was attended by sixty-nine participants, of which 19 were lecturers, coming from 14 different countries and two international organisations. The magnificent Tuscan setting provided an ideal background for two weeks of intensive, stimulating and wide-ranging discussions. This book contains the papers presented by the 19 invited lecturers, a number of selected short papers from the participants, abstracts of the other short papers, and the results of the panel discussions held during the meeting. This material has been edited to reduce overlaps and provide a consistent style. The order of the papers has been revised to present the reader with a more coherent structure. Although great care has been exercised by the editors, some errors and inconsistencies may still be present. We hope the reader will forgive that. Our main goal has been to edit and publish the book as quickly as possible so that it can truly serve its purpose as a snapshot of the current state of thinking.

We expect that the book will be of value to anyone who is interested in the problem of intelligent decision support systems in industrial applications. We hope that ideas and perspectives presented here can be used by the variety of engineers, plant operators, managers, and industrial and academic researchers from the relevant technical and social science disciplines that today collaborate to solve the problems in advanced process control.

Acknowledgements. We are deeply grateful to the *NATO Scientific Affairs Division, DELPHI, Intelli-Corp, Systems Designers Ltd,* and *TEMA* for having supported and financed the ASI on Intelligent Decision Aids in Process Environments, on which this book is based. We are also very grateful to Donald Norman, Jens Rasmussen, James Reason, and Giuseppe Volta who in their role as technical programme committee members provided us with ideas, advice and new perspectives from their considerable experience. We further wish to thank the OECD Halden Reactor Project (Halden, Norway), Computer Resources International A/S (Copenhagen, Denmark), the Commission of European Communities Joint Research Centre (Ispra, Italy), and the Westinghouse Research & Development Centre (Pittsburgh, USA) for having allowed us to spend part of our time to organise the ASI and edit this book, as well as for technical and organisational support. We are, of course, indebted to those who contributed to this book; not only did they provide the 'meat' of the ASI but their compliance with strict deadlines allowed us to produce the final manuscript in less than six months. Special thanks are finally due to the staff of 'I Cappuccini' at San Miniato for having offered an exceptionally warm environment for the meeting.

Copenhagen, December 1985

Erik Hollnagel
Giuseppe Mancini
David D. Woods

Introduction

The increasing complexity of technological systems has shifted the demands on human performance from a mechanical/physical level to a cognitive level. The role of the human in complex systems is to act as a supervisor of automated or semiautomated resources with goal setting, problem solving, and decision making as the primary tasks. While the negative consequences of possible malfunctioning have grown, the tolerances for performance variability have been reduced and the demands to accurate and safe performance increased, thereby leaving the human operator in a very difficult position. At the same time, advances in computer science (e.g. automated decision makers, object oriented programming, and expert systems) have provided new possibilities to support human performance that challenge established man-machine system design principles.

Decision making has emerged as the focal point for the often conflicting demands to human action. To ease the operator's task one must reduce these demands by incorporating basic intelligence functions in the man-machine interface. But it is essential that an intelligent interface supports the operator's decision making rather than replaces parts of it, i.e. that it is a tool rather than a prosthesis. The former may improve the task, while the latter will surely aggravate it. The logic behind decision making has been considerably extended in recent years, for instance by fuzzy-set theory and possibility theory. A clear distinction has also emerged between domain specific knowledge and decision making strategies. Decision making is no longer regarded as simply the strict following of a single set of rules or a strategy, but must also include the selection of the appropriate rules and the possible switch between them as the decision evolves, i.e. metalevel decisions.

The aim of this ASI was to identify the knowns and unknowns of intelligent systems that can support human decision making in process environments. The focus was the functional rather than the analytical aspects of such systems. The emphasis on and the development of systems that perform cognitive tasks require a corresponding shift in the multiple disciplines that support effective man-machine systems. It requires contributions from decision theory, theoretical and applied; from philosophy and logic; from process control theory and information science; from cognitive psychology and the study of human performance and human error; and from artificial intelligence and computer science. This was achieved by describing (1) the foundation provided by decision theory, (2) the problems of decision making in process environments, (3) the cognitive aspects of decision making, and (4) the possibilities of artificial intelligence and advanced computer applications. The ASI aimed to synthesise and integrate developments in all four areas.

As mentioned in the preface, this ASI joined a series of other meetings that all had dealt with problems of man-machine systems. The five major previous meetings were:

1976	Monitoring Behaviour and Supervisory Performance	Berchtesgaden, West Germany
1977	Theory and Measurement of Mental Load	Mati, Greece
1979	Changes in Nature and Quality of Working Life	Thessaloniki, Greece
1981	Human Detection and Diagnosis of System Failures	Roskilde, Denmark
1983	Human Error	Bellagio, Italy

The present ASI used the interdisciplinary cooperation and knowledge foundation established in the previous conferences to address the problems of cognitive man-machine systems. During the two weeks specialists from the above-mentioned fields worked together to advance the synthesis of knowledge needed for the development and evaluation of such joint cognitive systems, present the current unsolved problems, and point to possible solutions. The outcome is documented in this book.

Whereas most readers of this book probably will have a good understanding of what decision making and decision support systems are, it may be useful to describe briefly what is meant by process environments and how decision making in process environments differs from other types of decision making.

Process environments are first of all characterised by having short time constants. The decision maker must constantly pay attention to the process because the state of it changes dynamically; if he fails to do so he will lose control. If time was not that critical it would be possible to leave the process and look for information in other places, consult available experts and knowledge bases, etc. To use an example from medicine, if time is not critical one is dealing with a population phenomenon, for example the effects of smoking; if time is critical one is dealing with an individual phenomenon, for example toxification. If there is no time pressure one can always refer to data from the population and neglect the process without endangering it.

Formal decision theories do generally not contain any element of time (cf. the paper by Giuseppe Volta), hence fail to recognise an essential attribute of process environments. They are therefore not directly applicable but require further development and modification.

Another characteristic of process environments is the uncertainty about data, i.e. one is uncertain about what the next datum may be. There is less than complete predictability of what evidence will appear, both on the level of individual process parameters and on the level of the process as a whole. The uncertainty does not only arise from the imperfection or breakdown of single components as in normal reliability analysis or Probabilistic Risk Analysis. Far more important is the uncertainty of the evidence that comes from the dynamics and complexity of the process as such. This lack of certainty is in conflict with some of the main assumptions of both normative and descriptive decision theory and naturally makes decisions more difficult (cf. the comments by George Apostolakis).

Process environments are further characterised by being dynamic, by having multiple and possibly conflicting goals, and by having incomplete information. The latter may lead to tradeoffs between the speed and quality of the decision. Decision making in process environments normally means multiple decisions based on partial information, with little chance to consider all the alternatives or to make revisions. It is thus very different from the orderly world that is assumed by conventional decision theories – and provided by most of the experiments carried out in the behavioral sciences.

The presentations and discussions at the ASI, as well as the papers in this book, are organised in four sections corresponding to the four main aspects of Intelligent Decision Support systems. These are *decision theory, cognitive engineering, systems engineering,* and *artificial intelligence.* The development and application of Intelligent Decision Support systems must be based on a combination of these four aspects, as none of them is sufficient on its own. Each of them refer again to other scientific disciplines some of which, like cybernetics, are related to several aspects of Intelligent Decision Support. The presentations and discussions at this ASI clearly showed the insufficiency of any one of these aspects to account completely for the functioning of Intelligent Decision Support systems, but also demonstrated how interdisciplinary cooperation can bring about new ideas and solutions.

Decision theory, whether normative or descriptive, is a natural starting point for this book. It provides a basic understanding of the theoretical and practical issues and of the currently unresolved problems. Cognitive engineering gives the view of a recent combination of a number of disciplines that all focus on the cognitive aspects of system functioning, and in particular addresses the relation between knowledge and decision making. This view is closely related to the current developments in systems engineering which, together with decision theory, represent the more established views. The section on systems engineering summarises and identifies some of the shortcomings of a traditional engineering approach, and points to ways in which they can be solved. Finally, the section on artificial intelligence provides the experience from trying to incorporate intelligent capabilities in artificial systems. Artificial intelligence has produced a fascinating repertoire of ideas and tools which can be applied in developing Intelligent Decision Support systems, and the influence on cognitive engineering and systems engineering is obvious.

We believe that these four aspects are both necessary and sufficient to give a comprehensive description of Intelligent Decision Support systems, to point out the major problems, and indicate how the solutions can be achieved. It is also our strong belief that solutions can only be achieved by a combination of two or more of these disciplines. The insufficiency of an approach to Intelligent Decision Support systems based purely on either of them is demonstrated both in the papers and in the summarised panel discussion. Decision theory and systems engineering both focus on a description of the problem domain, decision making in process environments. Cognitive engineering and artificial intelligence provide an outline as well as examples of the methodological and functional solutions that can be applied on the problem domain. This ASI was an expression of the clearly recognised need for an interdisciplinary cooperation, and we trust that the book demonstrates the benefits of mixing minds and metaphors as a foundation for further advances into a new and challenging field.

Table of Contents

Section 5: Concluding Comments

Section 6: References

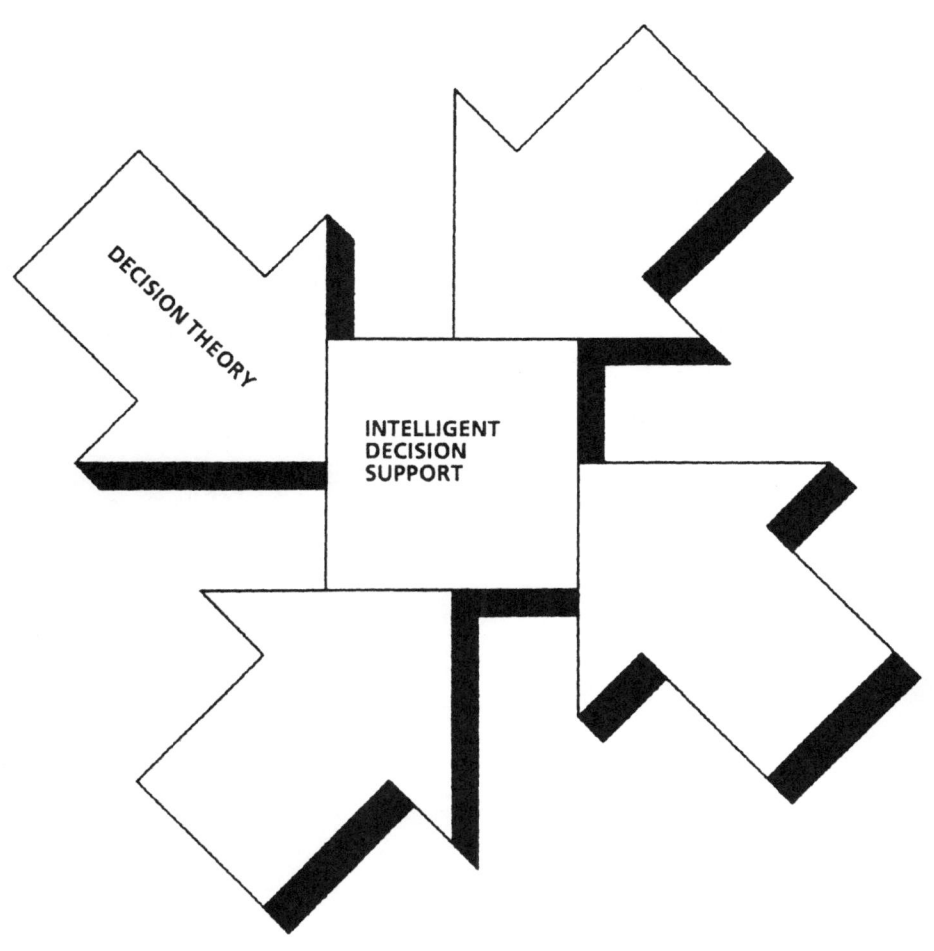

DECISION THEORY

INTELLIGENT
DECISION
SUPPORT

PHILOSOPHICAL FOUNDATIONS

RECENT MODELS OF UNCERTAINTY AND IMPRECISION AS A BASIS FOR DECISION THEORY: TOWARDS LESS NORMATIVE FRAMEWORKS

Didier Dubois and Henri Prade
Paul Sabatier University
Laboratory for Language and Information Systems
Toulouse, France

Abstract : Although a normative approach is necessary to provide sound foundations, utility theory has often been criticized as being too normative.
This paper intends to discuss various ways of relaxing the normative framework of subjective probability and utility theory, in order to account for
recent theories of uncertainty which encompass the probabilistic point of
view as a particular case. This attempt aims at providing new pieces of information in the debate between the normative point of view of decision theory and the (usually) descriptive point of view of deviant uncertainty theories, often used as a basis for approximate reasoning techniques in expert
systems.

INTRODUCTION

Utility theory relies for the most part upon probability theory, i.e.
the perception of uncertainty by individuals is assumed to be accurately
modeled by a probability assignment. This assumption, which looked natural
when probability theory was the only mathematical framework for the representation of uncertainty, can now be questioned, with the emergence of new
mathematical tools such as Zadeh (1965, 1978 a)'s fuzzy sets and possibility measures, Sugeno (1974)'s fuzzy measures, or Shafer (1976)'s belief functions, for instance. Any attempt to found a theory of decision upon deviant
measures of uncertainty judgments can be viewed as a fruitless mathematical
exercise as long as the need for other conceptual frameworks is not felt by
people in the field of decision theory itself. Interesting discussions providing careful accounts about applying Bayesian inference and utility theory to real decision problems appear in recent papers by Fischhoff and Beyth-
Marom (1983), Fishhoff, Goitein and Shapira (1982) respectively. In the

NATO ASI Series, Vol. F21
Intelligent Decision Support in Process Environments
Edited by E. Hollnagel et al.
© Springer-Verlag Berlin Heidelberg 1986

first paper the authors provide a listing of potential sources of bias in
Bayesian hypothesis evaluation, some of which can be tackled by enlarging
the normative framework of subjective probability. A similar discussion is
carried out for subjective expected utility (SEU) theory in the other paper.
Among their conclusions, we quote the following remarks : "Theoretically,
we need to go beyond simply trying to falsify SEU theory ; if only because
of the power of linear models, it will typically provide at least mediocre
predictions. Instead, we need to move on to sophisticated falsification,
finding theories that do what SEU does and at least a little more. The shape
of those theories might be new decision calculi with different primitives,
and combination rules...". The aim of this paper is to suggest new research
lines for building decision-making models, which take advantage of recent
models of subjective uncertainty. The emergence of new uncertainty measures
is directly linked to that of computers, and the search for knowledge representation techniques capable of handling information as provided by humans, i.e. not only is it structurally complex, but also pervaded with imprecision and uncertainty. Hence, among the ways to bring closer computer-
based decision support systems and decision theory, a study of how people
from Artificial Intelligence and related areas deal with uncertainty can
bring new ideas within SEU models.

The paper is organized in 2 parts, respectively devoted to quantitative uncertainty measures and the underlying qualitative assumptions, and
to potential applications of these uncertainty measures to decision theory.
Due to its brevity and also the present state of the art, the ambition of
the paper is limited ; it mainly discusses ideas and surveys material published elsewhere in scattered journals, and draws from the dissertation of
one of the authors (Dubois, 1983).

1 - QUANTITATIVE MEASURES OF UNCERTAINTY : A UNIFIED FRAMEWORK

Before proceeding forward it is crucial to distinguish between uncertainty and imprecision. Roughly speaking, a proposition (or equivalently
an event) is said to be imprecise when it is not elementary, but composed
of a cluster of elementary propositions. An imprecise proposition is said
to be vague if the set of elementary propositions it refers to has no precise boundaries. On the other hand, an uncertain proposition is one the
truth of which is not completely known. Uncertainty can be expressed by as-

signing a grade of confidence to the assumption that an imprecise proposition is true and a grade of confidence to the converse assumption (since an imprecise proposition is either true or false). Note that for a given state of knowledge, increasing the imprecision of a statement leads to increase its certainty, so that although being different notions, uncertainty and imprecision are not unrelated. Lastly, the case of vague propositions, which often occur in human-originated information, is special, in that even in a deterministic environment a vague proposition can be neither true nor false (but in-between), if the actual fact falls among the borderline events encompassed by the proposition. In the following we investigate various proposals for building grades of confidence in imprecise propositions, starting with a discussion of the probabilistic approach.

1.1 - Limitations of the probabilistic model of subjective uncertainty

Let P(p) be the grade of confidence in the truth of proposition p, considered as a probability. The relevance of probabilities as modeling subjective judgments has been questioned from several points of view : limited expressive power, strong underlying assumptions regarding the available evidence, assessment difficulties.

First, to allocate a probability a to proposition p (P(p) = a) requires a probability 1-a to be allocated to proposition ¬p ('not p'). Hence in the probabilistic framework the grade of belief in p (P(p)) and the grade of doubt in p (P(¬p)) are closely linked. As a consequence it is not possible to conveniently represent a state of total ignorance in which a complete lack of belief in p goes along with a complete lack of belief in ¬p. Nothing indicates that the state of knowledge of individuals is enough advanced to be liable of a probabilistic representation, where a single number characterizes a state of uncertainty about a statement p [1].

Another problem with the probabilistic framework is the necessity to provide an exhaustive set of mutually exclusive alternatives, which act as elementary propositions. Once again individuals may be aware of only part of the consequences of decisions, and may not be able to describe them precisely enough, so that the mutual exclusiveness property may fail to hold. [2]

Moreover the practical application of probabilistic models often leads to an extensive use of independence assumptions ($P(p \wedge q) = P(p).P(q)$), which underlie random compensation of errors. It is often hard to justify the classical definition of independence for subjective probabilities.

The two main attempts at supporting a probabilistic model outside the frequentistic school, are the betting behavior interpretation, and the qualitative approach. In the theory of betting behavior, $P(p)$ reflects the amount of money one accepts to bet upon the truth of p ; $1-P(p)$, the remaining amount, is implicitly bet upon the truth of $\neg p$. In this interpretation, individuals are obliged to bet, and they are not allowed to refrain from putting the complete amount of money into the game (another view of forbidding total ignorance). It is difficult to support the idea according to which, in every situation, the gathering of information from individuals, should be based upon a betting procedure. The grades of confidence one obtains are biased by the fear of loosing money.

Besides, several researchers, among whom Savage (1972), have proposed qualitative axioms to represent intuitive properties of a relation "at least as probable as" ($\geq$) on a set of events (or propositions). Namely, denoting $\mathbb{1}$ and $\mathbb{0}$ the sure and impossible events respectively :

P0) $\mathbb{1} > \mathbb{0}$ (non triviality)
P1) $\geq$ is a simple ordering (comparability and transitivity)
P2) $p \geq \mathbb{0}$
P3) $p \wedge (q \vee r) = \mathbb{0} \rightarrow (q \geq r \leftrightarrow p \vee q \geq p \vee r)$

where $\wedge$ and $\vee$ stand for conjunction and disjunction, $>$ is the strict ordering based on $\geq$. As such, P0-P3 are not enough to ensure that the only quantitative counterpart of $\geq$ is a probability measure. Several authors have proposed extra axioms which enable this property to be valid, but these axioms are usually hard to justify from an intuitive point of view (See Fine, 1973, for a discussion).

Lastly, even if a probability measure is a good model of uncertainty judgment, the probability values supplied by individuals cannot be taken for granted. One must admit that they are not perfect measurement devices, and that imprecision pervades the elicitation techniques. A study of error pro-

pagation in long inference chains may invalidate the conclusions of hypothesis evaluation procedures using Bayesian inference, because error intervals
have grown too large.

1.2 - Decomposable measures of uncertainty

Any quantitative representation $g(p)$ of a grade of confidence in the
truth of p, taking its values on $[0,1]$ should obey the following consistency
axiom (Sugeno, 1974) :

$$\text{if p entails q (i.e. } \lnot p \lor q = 1 \text{), then } g(p) \leq g(q) \tag{1}$$

because any piece of evidence supporting p also supports q. Besides $g(0)=0$
and $g(1)=1$ are taken for granted. Although intuitively satisfactory, such a
framework is not very easy to use. Procedures to build such uncertainty measures are needed, and only a subclass of computationally attractive functions
are likely to be used. A first idea to delimit such a subclass consists in
generalizing the additivity axiom of probability into :

$$\text{if } p \land q = 0 \text{, then } g(p \lor q) = g(p) * g(q) \tag{2}$$

This is the decomposability axiom. The set of candidate operations $*$ is described in (Dubois and Prade, 1982a). Such operations are semi-groups on $[0,1]$
with identity 0, known under the name of "triangular co-norms" (Schweizer
and Sklar, 1963). The following property obviously holds, as a consequence
of (1) :

$$g(p \lor q) \geq \max(g(p),g(q))$$

Probability measures are recovered when $* = +$. Another interesting
case is when $* = \max$. Then Zadeh (1978a)'s possibility measures are obtained.
More generally, the set of uncertainty measures derived from the decomposability axiom can be split into two families : the functions g such that $\forall p$,
$g(\lnot p)$ is completely determined by $g(p)$, and those such that $g(\lnot p)$ does not
contain the same amount of information as $g(p)$. The first class corresponds
to probability measures and isomorphic set-functions. The second class includes possibility measures and set-functions such that $\max(g(p),q(\lnot p))=1$. For
instance if g is a possibility measure, $g(p)$ is the grade of possibility of
p, and $1-g(\lnot p)$ represents a grade of necessity in the same proposition. A
noticeable feature of this class of uncertainty measures is its ability to
capture the concept of total ignorance by stating $g(p) = g(\lnot p) = 1$ (which
means equal possibility for p and $\lnot p$). It provides a mathematical model of
concepts of "plausibility" or "surprise" extensively described by Shackle

(1961) who argued against probability as the universal approach to uncertainty in economics. See Hamblin (1959) for a discussion of Shackle's ideas in terms of possibility measures of logical truth, as compared to epistemic probabilities. A similar approach is that of Rescher (1976).

Decomposable measures are easy to define since, just as probability measures, it is enough to assess a set of numerical values representing the weight of each elementary proposition, i.e. an equivalent of the "density" is available. Decomposable measures can be put into a qualitative setting, since P0-P3 imply the existence of $*$ such that (2) holds (Fine, 1973). However, as pointed out by Dubois (1984), P0-P3 cannot describe possibility measures. Actually the set of decomposable measures can be attained if we relax P3 into

P4) $p \wedge (q \vee r) = \mathbb{0} \to (q \geq r \to p \vee q \geq p \vee r)$

Moreover possibility measures ($*$=max) can be characterized by a strong form of (P4) defined as follows (see Dubois, 1984)

P5) $\forall \, p, (q \geq r \to p \vee q \geq p \vee r)$

(an axiom which does not hold for probability !).

These results clearly indicate that it is possible to go beyond the probabilistic framework in a meaningful way in the setting of qualitative probability relations.

1.3 - Plausibility and credibility measures

Instead of relaxing the additivity axiom of probability, another way of going beyond the probabilistic framework is to do away with the exhaustivity and mutual exclusion of the available evidence. Then, following Shafer (1976), total certainty is shared among a set F of (possibly not elementary) propositions, called focal propositions about which some knowledge is available. This is done under the form of a basic assignment m such that

$$\forall \, p_i \in F, \; m(p_i) > 0 \; ; \; m(\mathbb{0}) = 0 \qquad (3)$$
$$\Sigma\{m(p_i) \mid p_i \in F\} = 1.$$

The probabilistic setting is easily recovered by assuming that the members of F are elementary propositions ; m is then a standard allocation of probability. But in the general case, focal propositions are no longer mutually exclusive, and explicit exhaustiveness can be obviated by including $\mathbb{1}$ in the focal propositions, $m(\mathbb{1})$ being the weight committed to the total ignorance.

The truth status of propositions can no longer be evaluated in terms of probability. Because the propositions conveying the available evidence are imprecise, only bounds on probability values can be obtained, namely, for each proposition p

- a lower bound, denoted Cr(p), cumulates the weights of focal propositions which entail p. Cr(p) is called the credibility (or belief) in p because it reflects the evidence which supports p.
- an upper bound, denoted Pl(p), cumulates the weights of focal propositions which are not exclusive with p, i.e. which make p possible. Pl(p) is called the plausibility of p because it reflects the evidence which does not support ⌐p. Indeed, $Cr(\lnot p) = 1-Pl(p)$.

These set-functions have been originally suggested by Dempster (1967), with a statistical point of view. If the weight $m(p_i)$ of each focal proposition p which is not elementary, is shared among its elementary implicants, then the obtained allocation of probability is such that $\forall p, Cr(p) \leq P(p) \leq Pl(p)$. When the focal propositions are imprecise but consonant , i.e., it is possible to order them so that $p_i \to p_{i+1} = 1\!\!1, p_i, p_{i+1} \in F$, then the credibility and plausibility measures satisfy the decomposition axiom of necessity and possibility measures : $Cr(p\land q) = min(Cr(p),Cr(q))$, and $Pl(p\lor q) = max(Pl(p), Pl(q))$. Hence Zadeh (1978a)'s possibility measures can be interpreted in this probabilistic-based framework [3].

1.4 - Upper_and_lower_probabilities

Other extensions of the probabilistic framework have been suggested in the literature ; authors like Good (1962), Smith (1961) have proposed their own views of upper and lower probabilities which differ from Shafer's proposal. Good introduces the following axiom

$$p\land q = \emptyset \Rightarrow P_*(p)+P_*(q) \leq P_*(p\lor q) \leq P_*(p)+P^*(q) \leq P^*(p\lor q) \leq P^*(p)+P^*(q) \quad (4)$$

Suppes (1974) proposes qualitative axioms for belief structures which account for this axiom. Dempster (1967) proves that plausibility and credibility measures satisfy (4), i.e. Shafer's framework is more restricted than Good's.

Giles (1982) recently suggested a relaxed version of (4), in the spirit of betting behavior interpretation. Namely, $P^*(p)$ is viewed as the least amount of money such that for a fee of $\$P^*(p)$, an individual will agree to pay \$1 if p turns out to be true. Giles departs from the Bayesian point of view because he admits that $P^*(p)+P^*(\lnot p) \geq 1$. More specifically, the axioms of this set-function are supposed to be (1) and

$$\forall\ p,\ q,\quad P^*(p \vee q) \leq P^*(p) + P^*(q) \tag{5}$$

This inequality stems from the fact that asserting $p \vee q$ involves a weaker commitment than asserting both p and q separately, so that the required fee is smaller in the first case. It is easy to see that this framework encompasses Shafer's plausibility measures, Zadeh's possibility measures (they provide the lower bound for $P^*(p \vee q)$), and also decomposable measures such that the decomposability operation $*$ satisfies $a*b \leq a+b\ \forall\ a,b \in [0,1]$. Probability measures are recovered for Bayesian agents, i.e. for whom $P^*(p)+P^*(\neg p)=1$.

Giles (1982) introduces another axiom which encompasses both (5) and (1) as follows : If $p_1 \ldots p_n$ are propositions, r out of which hold true, and p_0 is another proposition such that if p_0 is true then s more propositions among $p_1 \ldots p_n$ hold true, the following inequality holds

$$\sum_{i=1}^{n} P^*(p_i) \geq r + sP^*(p_0) \tag{6}$$

Indeed, because r propositions are known to hold true, the total fee must exceed \$r ; the amount \$$sP^*(p_0)$ accounts for the uncertainty on the truth of p_0.

Once more Zadeh's possibility measures as well as probability measures satisfy (6).

Giles (1982) proves that set-functions satisfying (6) are indeed upper probabilities in the sense that there is a set P of probability measures such that $\forall\ p,\ P^*(p) = \sup\{P(p) \mid P \in P\}$ and conversely, given a set P of probability measures, this formula yields a set-function that satisfies (6). The latter point is obvious since given any family of set-functions P_i^*, $i \in I$, satisfying (6), $P^* = \sup_{i \in I} P_i^*$ also satisfies (6).

Hence Giles's concept of uncertainty measure comes close to Smith (1961)'s idea of modeling uncertainty judgments by means of interval-valued probabilities, from which upper (and lower) probabilities are obtained. Similarly, plausibility measures, being upper probabilities, also satisfy (6).

1.5 - Summary_and_open_problems

There are two basic concepts of uncertainty measures which are obtained by taking the probabilistic framework as the reference to go beyond. First the idea of "distorted" probabilities which are embodied in decomposable measures. This means that the human mind sort of more or less falsifies the additivity rule in some way, sometimes leading to models which are no longer isomorphic to probabilities. The other idea is that the (ideally) right probability is out of reach, and only a set of probability measures can be characterized in some way. Shafer, Good and Smith have their own view of procedures to achieve this characterization. Zadeh's possibility measures have some special status regarding this dichotomy since they appear in both frameworks. This is also true for a few other decomposable measures (see Dubois & Prade, 1982a); but possibility measures, because of the use of the maximum operation, are qualitative in essence, hence rather well-adapted to model subjective grades of uncertainty. Moreover, in terms of betting behavior, and interpreted in Giles (1982) framework, they model cautious gamblers. The fact that a possibility measure can be viewed as a weighted set of nested sets enables it to capture the notion of a vague proposition translated as a fuzzy set (Zadeh, 1965) of possible values of some variable. The characteristic function of the fuzzy set is the "density" on which the decomposable measure is built. It then provides a powerful tool to model vague statements uttered in natural languages (Zadeh, 1978b), and also is the basis for computational techniques with vaguely-defined numbers, consistently with sensitivity analysis (Dubois & Prade, 1980, 1985a).

Although broadening the set of mathematical models at hand, these various attempts to go beyond the classical probabilistic framework (which have fore-runners in the 17th and 18th centuries[4] (Shafer, (1978)) also leave room for many open problems. Indeed the power of probability theory is basically due to Bayes theorem which founds Bayesian inference, and the concept of expectation which is at the root of utility theory and of models of trade-off in decision-making. In order to do something with uncertainty measures, we need some canonical counterparts of these notions. The reader will not be surprised to hear that extensions of conditioning, independence, and expectation to uncertainty measures are not unique. Moreover the problem is far from being solved, and only bits of a future extensive setting are available.

a) Imprecise conditioning and expectation

On the side of imprecise probability, i.e. upper and lower probabilities, it is natural to define upper and lower expectations. This has been done by Dempster (1967), Smith (1961), and the authors have used this concept on possibility measures to define the mean value of a fuzzy number (1985c). Conditioning is slightly more difficult. Smith (1961) defines upper conditional probability $P^*(p|q)$ as

$$P^*(p|q) = \text{Sup}\{P(p|q) \mid p \in P\}$$

But Dempster (1961) suggests the quantity $P^*(p|q)/P^*(q)$ and proves that it provides tighter bounds on unknown probability values. Behind conditioning lies the idea of independence. This concept can be studied in a qualitative belief structure (Fine, 1973), and in the case of probability measures this approach leads to define p as independent from q as soon as there is some operation * such that

$$p(p \wedge q) = P(p) * P(q) \tag{7}$$

With this definition, p and ⌐q are independent in the sense of another operation, since $P(p \wedge \neg q) = P(p) - P(p)*P(q) \neq P(p)*(1-P(q))$. The choice of * is not arbitrary, although * = product is not the only solution. Dubois (1983) has shown that * must be choosen in a family of triangular norms (Schweizer & Sklar, 1963) which depends on a single parameter. This family was characterized by Frank (1979) as being the only possible associative function * such that a*b and a+b − a*b are both associative. The extension of these results to upper and lower probabilities is among the open problems. Defining independence with a statistical view point, when frequencies are obtained through random experiments yielding imprecise outcomes (Dubois & Prade, 1985a) is even not trivial anyway.

These notions must be clarified in order to generalize Bayesian inference in a proper way. Particularly, Dempster (1967) rule is not an extension of Bayes theorem, but extends the concept of conditioning to the combination of several plausibility measures. The question of updating prior uncertainty measures from the knowledge of (generalized) likelihood ratios is in order. A preliminary treatment of this question appears in Wierzchoń (1985), in the setting of plausibility measures, and for a particular family of such functions.

The introduction of imprecision in probability theory brings logical connectives into the measure-theoretic arena, and also new information measures [5].

b) Distorted conditioning and expectation

The approach by decomposable measures leads to a different kind of extension of conditioning and expectation. Viewing an uncertainty measure as a "distorted" probability, one gets the intuition for a "distorted" expectation, and another view of conditioning.

For instance if g is isomorphic to a probability measure, i.e. there is an increasing bijection φ of the unit interval such that (see Dubois & Prade, 1982a) $\varphi(g(\mathbb{1})) = 1$, $\varphi(g(\mathbb{0})) = 0$,

$$p \wedge q = \mathbb{0} \Rightarrow \varphi(g(p \vee q)) = \varphi(g(p)) + \varphi(g(q)) \tag{8}$$

since $\varphi \circ g$ is a probability measure, it may look natural to define $g(p|q)$ as follows (Dubois, 1983) :

$$g(p|q) = \varphi^{-1}(\frac{\varphi(g(p \wedge q))}{\varphi(g(q))}) \tag{9}$$

This view yields a "distorted" definition of independence, close to (7). The concept of "distorted" expectation is not uniquely defined. See Weber (1984), Schwyla (1980), Dubois (1983) for different definitions. Contrastedly, Zadeh (1978a)'s concept of the possibility of a fuzzy event, is not related to expectation in the usual sense, but to the operation 'median', as in Sugeno (1974)'s "fuzzy integral" for a monotonic set-function. Anyway, expectation (i.e. averaging) and median are similar notions, and the concept of "distorted" expectation should enable both notions to be involved in the same setting, since triangular norms and conorms contain linear operations as well as the minimum and the maximum, used in possibility theory.

The fact than some decomposable measures other than possibility and probability measures can be viewed as upper or lower probabilities (see Dubois & Prade, 1982a) even complicates the question of finding the proper notions of expectation, conditioning and inference.

2 - APPLICATIONS TO DECISION THEORY

The existence of two points of view on uncertainty measures, one reflecting a distortion of probability axioms by the mind, the other reflecting the imprecision pervading subjective assessments lead to two potential extensions of utility theory, which do not really compete each other since

one can combine (at least conceptually) these two views by claiming that subjective judgments are both distorted and imprecise. However it is good to
keep the points of view separate. Following one path, we look for new types
of utility functions which are no longer linear, stricto sensu. Following the
other path, we introduce some new kinds of sensitivity analysis in standard
utility theory, thus converging towards procedures already used by utility
theory tenants. These extensions are only partially (and sometimes poorly)
developed. However the discovery of possibility measures, as based on fuzzy
sets, has improved our capability to model imprecision consistently with
standard interval analysis. Namely, via the use of the concept of expectation
it is possible to define the probability of a fuzzy event as the expectation
of its membership function (Zadeh, 1968) ; this fuzzy event models a vague
proposition. Moreover ill-assessed probability values can be modeled by possibility distributions, i.e. fuzzy numbers in the unit interval. Lastly, a
fuzzy set is similar to a value function, and fuzzy set-theoretic operations
can be useful in multiple criteria decision problems.

2.1 - Towards distorted utility functions

It is possible to reconsider the axioms of Von Neumann and Morgenstern (1944)'s utility theory with decomposable measures or upper and lower
probabilities as a model of subjective uncertainty. Let X be the set of possible consequences of a decision ; (x,p,y) denotes the case when the probability of consequence x is p while that of y is 1-p. The set of uncertain con-
sequences on X is then $\underline{X}$, constructively defined by the rules $X \subseteq \underline{X}$ and
$(x,y \in \underline{X} \Rightarrow (x,p,y) \in \underline{X} \ \forall \ p \in [0,1])$. A simple ordering $\geq$, reflecting the decider's preferences over $\underline{X}$ is then built. Among natural axioms it should sa-
tisfy, Von Neumann and Morgenstern suggested

$$((x,p,y),q,y) \sim (x,pq,y) \tag{10}$$

where $\sim$ is the equivalence relation associated to $\geq$.
This axiom relies on an independence assumption between consequences of suc-
cessive decisions. A first idea to extend (10) is to relax this independence
assumption using (7). This investigation is made in (Dubois, 1985) who exhibits a generalized utility function u_s which depends upon a single parameter s (the one which appears in Frank (1979)'s triangular norms family) ranging over $[0,+\infty)$. For s=1, $u_s(x,p,y) = u_s(x)p+(1-p)u_s(y)$, i.e. the standard
utility function is recovered. Moreover, if u_s is normalized so as to range
on $[0,1]$, then for $s = +\infty$ (i.e. $* = $ minimum in (7)), the utility function is

a median : $u_s(x,p,y)$ = med$(u(x),p,u(y))$ which violates the following axiom

$x > y \rightarrow \forall p \in (0,1), x > (x,p,y) > y.$

More generally a set of uncertain consequences could be defined from elements of the form $(x,g(x),y,g(y))$ where g is some uncertainty measure over $\{x,y\}$. A reformulation of Von Neumann and Morgenstern axioms in this setting would lead to more general families of utility functions displaying various behaviors which could be interpreted under the light of the underlying axioms. Of course, the authors are aware that an even more convincing approach would be to start from purely qualitative arguments to build the uncertainty measure and the utility function conjointly as Krantz et al. (1971) in the classical setting. However there is still a long way to go, in order to reach this goal.

The interest of such types of generalized utility functions lies in their potential to overcome paradoxical behaviors of classical utility functions such as those mentioned by Allais (1953), by choosing the appropriate value of the parameter, depending upon the decider's attitude in front of risk. Note that Kahneman and Tversky (1979) proposed a notion of distorted probability to solve such kinds of paradoxes, in their prospect theory. They use a bilinear expression of expected utility where probabilities are changed into "decision weights" which can be viewed as defining some kind of decomposable measure of uncertainty.

2.2 - Using fuzzy numbers in utility-based decision models

Although mathematically sound, the S.E.U. model has proved difficult to use in real applications, one of the reasons being that the required knowledge is beyond what an individual can supply. The possible states of the world are only roughly perceived and the knowledge of the actual state is sometimes too weak to derive precise probability values expressing the uncertainty pervading decisions.

In order to cope with this lack of knowledge, one idea is to perform some sort of sensitivity analysis in the evaluation of the expected utility. Then the state-to-consequence mapping becomes a multi-valued mapping, and the probabilities of the consequences are upper and lower probabilities. An extensive survey of sensitivity analysis in decision-support systems is in Sage (1981).

Consider first the case when the possible consequences of a decision are imprecisely described, i.e. they are viewed as subsets $S_1...S_p$ of the set X. If precise probabilities of $S_1...S_p$ are available, they induce a pair of credibility/plausibility measures on X, in the sense of Shafer (1976), and the utility of the decision is obtained by means of upper and lower expectations [6].

If probabilities only are imprecise, then let $[p_\star^i, p_i^\star]$ be the interval containing the probability of consequence x_i. The expected utility E is an interval defined by

$$E = \left\{ \sum_{i=1}^{p} u(x_i)p_i \mid \forall\, i,\ p_i \in [p_\star^i, p_i^\star],\ \sum_{i=1}^{n} p_i = 1 \right\}.$$

It is bounded by upper and lower expectations in the sense of Smith (1961), which can be calculated by a method described in Dubois-Prade (1981).

Proceeding further along this line, some researchers have suggested the use of fuzzy sets for integrating imprecision and uncertainty in decision-support systems, in a way to get closer to the actual information provided by individuals. This trend can be examplified by papers of Watson et al. (1979), Freeling (1980) among others. Fuzzy consequences of decisions and fuzzy probabilities of these consequences only are available, usually described in linguistic terms. A fuzzy probability value is similar to a second-order probability, but is a possibility measure on the unit interval. The interesting point is that although carrying more information than interval-valued probabilities, fuzzy probabilities lead to an amount of computation which is not significantly larger. A survey of computational methods for fuzzy decision analysis is in Dubois and Prade (1982b). See also (Dubois and Prade, 1985c) for complementary material, in relationship with plausibility and credibility measures.

2.3 - Multiattribute evaluation using fuzzy sets

This topic is an application of possibility measures basically. A possibility measure is a generalization of a set, contrastedly with a probability measure. Indeed the "density" of an all-or-nothing possibility measure is exactly the characteristic function of a set, while an all-or-nothing probability measure focuses on a point. In the general case the "density" of a possibility measure can thus be interpreted as the membership function of a fuzzy set. This membership function classifies objects to which it applies in terms of their compatibility with a vague proposition described by the fuzzy set. Here this proposition refers to a fuzzy goal G and the membership grade $\mu_G(\omega)$ expresses the grade of compatibility of object ω with respect to goal G. In this context, the membership function is by nature a value function (Krantz et al., 1971), normalized in such a way that $\mu_G(\omega) = 1$ means total compatibility of ω with G and $\mu_G(\omega) = 0$ means total incompatibility of ω with G. The questions of elicitation procedures for membership functions, based on fundamental measurement theory, are discussed in Norwich and Turksen (1982).

In the case of multiple attribute evaluation, a methodology has been suggested by Zimmermann and Zysno (1983) and the authors (Dubois and Prade, 1984). The idea is to identify the membership function μ_{G_i} along each attribute i. The overall evaluation is described in terms of a hierarchy of partial evaluations, the leaves of which correspond to elementary goals G_i. The aggregation problem is then to find a connective h : $[0,1]^q \rightarrow [0,1]$ where q is the number of elementary goals, such that the overall evaluation function μ_D is defined by

$$\mu_D = h(\mu_{G_1}, \mu_{G_2} \ldots \mu_{G_q})$$

h accounts for the various interactions between elementary evaluations, and should satisfy natural requirements such as $h(0,0,\ldots,0)=0$, $h(1,1,\ldots,1)=1$, and consistency with vector-ordering in $[0,1]^q$.

The actual derivation of h is made by interpreting it as a fuzzy set-theoretic operation. The class of fuzzy-set-theoretic operations is much richer than the class of set-theoretic operations, and has been significantly investigated to date, via the axiomatic approach (see Dubois and Prade, 1985d for a recent survey). The fuzzy set-theoretic framework enables three basic behaviors in front of aggregation to be modeled : the conjunctive behavior (simultaneous attainment of goals) expressed by fuzzy set-intersections, (h ≤ min), the disjunctive behavior (redundancy of goals) expressed by fuzzy set-unions (h ≥ max) and the trade-off behaviors expressed by mean operations (min < h < max). As noticed by Fishhoff et al. (1982) multiattribute utility theory is interested only in the third type of behavior. Of course the actual behavior of a decider does not necessarily matches one of these extreme cases. One may observe hybrid behaviors depending upon the levels of elementary evaluations. The fuzzy set-theoretic framework is capable of handling more complex aggregation schemes than the pure conjunctive, disjunctive, and trade-off rules, including attributes of variable importance (Dubois and Prade, 1985b). It could also deal with value-functions expressing aggregation of the form "either this or that but not both" mentioned in Fishhoff et al (1982), which violate requirement $h(1,1)=1$. For this purpose one may use a fuzzy symmetrical difference operation (see Dubois and Prade, 1980 for instance).

A comparison of axiom systems underlying fuzzy set theoretic operations (e.g. Dubois, Prade, 1985d) and those of conjoint measurement theory (e.g. Kranz et al., 1971) would be worth-while investigating. In both cases, underlying additive structures exist, induced by associativity properties in the fuzzy framework (e.g. a set-intersection is associative) and by attribute-independence in the other framework. However value functions in the fuzzy framework are not assumed to be strictly monotonic with respect to qualitative preference relations, so that operations such as minimum and maximum are not ruled out. Lastly there are some interesting mathematical connections between decomposable measures and multiattribute utility theory, especially multiplicative forms of utility functions (Keeney, 1974 ; see Dubois, 1983, Dubois-Prade, 1985b, for details). The basic idea is that the concept of attribute importance can be modelled by an uncertainty measure on the set of attributes ; it is a probability measure for additive utility function, and a special parametered family of decomposable measures studied by Sugeno (1974) for multiplicative utility functions.

CONCLUSION

The aims of this paper are twofold : first to provide an overview of recent proposals for modeling uncertainty and imprecision in a way that includes the probabilistic approach as a particular case ; next, to indicate some research lines for the development of decision-making models which are based on these new uncertainty measures. These new tools have potentials to deal with some biases in Bayesian hypothesis evaluation and Subjective Expected Utility such as non exclusive, non exhaustive evidence, deviant combination rules, cautious behavior in front of risk, non-compensatory criteria. Of course, uncertainty measures will not solve all the problems. For instance, even based on uncertainty measures, decision models remain static. But time may be an important aspect of a decision process (see Volta, 1985). Similarly, the representation of uncertain judgments outside the probabilistic framework must still face biases such as miscalibration, incoherence, base-rate fallacy, premature conviction, etc ...

It is our belief that decision-making models should be developed as a trade-off between the normative point of view of classical utility theory and the descriptive point of view of artificial intelligence[7]. Indeed the latter is meant to simulate human behavior when solving problems, while the former prescribes what a rational decision should be. But both are concerned with reasoning processes in uncertain environments. The introduction of rules of rationality in A.I. models may improve the performance of empirical approximate reasoning techniques as found in existing expert systems (e.g. MYCIN (Shortliffe and Buchanan, 1975), etc ...). On the other hand, the relaxation of axioms which underlie SEU models may bring them closer to the kind of information individuals can actually supply, while preserving some normative foundations.

Additional notes (after discussions at the Institute) : Notes are referred to in the main text by upper-script numbers in parentheses.

1 - A single number is not always sufficient to assess uncertainty

Consider a (fake) experiment in cognitive psychology, similar to those described in Wagenaar & Keren (1985) paper. Students have followed a skating contest all sunday long. On the following day they are asked to give their confidence rating on statements pertaining to the color of the winner's shirt. The two statements are
  S1 : the winner's shirt was black (B)
  S2 : the winner's shirt was dark (D)
where 'dark' includes black, brown, ..., and other colors. The students answer consists in two numbers P(B) and P(D) which characterize their state of uncertainty about the statements. P(B) = 1 means "I am sure it was black", P(B) = 0 means "I am sure it was not black", i.e. P(B)+P(notB) = 1, as usual with probabilities. A natural consistency condition is P(B)≤P(D) since black implies dark. (See equation (1)).

Now consider the case of a student who followed the race on the radio then he has no idea about the color of the shirt. His only way of expressing his ignorance about S1 and S2 is to put on each statement P(B) = P(D) = 1/2, since 1/2 is half way between positive and negative certainty (1 and 0 respectively, in the model). Assuming that P(B) and P(D) are probabilities leads to conclude that

$$P(D \text{ and not } B) = P(D)-P(B) = 0 \text{ (since } B \subset D)$$

Hence we conclude that the ignorant student is sure that the shirt is not brown, which is a paradox in case of total ignorance! What is wrong in the model here is to assume that one number can represent the available evidence on both a statement and its negation, when two descriptions of the world are simultaneously used (the partitioning induced on the set of colours by the predicates "black" and "dark" respectively). This situation is commonly encountered in expert systems applications where uncertain knowledge from several sources with different points of view on the world is available, must be consistently represented and combined.

N.B. In case of total ignorance, people may just express the base-rate of similar events. This attitude, which is hard to detect in individuals is not always satisfactory. Indeed, it depends upon the extent of the similarity between standard events and the one under concern. Moreover people may be unaware of the base-rate itself if they are not related to the kinds of events they are asked about.

A way out of this paradoxical behavior is to adopt the following conventions. If the question involves the truth of a statement p, ask for two numbers N(p), N(not p) with interpretation as follows

$N(p) = 1$; $N(\neg p) = 0$ means certainty of p being true
$N(p) = 0$; $N(\neg p) = 0$ means total ignorance
$N(p) = 0$; $N(\neg p) = 1$ means certainty of p being false.

The rules associated to these numbers are :

$N(p) > 0 \Rightarrow N(\neg p) = 0$ (certainty in favor of p rules out certainty in its negation)
$N(p$ and $q) = min(N(p),N(q))$
if p implies q, then $N(p) \leq N(q)$

These rules are those of possibility theory (Zadeh (1978a)) viewing $1-N(\neg p)$ as the grade of possibility of p. The reader can check that confidence ratings following these rules obviate the paradoxical behavior observed with rules of probability theory.

2 - The paper by Garbolino (1985) tends to indicate that even if the states of the world could be described with infinite precision, such descriptions may turn out to be useless due to the complexity of the corresponding decisional Bayesian machine which would forbid any computation of the optimal decision in a reasonable amount of time. This remark brings forward a limitation of Bayesian models which questions the quest for unlimited precision. This necessary trade-off between precision and complexity in decision analysis is quite similar to the trade-off between precision and meaningfulness in large systems, as pointed out by Zadeh (1973).

3 - Shafer's theory of evidence and possibility measures

A simple example can give more intuition about the meaning of credibility, plausibility and other confidence measures.

Suppose we want to get some idea about the age of the president of the country, and that the only available way of getting some information is to ask people around. They are not supposed to give precise estimations if they cannot provide them, but only a range contained in [0, 100] years. Assume we get the following answers

Age range	proportion of answers
F_1 = {65}	$m(F_1)$ = 0.1
F_2 = [60, 70]	$m(F_2)$ = 0.5
F_3 = [0, 100]	$m(F_3)$ = 0.2
F_4 = [55, 75]	$m(F_4)$ = 0.1
F_5 = [71, 80]	$m(F_5)$ = 0.1

The focal propositions make the set of ranges $\{F_i | i=1,5\}$. Note that F_3 = [0, 100] expresses ignorance. The basic assignment simply reflects frequencies here. Now consider the proposition P(p) of people who believe that the president is at least 55 years old and 70 years old at most. Because of the imprecision of the answers, this proportion is itself imprecise namely $p \triangleq$ Age $\in$ [55, 70], and

$$Cr(p) = \sum_{F_i \subseteq [55,70]} m(F_i) \leq P(p) \leq \sum_{F_i \cap [55,70] \neq \emptyset} m(F_i) = Pl(p)$$

i.e. $m(F_1)+m(F_2) \leq P([55,70]) \leq m(F_1)+m(F_2)+m(F_3)+m(F_4)$

i.e. $P(p) \in$ [0.6, 0.9]. The lowerbound is interpreted as a grade of belief or credibility because it gathers the positive evidence in favor of [55,70]. The upperbound is interpreted as a grade of plausibility because it gathers what is left when evidence against [55,70] has been set aside. If p = [55,70], the negation $\neg p$ = [0.54] $\cup$ [71, 100]. It is easy to check that $Pl(\neg p)$ = 1-Cr(p) = 0.4, $Cr(\neg p)$ = 1-Pl(p) = 0.1.

Now assume that F_5 = [55, 80] instead of [71, 80]. As a consequence the focal propositions are nested ($F_1 \subseteq F_2 \subseteq F_4 \subseteq F_5 \subseteq F_3$). It can be checked that if p = q$\vee$r is a disjunction of propositions then

Pl(p) = max(Pl(q),Pl(r)).

For instance Pl([55, 70]) = 1 (instead of 0.9) = max(Pl[55, 63], Pl([60, 70])) since Pl([60, 70]) = 1. Pl is then called possibility measure, as coined by Zadeh (1978). See also Dubois-Prade (1980, 1982c) and Shafer (1985). The associated credibilities 1-Pl($\neg$p) are called necessities or certainties (Dubois & Prade (1980, 1985b), Zadeh (1979a)), and satisfy the dual decomposability axiom as mentioned in the main text.

A very important consequence of this decomposability property is the ability to characterize Pl and Cr in terms of an allocation of weights on the reference set Ω (here on [0, 100]). Namely, in the finite case this allocation is defined by $\pi(\omega)$ = Pl($\{\omega\}$) $\forall \omega \in \Omega$. In our example ω is an age and

$\pi(\omega) = \Sigma\{m(F_i) \mid \omega \in F_i\}$

which yields the following figure in the modified example

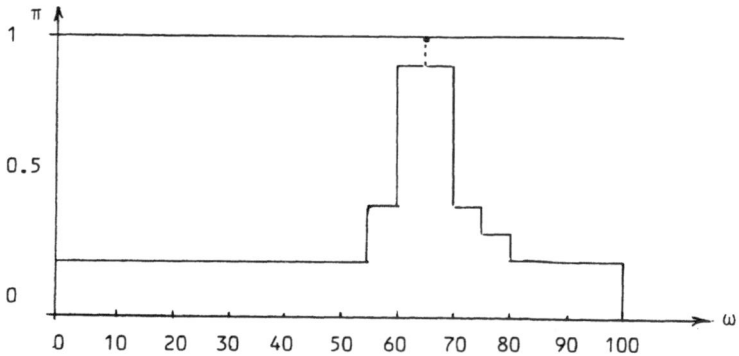

And from the decomposability property we get for any proposition p of the form "age ∈ A", A being a subset of Ω,

$$PL(p) \triangleq PL(A) = \max_{\omega \in A} \pi(\omega)$$

$$Cr(p) = 1 - PL(\neg p) = \min_{\omega \notin A} 1 - \pi(\omega).$$

The possibility distribution π can be viewed as the membership function μ_F of the fuzzy set F of possible values of the age of the president. Zadeh (1965)'s fuzzy sets can be viewed as convex sums of nested sets. These remarks enable fuzzy sets to be clearly distinguished from probability measures (i.e. convex sums of (disjoint) points), and lead to a statistical interpretation of fuzzy sets out of imprecise observations (Dubois and Prade, 1985a). This statistical interpretation can live along with subjectivist ones, of course. Another discussion of Shafer's framework (it inspired the present one) is in Zadeh (1984).

4 - Volta (1985) notes that after the Medieval Ages, logic lost the concept of time, imbedded in the categories of "possible" and "necessary" (ontological time). It is interesting to notice that quite at the same epoch, the concept of additive probability became more and more "natural", so much so as Lambert's works on non-additive probability in the XVIIIth century were unintelligible to scholars of his time, as Shafer (1978) points out. Actually in both cases similar concepts of possibility = plausibility, and certainty = necessity were forgotten in different areas.

5 - From example 3, it is clear that Shafer's theory deals with convex combination of sets (nested or not) - the F_i's weighted by the $m(F_i)$'s. Hence plausibility measures (or belief functions) can be viewed as generalized sets, and can be combined by means of the connectives of logic, such as union, intersection, complementation and so on. For instance given two bodies of evidence (F_1, m_1) and (F_2, m_2) where F_i is a set of focal subsets F_{ij}, $j = 1, k_i$ and m_i is a basic assignment, given a logical operation $*$ on sets, the logical combination of (F_1, m_1) and (F_2, m_2) via $*$ yields (F_*, m_*) defined by $F_* = \{A \subseteq \Omega \mid m_*(A) > 0\}$ and

$$\forall A \subseteq \Omega, \quad m_*(A) = \Sigma\{m_1(B).m_2(C) \mid A = B*C\}.$$

This expression views the basic assignment as defining a probability measure on the power set 2^Ω, and assumes (F_1, m_1) and (F_2, m_2) are stochastically independent. Note that this view is valid on 2^Ω because $\forall\ A,\ A' \in F_1$, $\{A\} \cap \{A'\} = \emptyset$ even if $A \cap A' \neq \emptyset$ in Ω ! When $* = $ intersection the above definition yields Dempster (1967) rule of combination of evidence, provided that m^* is normalized (this is done dividing m_* by the constant $1-m_*(\emptyset)$, which gets rid of the discrepancies between F_1 and F_2). More information on Dempster rule is in Shafer (1976, 1985), Dubois-Prade (1982c), Zadeh (1979b, 1984). For other logical connectives see Oblow (1985), Yager (1985a), Dubois-Prade (1986a). The logical connectives for fuzzy sets do not follow the above rule (see e.g. Dubois-Prade, 1985d).

Viewing the cardinality of a set A as a measure of the imprecision of the statement $v \in A$ where v is a variable, the view of a plausibility measure as a generalized set leads to measures of the imprecision of a body of evidence (F, m) such as

$$HI(m) = \sum_{A \subseteq \Omega} m(A) . \text{Log}_2\ |A|$$

where $|A|$ denotes the cardinal of A. This measure is an extension of one proposed by Higashi and Klir (1983) in the setting of fuzzy sets. Another proposal for imprecision measures is in Yager (1982) who also generalizes Shannon's entropy into a measure of dispersion

$$HD(m) = - \sum_{A \subseteq \Omega} m(A) . \text{Log}_2\ Pl(A)$$

HI = 0 for probability measures (which imbed precise information : $m(A) > 0 \Rightarrow |A| = 1$), while HD = 0 for possibility measures (which imbed consonant information $m(A) > 0 \Rightarrow Pl(A) = 1$). More details can be found in Higashi and Klir (1983), Yager (1982), Dubois and Prade (1985f, 1986b).

6 - This approach enables the maximin, maximax and expected utility rules in decision analysis to be cast in a unique framework. Let X be a set of elementary consequences, $S_1, ..., S_p$ a set of perceived consequences with probabilities $p(S_i)$ such that $\sum_{i=1}^{p} p(S_i) = 1$. The S_i's are subsets of X. Let $u(x)$ be the utility of an elementary consequence, the upper and lower expectations of utility are respectively (e.g. Shafer, 1981)

$$E^* = \sum_{i=1}^{p} p(S_i) . \max_{x \in S_i} u(x)$$

$$E_* = \sum_{i=1}^{p} p(S_i) . \min_{x \in S_i} u(x)$$

When the S_i's are precise $E_* = E^* = $ the usual expected utility. When a single imprecise consequence S is perceived with certainty, the max-min (resp. max-max) rule is obtained as a result of finding the decision which maximizes E_* (resp. E^*). See also Bolaños et al. (1985), Yager (1985b), and Dubois and Prade (1985c) when the S_i's are fuzzy sets.

7 - See Prade (1985) for an extensive survey of approximate reasoning techniques in Artificial Intelligence, with a descriptive point of view.

DECISION COMPLEXITY AND INFORMATION MEASURES

Paolo Garbolino
Scuola Normale Superiore
Pisa, Italy

In this paper I shall discuss the problem of complexity of deci=
sion rules in the framework of machine complexity. In this way,
we can actually have a conceptually well-structured discussion
of what one might mean by "decision complexity" and we can
study the bearing the issue of complexity might have on the
foundational problems of Decision Theory. I shall take a deci=
sion rule to be a Finite State Machine (FSM), and I shall define
an ideal Bayesian Decision Machine (BDM). I would like to stress
that I am thinking of a "machine" as a mathematical object. I
shall show how BDM will meet with complexity troubles and I
shall argue that it is possible to deal with them by defining
decision machines endowed with "information processors" which
differ from each other according to different organizations of
Long Term Memory (LTM). My main claim will be that "information
processors" are procedures or algorithms which denote abstract
objects, namely functions, and that, whatever the underlying
architecture of LTM is, it must satisfy certain very general
and formal conditions to guarantee the existence of "information
processing" functions on it.

I

The most recent brand of Bayesian Theory put forward by Jeffrey
(1983) is the unified theory of preference that attributes
probabilities and utilities to the same objects, that is propo=
sitions. By a "proposition" Jeffrey means, drawing upon Carnap's
(1947,1950) semantic theory, the following: a non-empty set of

NATO ASI Series, Vol. F21
Intelligent Decision Support in Process Environments
Edited by E. Hollnagel et al.
© Springer-Verlag Berlin Heidelberg 1986

sentences in the decision maker's language is a "state descrip=
tion" if it is a maximally consistent set of sentences, i.e. a
set that does not contain sentences which logically contradict
each other and such that it contains exactly one of each contra=
dictory pair of sentences in the language. A state description
is a "possible world". According to the semantics of natural
languages developed by Montague (1974) and Kripke (1963,1972),
the meaning of a sentence can be thought of as the set of state
descriptions or possible worlds in which it would be true. In
this way we can define a "proposition" as a set of state descrip=
tions or possible worlds and say that each sentence of the
decision maker's language is to be assigned a proposition. The
objects of desires and beliefs are propositions and not sen=
tences in order to avoid the possibility of fooling the decision
maker just by offering him a bet on different sentences expres=
sing the same proposition. In other words, the prerequisite of
rational decision making according to Bayesian Theory is the
ability to understand the language in a very strong sense: as
a matter of fact, requiring that the decision maker must bet
on propositions and not sentences, amounts to say that he ought
to recognize logically equivalent sentences, for logically
equivalent sentences in Montague's semantics denote the same
proposition, and only one. I identify, as usually, state
descriptions or possible worlds with "elementary events" in
mathematical Probability Theory, and propositions with "events".
To the set Ω of elementary events and to the Boolean algebra β
of subsets of Ω, we add as another primitive term the prefer=
ence relation " $\leqslant$ " and we get the relational system (r.s. for
short) $\langle \Omega, \beta, \leqslant \rangle$. The question whether,it does exist an expected
utility function representing preferences on it,it has been
answered in the positive by Bolker (1967), who formulated
sufficient conditions when β is a complete and atomless Boolean

algebra, by Jeffrey himself (1978), who adapted Bolker's condi=
tions to the case in which ß is a Boolean σ-field of sets, and
by Domotor (1978), who gave necessary as well sufficient condi=
tions to represent preferences on finite and atomic Boolean
algebras. The choice of propositions as primitives is not
common practice among decision theorists. I did it because one
of the aims of this paper is to define BDM: it receives as
inputs finite strings of sentences and its task is to print on
the output tape the same sentences transformed in a certain
way we shall see in a while, ordered according to the expected
utilities of the propositions expressed by them. Therefore, it
is nice that the primitives of the decision rule, the inputs,
and the outputs are in the same format. Furthermore, Jeffrey's
theory is compatible with the general theory of Conditional
Decisions put forward by Luce and Krantz (1971a,1971b). The
fact that the algebra ß can be finite and atomic allows us to
take the finite set Ω as the set of database stored in LTM.
Probability and utility functions are thus determined by the
preferece ranking of propositions and the desirability of a
proposition is a weighted average of the desirabilities of all
the state descriptions or possible worlds in which it is true,
where the weights are proportional to the probabilities of the
state descriptions. The probabilities of state descriptions
are, on their turn, determined by the preference ordering " $\lesssim$ "
among them. This feature of the theory is important because
BDM is then fully described by a preference ordering of all the
state descriptions or possible worlds stored in memory. The
idea of attributing preferences to FSM's is due originally to
Putnam (1967), who associated with each machine a preference
function assigning a "utility" to possible worlds, and a "degree
of confirmation" function (in terms of Carnap's system of
inductive logic), the two functions together completely deter=

mining the machine's behavior. Now we know that all we need is just a preference ordering of possible worlds.

II

BDM is a quintuple $\langle X,Y,Z,\lambda,\delta \rangle$ where the set X of inputs consists of stringsof sentences, the set Y of outputs consists of ordered strings of sentences, Z is the set of internal states, λ is the state transition function, and δ is the output function. The machine must fulfill two sub-goals: the first one is to "under= stand" the inputs by mapping them into their corresponding propositions as defined above; the second one is to calculate the expected utilities of these propositions. Thus, it is possi= ble to represent BDM by a serial decomposition into machines M_1 and M_2 to generate the machine $M_1 \otimes M_2$: the sub-machine M_1 checks, for any sentence it receives and for all the state descriptions stored in its LTM, if that sentence is true in that state description. Let's suppose that M_1 has a parallel- -processing capacity, so that all the input sentences can be checked in a parallel manner: therefore, it is realized by parallel decomposition into component machines to generate $M_1 = M_{11} \oplus \ldots \oplus M_{1n}$. Any output from M_{1i} is a vector $A = \Sigma A^C(\omega)\omega$, where $A^C: \Omega \to \{1,0\}$ is the characteristic function for any $A \in \beta$, and $A^C(\omega)=1$ if $\omega \in A$, else 0, for all $\omega \in \Omega$. The output of M_1 is a set of vectors and the set Y_1 of outputs from M_1 is a finite dimensional real linear vector space. As Domotor has proved , using a theorem first proved by Scott (1964), if certain ine= qualities are satisfied in this space, then probability and utility functions exist that take their values in fields of non-standard real numbers, e.g. propositions can have infinites= imal probabilities. The set of vectors produced by M_1 is then fed as input to the sub-machines M_2 that checks the existence of those inequalities and computes the expected utilities of the vectors. We can now use the Complexity Axioms given by

Gottinger (1983) to obtain the complexity of the Bayesian deci= sion rule as the complexity number of BDM. The axioms say that the complexity of a machine realized by a parallel decomposition is equal to the max of the set of complexity functions of the component machines; the complexity of a machine realized by serial decomposition is not greater than the sum of the complex= ity functions of the component machines. The last axiom says that a "flip-flop" machine, that is an identity reset machine which corresponds to a combinatorial semigroup in the Mathemat= ics of Machines, has zero complexity. Let's denote by μ the machine complexity function, by γ_G the group complexity function (recall that to a machine realized by serial-parallel decomposi= tion corresponds a semigroup and that "finite semigroup theory is finite group theory plus the 'flip-flop'"; for reference sources see Arbib (1968) or Eilenberg (1976)). Finally, let $\underline{f}^S$ be the semi-group of a machine $\underline{f}$ realized, as BDM, by serial- -parallel decomposition. Then,

$$\mu(\underline{f}) = \gamma_G(\underline{f}^S).$$

Let $\mu(BDM) = \gamma_G(BDM^S)$ be the design complexity, understanding by it the complexity number associated with the transformation process in which full use of the system potential is made. As it is well known, the upper bound for design complexity follows from the machine inequality :

$$\gamma_G(\underline{f}^S) = \mu(\underline{f}) \leqslant |Z| - 1.$$

The design complexity of BDM depends in a critical way upon the design complexity of M_1. A simple model for M_1 would be a system composed of primitive computing elements or units connec= ted to each other. Each unit is a two-state automaton: when the i-th unit is on, the machine prints 1 on the output tape, and when the same unit is off, the machine prints 0. Then, the computational complexity is the number of links between units,

times the number of interactions required to compute a string
$A = \Sigma A^c (\omega) \omega$. By structural complexity I mean, following again
Gottinger (1983), the complexity of the units that hooked to=
gether realize the system. We see that in the machine M_1 as
sketched above the computational burden falls entirely on the
side of computational complexity. Now suppose that the language
of BDM contains $\underline{n}$ atomic sentences which can be used to form
2^n possible state descriptions; hence, there are 2^n units to be
scanned and M_1 has 2^n states. Therefore, we are able to set a
lower bound for the design complexity of M_1: given a finite
number $\underline{n}$ of atomic sentences, it must hold that

$$2^n \leqslant |z_1|.$$

This very fact has far-reaching consequences: with as few as
300 atomic sentences, the number of possible state descriptions,
as Harman (1980) pointed out, would be a lot larger than the
number of molecules in the universe. The "combinatorial explo=
sion" argument, raised by Suppes (1974) as well in arguing
against the plausibility of Scott's 1964 axioms as a theory of
belief, is not a decisive argument but it is indeed a strong
incentive to depart from viewing elicitation of probability
judgments as a matter of introspecting pre-existing probability
distributions and to move toward thinking of such an elicitation
as a matter of working up such functions "on the spot", as occa=
sions demand. Jeffrey (1984) himself has recently argued for
such a move. If we begin to think about the problem in this way,
we are able to identify a common area of interest for Decision
Theory and Artificial Intelligence.

III

Maybe the reader has already realized that the core of the
problem is very similar to chess playing problems, a typical
paradigm for early researches in Artificial Intelligence. We

know that chess players, both humans and computers, do not
consider all possible board positions, but they generate and
examine a small number of them. The same thing we can try to do
with BDM, trading-off computational complexity with structural
complexity, that means endowing BDM with some degree of problem
solving power. The need for supplying BDM with heuristic power
to operate searching procedures in LTM comes from the seeming
impossibility to take any quicker way to reduce complexity.
Hacking (1967) proposed to take sentences as objects of belief
instead of propositions. This move runs immediately into a
major drawback: either the set of sentences belonging to the
"personal language" of BDM does not exhibit any interesting
algebraic structure, and by consequence no representation
theorem is possible to be proved for it and thus no measure of
belief whatsoever it is possible to assign to sentences, or, if
it does exhibit some algebraic structure so that the set of
sentences is deductively closed under some suitable entailment
relation, then this structure is not less complex than the
propositional one, as showed by Mondadori (1984). The way out
from such an unpleasant situation is to notice that in all
choice situations the decision maker does not deal with the set
of all possible worlds, but with that entity that Savage (1954)
called a "small world". A "small world" is a partition of the
"grand world" (in Savage's own term) which is the set Ω of all
the elementary events or state descriptions given the decision
maker's language. "Small world events" are sets of elementary
events and any "small world event" generates a Boolean sub-alge=
bra of the Boolean algebra β in the above mentioned r.s.
$< \Omega, \beta, \leqslant >$. If the preference relation "$\leqslant$" is closed under sub-
-models, that means the axioms for "$\leqslant$" which hold in $< \Omega, \beta, \leqslant >$
still hold for any non-empty sub-algebra of β, then we can get
a representation theorem for expected utilities on the sub-alge=

bra in Domotor's (1978) fashion, and we don't have to bother about the "grand world" to get a probability distribution on the smaller world. Savage (1954) limited himself to say that

the selection of small worlds to be used in a given context

is a matter of judgment and experience about which it is impossible to enunciate general principles. Sure, in choosing certain small worlds descriptions of the decision situation at hands, the decision maker implements some sort of heuristic power. The basic problem here is to define a procedure which, starting from a given knowledge base in LTM, generates "small worlds events". Even if we are not able to define such a proce= dure, some recent developments in the Theory of Programming Languages allow us to state some very general principles, after all, which might be useful, at least, to give a clear formal framework to the problem, in which further work could be done, hopefully. Before going farther to see those developments, some other remarks about "small worlds" are proper at this point. Sometimes it happens that the choice of the "small world" de= scription is clearly determined by the context. Let's take a simplified version of Savage's example about "eggs worlds": one might be uncertain about whether the unique brown egg in the pair of eggs is rotten or about which in the pair are rotten, if any. In the first case we are just satisfied with two state descriptions, "the brown egg is rotten" and its negation; in the second case we have 2^2 of them: both the eggs are rotten or the brown one is rotten and not the white one or viceversa or them both are good. The sentence "the brown egg is rotten" describes the smaller world completely and corresponds to a set of 2^1 states of the larger world: that one in which both the eggs are rotten and that one in which the brown egg but not the white one is rotten. The former sentence is true if and only if the logical **disjunction** of the latter two, let's call it ϕ, is true. On the

other hand, the sentence "there is at least one rotten egg" is
true if and only if the logical disjunction of three state de=
scriptions, call it ψ, is true, as the reader can easily check
by himself. He can immediately see also that the truth of ϕ
implies the truth of ψ. Sentences in this form can be ordered
by the Boolean inclusion relation "$\subseteq$": in particular, $\phi \subseteq \psi$.
This means that they behave like Gottinger's (1973,1974) "stan=
dardized informative propositions": let Φ be the set of all
propositions of the language of BDM in such a disjunctive form
and let Φ be endowed with a relation "$\leqslant$:"; Φ contains also a
unit element ∇ (the tautological disjunction) and a zero element
Δ (the impossible or self-contradictory state description). Then
define

$$\text{if } \Delta \leqslant: \phi \leqslant: \psi \leqslant: \nabla \text{ then } \Delta \subseteq \phi \subseteq \psi \subseteq \nabla$$

holding for any $\phi, \psi, \ldots \varepsilon \Phi$. Gottinger showed that the r.s.
$\langle \nabla, \Phi, \leqslant: \rangle$, called a Qualitative Standardized Information Structure
(QSIS) is a Boolean algebra; therefore, there exists a Boolean
homomorphism H between QSIS ans a Qualitative Probability Space
$\langle \Omega, \beta, \leqslant \rangle$ such that $\phi \leqslant: \psi \overset{H}{\underset{\rightarrow}{}} A \leqslant B$. Hence, if $\langle \Omega, \beta, P \rangle$ is a Probabil=
ity Space that agrees with the qualitative probability "$\leqslant$", we
have that if $\phi \leqslant: \psi$, then $P(A) \leqslant P(B)$. Considering that, in this
context, P is a logical probability, it is immediately seen that
Gottinger's "information" is the reverse of the measure of
"semantical information" proposed by Carnap and Bar-Hillel
(1953; Bar-Hillel,1964). This "semantical information" is called
also the "content" of a sentence and it is defined as $\underline{cont}(\phi) =$
$= 1 - P(\phi)$, where P is just the logical probability of a sentence
ϕ. The interesting point here is that we are able to account for
the intuitive advantage of being in a "small world", compared
with a larger world, in terms of Carnap's semantic information.
The relative advantage of betting on a sentence that will come
out to be true in a small world, over a sentence that will come

out to be true in a larger world consists of the fact that the
first one carries more semantic information than the second one.
This measure of "semantic information" is related to the logical
probability of the sentence and to its logical "content"; this
is not exactly what we need. What we need is a measure of the
"semantic information" related to the personal probability of a
sentence, and to the logical "content" of the "small world" in
which the decision maker has framed his decision problem. Is it
possible to define a measure of the "information" contained in
"small worlds", and to define it independently from the existen=
ce of a probability distribution over "small worlds events"?
The latter requirement is imposed by the new approach suggested
above that personal probability is not elicited once for all
but it is elicited each time a "small world" is newly called
"on the screen". In the next and last paragraph we shall show
how it is possible to answer to this question in the positive.

IV

The problem the decision maker is facing is to obtain in a con=
structive way a "description" of an unknown elementary event
$\omega \in \Omega$ in $<\Omega, \beta, \preccurlyeq >$; possible bodies of information about an -
-otherwise indetermined - element of Ω amount to the assertions
that this unknown element of Ω possesses certain specified
features. The logical conjunction of the bodies of information
about an unknown elementary event is a "description" of it. It
can be a "total description" if it is equal to the elementary
event itself; otherwise, it is a "partial description". Let's
suppose that bodies of information are objects which are sets:
then, they are ordered, in a natural way, by the set-theoretic
inclusion relation; for ex., if A and B are bodies of informa=
tion about a certain ω, then $\{A\} \subseteq \{A,B\}$ and they generate two
"partial descriptions" of ω which are two small worlds events
such that $(A \& B) \subseteq (A)$. Now, take the set Π of all possible
partitions of the set Ω. A partition $\pi \in \Pi$ is a system of pairwise

mutually exclusive "descriptions" -concerning an unknown element of Ω- none of which is logically inconsistent, for ex. $\pi_1=\{A\&B,$ $,A\&\neg B,\neg A\&B,\neg A\&\neg B\}$ and $\pi_2=\{A,\neg A\}$. As it is well known, we can define an ordering " $<$ " (...is finer than...) between partitions in the following way (Domotor,1970):

$\pi_1 < \pi_2$ iff for all $A^{\epsilon}\pi_1$, there is a $B\epsilon\pi_2$ s.t. $A\subseteq B$. The inclusion relation between "descriptions" induces a refine= ment relation between partitions, and the r.s. $<\Omega,\Pi,<>$ is a complete lattice with minimum $\Delta=\{\Omega,\emptyset\}$ and maximum $\nabla=\{\{\omega\}|\omega\epsilon\Omega\}$ $\{\emptyset\}$, that is, the partition whose elements are the elementary events themselves or, in other words, "total descriptions" or "grand world events". This structure is isomorphic to a complete lattice of Boolean subalgebras of the Boolean algebra ß on Ω: therefore, it is a partial ordering, according to the refinement relation, of the set of Savage's "small worlds". Forte and Pintacuda (1968a,1968b) gave an axiomatic definition of the measure of information $H(\pi)$ associated to a partition $\pi\epsilon\Psi$, where $\Psi \subseteq \Pi$ and it is the subset of "independent" subalgebras of ß, according to a purely algebraic definition of "independence" (Marczewski,1948; Sikorski,1950). The information given by a partition $\pi\epsilon\Pi$ is a function $H:\Pi \rightarrow Re^+$ satisfying, among others, the following conditions:

(1) for $\pi_1,\pi_2\epsilon\Pi$, if $\pi_1<\pi_2$ then $H(\pi_1)>H(\pi_2)$;

(2) for $\pi_1,\pi_2\epsilon\Psi$, if $\pi_1\cup\pi_2\epsilon\Pi$ then $H(\pi_1\cup\pi_2)=H(\pi_1)+$ $+ H(\pi_2)$.

Thanks to the isomorphism noticed above, we get in this way a measure of "information" associated to the lattice of Boolean subalgebras of ß, and, furthermore, this measure is defined independently from the existence of personal probability distri= butions on these subalgebras, given that one can define "inde= pendence" without any reference to probabilistic "independence", even if the underlying subalgebras are probability algebras (Kappos,1960;1969). The problem now is to give an interpretation

of the objects we have called "bodies of information" such that
they are set-theoretic objects and, moreover, they are compati=
ble with that abstract object which a BDM is. "Bodies of infor=
mation" are the stuff LTM is made of. They could consist of
simple sentences or "facts" and of chunks of knowledge called
"frames" (Minsky,1975) describing something in terms of its
properties. A frame is thus a "description" and each of the
properties that constitute the description is conveyed by a
slot-value combination. Each frame is a "description" in itself
and can generate other "descriptions" which are subsets of the
former. Let's take a simple example :

FRAME NAME	SLOT	VALUE
Car	A-kind-of	Machine (M)
	Operation	Faulty (F)
		Not-Faulty (F)
	Color	Red(R)

From this very simple frame one can get an ordered set of "small
worlds": $\{M\&F\&R, M\&F\& R, M\& F\&R, M\& F\& R\} < \{M\&F, M\& F\}$ or $\{M\&R, M\& R\}$.
The minimum of bottom of the set would be $\{\Omega, \emptyset\}$ that could stand
for "Something or Nothing". Frames generate "descriptions" and
Boolean subalgebras of ß. So, let's suppose that BDM's LTM is
a "semantic network" of simple "facts" and "frames" over which
an "inheritance relation" (an IS-A or a A-KIND-OF relation) is
defined. Do the nodes of this "semantic network" denote objects
which are sets, so that the "inheritance relation" can be inter=
preted as a Boolean inclusion relation and, by consequence, one
can associate to a particular subset of it an H-Information
measure? The theory of Denotational Semantics for Programming
Languages (Stoy,1977) allows us to answer YES to the question.
Here are the reasons for that, briefly stated, taken from a
brand-new exposition of the theory by Scott (1982): an "informa=
tion system" is a r.s. $I = \langle X, \underline{Con}, \vdash \rangle$, where X is a set of data

objects or propositions, _Con_ is a set of finite subsets of X
(the consistent sets of objects), and "$\vdash$" is the entailment
relation defined as follows:

> for all $\underline{a} \subseteq X$ and for all $x \epsilon X$, $\underline{a} \vdash x$ iff either
> $x \epsilon \underline{a}$ or $x = \Delta$, where Δ is the less defined object
> (for ex., "Something").

Scott offers an axiomatic definition of I , giving a list of
axioms for it, for which I refer to his paper. Among them, that
one that says that if $x \epsilon X$, then $x \epsilon$ _Con_ (i.e. _Con_ contains all
singletons) is particularly important. "Elements" of the Domain
D of semantic values are sets of data objects; an "element" is
the set of data objects which are true of it:

$$\alpha = \{ x \epsilon X \mid \alpha^c(x) = 1 \}.$$

The Domain of semantic values is a r.s. $D = <X, E, \subseteq>$, where $E \subseteq$ _Con_
is the set of those finite consistent subset of X which are
elements, for which Scott gives the following axioms:

> (1) all finite subsets of α are in _Con_;
> (2) if $\underline{a} \subseteq \alpha$ and $\underline{a} \vdash x$, then $x \epsilon \alpha$ (closure of α
>   under entailment).

The set E is partially ordered by the set-theoretic inclusion
relation in quite a natural way, for "elements" $\alpha \epsilon E$ _are_ sets.
This partial ordering has a bottom element which is {X}, the
singleton of X itself, and, in case all the finite subsets of X
are consistent, it has also a top element, which is the set X
itself. If D has both a top and a bottom element, then it is a
complete lattice under "$\subseteq$". Any set $\underline{a} \epsilon$ _Con_ generates an element:

$$\overline{\underline{a}} = \{ x \epsilon X \mid \underline{a} \vdash x \},$$

and any $\alpha \epsilon E$ is the limit of its finite approximations:

$$\alpha = \bigcup \{ \overline{\underline{a}} \mid \underline{a} \epsilon \text{_Con_} \ \& \ \underline{a} \subseteq \alpha \}.$$

The theory of Domains offers a set-theoretic semantics for our
"semantic network": single data objects denote singletons (re=
member that all singletons belong to _Con_), and frames denote

sets of data objects: they both denote "elements", singletons
or sets. "Descriptions" are sets of data objects as well, and
they are generated by frames, so that "descriptions" are "ele=
ments" generated by other "elements". In Denotational Semantics
we are able to assign a "meaning" to the procedure, if it does
exist, which generates "small worlds" starting from a given
"information system": if this procedure can be defined as an
algorithm, then it denotes a recursive function that is amapping
from D onto itself; we can give recursive definitions of domains
of this kind:

$$D = [D \rightarrow D].$$

The "elements" of these recursive or "reflexive" domains are
the approximable functions $\phi \epsilon |D \rightarrow D|$. These functions have a
least fixed point which is taken, in Denotational Semantics, to
be their "meaning". Therefore, we can endow BDM with an "infor=
mation processor" that is an algorithm that denotes a function
$\phi \epsilon |D \rightarrow D|$. BDM is to be represented now by a serial decomposition
into machines M_0, M_1, and M_2. The sub-machine M_0 receives as an
input a sentence, it generates as an output a set of "descrip=
tions" and this set is fed to sub-machine M_1 that checks, as
before, if that sentence is true in those "partial state descrip=
tions". BDM is now a machine M such that $M = M_0 \otimes M_1 \otimes M_2 = <X, Y, Z_0$
$Z_0 \times Z_1 \times Z_2, \lambda, \delta>$.
It is worth while to do a last remark: the H-information is a
measure associated to the "relative complexity" of choice sit=
uations. When two sentences representing alternative choice
options are fed to BDM, the difference in the measure H asso=
ciated to the two subalgebras generated by the "information
processor" could be a measure of their comparative "complexity".
This possibly introduces a new parameter to be considered in
making a decision, besides expected utility .

THE USE OF WEAK INFORMATION STRUCTURES IN RISKY DECISIONS

Sergio F. Garriba
Politecnico di Milano
Milano, Italy

Abstract: Decision making in presence of weak information raises the problem of selecting the best decision model. The choice ought to be made according to pragmatic criteria. Specifically, the process of risk assessment entails various decision steps where fuzzy algebra with related generalized measures of uncertainty and preference ordering may be helpful. Two simple examples are given to show how these decision models can find applications.

Keywords: Decision model, Fuzzy algebra, Possibility measure, Risk, Non-destructive testing, Environmental damage.

1.   Premise. A risky decision means in the following that a solution is looked for, while economic and social costs are at stake. Accordingly, there will be a pragmatic equilibrium between what we can and what we wish to do in decision making. This equilibrium will be influenced first by the magnitude of the cost vs. benefit balance associated with the industrial process or any other activity, second by the cognitive structure of the problem.

It has been variously argued that decisions cannot base upon complete information regarding the facts they address. Every information which is added makes use of further explanatory considerations (data and models), which serve as justifications and are left unexplained. Therefore decisions are always conditional upon a state of knowledge. We are led to recognize that there is no reason to expect that technical perfectibility is possible, even in principle, and even if we have achieved it, we would not be able to claim success with warranted confidence.

Prescriptive and descriptive decision making strategies are generally taken to be contrasting. They share, however, the assumption that normative decision models explain what subjects ought to be doing and what they would be doing, where it not for cognitive limitations. In practice the more normative or informative a decision model is, the less secure it is. Conversely, the less normative, the more secure it appears.

All other things being equal, a degradation relation could describe the domain of feasible decisions. The security of decision models we can adopt declines with the sharpness to which we aspire (Fig.1). If one insists in making decisions with high resolution, one's confidence in the outcome should be drastically reduced. The crucial point is that we want to operate on the right-hand side of the diagram of the figure, since we always strive for the maximal achievable definiteness in decision making. The ever-continuing pursuit of increasing

NATO ASI Series. Vol. F21
Intelligent Decision Support in Process Environments
Edited by E. Hollnagel et al.
© Springer-Verlag Berlin Heidelberg 1986

sharpness, widened comprehensiveness while preserving security, is the motive force behing research of the best decision model.

2. The selection of the decision model. If information is abundant and external requirements are low, the solution to the problem of metadecision, in other words the choice of the best decision model could be straightforward. This may not be the case when available information is inherently qualitative and sparse, interpretative models are shaky, new data do not just supplement but generally upset knowledge-in-hand. We refer to this occurrence as a case of weak information.

In principle, a formal decision model relies upon two components, i.e. some kind of evidence logic and a theory of preferences. In its turn, evidence logic contains an algebra to represent events and a measure of uncertainty. In the probabilistic decision model, events are expressed in terms of crisp algebra, uncertainty is captured by subjective probability, and preference ordering can be variously expressed by means of utility theory (Keeney and Raiffa, 1976). The situation looks more complicated in the case of weak information where events have imprecise boundaries. Fuzzy algebra would allow to represent ill-specified and not distinct collections of objects with uncrisp boundaries in which the transition from membership to non-membership in a subset of a reference set is gradual rather than abrupt. Clearly, probability measure and utility theory no longer apply and should be substituted by other types of (generalized) measures of uncertainty and preference classification. A similar situation would occur when probability and/or utility functions cannot express the subjective judgment about uncertainty and/or preferences, respectively (Prade, 1985). Intriguing problems are encountered when events besides being imprecise, are vague in the sense that the truth of events cannot be expressed by the value true or false, but other values are allowed. If this is the case, it can be suggested to extend the concepts of fuzzy algebra to modal logic and possibly to emphasize the fuzzy temporal dimension entailed by the "open texture of language" (Negoita, 1985).

A prospect of unending decision models escalation lies before us. When facing a problem characterized by a weak, often qualitative information structure there will be a variety of applicable models. The decision maker would have to select the most appropriate one according to some subjective judgment in which weighing cost and resources has a great part. On the other side, it will be the task of decision analysts to show how the different models compare and relate each other.

3. Risk assessment as a decision process. Given an industrial plant or any other type of activity bearing a significant impact on the environment, risk assessment tends to reflect the goals expressed by decision maker. Four inter-related steps can be separated, namely system representation, risk

estimation, risk evaluation and risk management (Fig. 2).

System representation entails the functional and causal explanations of the interactions among the plant, its components, and the environment. Risk estimation refers to the identification of planned and unplanned events that may occur in the plant. The possible consequences on the environment, their magnitude in space and time, and associated measures of uncertainty are computed. The next step consists of risk evaluation, that is the determination of the meaning, or values of combinations of estimated consequences and uncertainties to those affected, individuals, local communities and society. Evaluation may be thought of as a method for ranking combinations of consequences and uncertainties, which means risks so that they can be compared. This method for ranking allows to set levels and to define the acceptability of risk according to some decision goals. Finally, risk management or risk control refers to actions one may take, given the evaluation of risk and the corresponding attitudes of those affected. Conflicts would be resolved by changes in the plant, in the decision process, or in the environment. The last aspect asks for special forms of participation and social choice which are beyond our consideration.

Two simple examples are now given to show how some decisions entailed by this complex process of risk assessment can be made in presence of weak information.

4.   Structural safety of a pressure vessel. The damage of a pressure vessel due to a specific crack can be described within the framework of fracture mechanics. A catastrophic failure is predicted when the size of a crack exceeds certain critical crack size C_0 depending on material properties, the applied load and the geometry of the component under consideration. Probabilistic evidence logic has been applied to predict the failure probability of the pressure vessel. The crack size distribution entering into these calculations in the form of a probability density function is obtained from histograms of cracks detected during the non- destructive testing. Due to the fact that small cracks cannot be detected, there is a lower bound in the histograms; due to the limited sample size, there is an upper bound; due the various sources of error in the measurement process the definition of detected events may become imprecise. Thus, the assessment of safety (i.e. absence of rupture) of the pressure vessel may induce high uncertainty of the results predicted.

These uncertainties, which refer to system representation and risk estimation in our nomenclature, should be looked at more closely. A descriptive measure of performance is planned by decision maker to evaluate the detection process. Specifically, it is admitted that different teams ($k = 1,...,k$), all using the same procedures and techniques in non-destructive testing are confronted with the same population of defects. The distribution of these defects is fully known to the decision maker. Interest lies in selecting the best team who has the ability of detecting defects with minimum errors.

Cracks are identified by measuring their length (X) and

location (Y). Cracks, however, have a rather complex geometry and orientation. Therefore crack detection event is imprecise and should be treated as a fuzzy event. Errors made by each team can be given in terms of possibility and necessity distributions, say $\Pi_k(X)$, $\Pi_k(Y)$, and $N_k(X)$, $N_k(Y)$, respectively. Use of possibility and necessity logic seems particularly workable in this situation (Dubois and Prade, 1985). Intersection of the single team possibility distributions Π_k allows to express imprecision connected with measurement chain, $\Pi_M(X) = \min \{\Pi (X_k)\}$ $\Pi_M(Y) = \min \{\Pi (Y_k)\}$. Conversely, union of Π_k would signify imprecision due to human operators, $\Pi_H(X) = \max \{\Pi (X_k) \}$, $\Pi_H(Y) = \max \{\Pi (Y_k) \}$. The best detection team can be selected by a proper weighting of the difference $\Pi_k - \Pi_M$. To this aim it is observed that risk aversion may play a role. Errors concerning X and Y have a different meaning and large errors may induce unbearable hazards. A fuzzy two-attribute utility function can be elicited according to the additive form $u(X,Y) =$ $= r\ u(X) + (1-r)\ u(Y)$ with fuzzy scaling constant r (Seo and Sakawa, 1985). If decision relies upon the maximum (subjective) expected preference, one has to look for $\max_k \{E (u)\} =$ $\int u \cdot (\Pi_k - \Pi_M) \, dXdY$.

As soon as defect population is known in terms of possibility and necessity distributions, it would be feasible to estimate the rupture hazard by adopting a crack propagation model. A fuzzy random process ensues and C_o-level upcrossing would be determined by means of associated possibility and necessity process outcomes, with reference to the time interval of interest in the life of the pressure vessel. It is obvious that crack propagation model itself is affected by uncertainties which ought to be included in the final estimate.

5.   Environmental damage produced by acid rain.

Decision analys methods can be applied to determine the impact of acid rain as the risk evaluation step of a comprehensive control policy involving a number of different countries. In this case, the decision analysis methodology should permit the evaluation of policy options according to different decision criteria, thus reflecting the variations among state governments.

A model can be built basing upon a receptor-oriented approach. Aiming at the reduction of the problem of multimensional environmental impact of acid rain into manageable components, it is chosen to focus on forest damage as a single receptor type. Some state governments, however, feel secondary "acid rain" concerns, other than forest damage. Attention of some countries focuses indeed on aquatic ecosystems, agriculture, ancient monuments and buildings. Furthermore, the perception of the size of damage changes in time and the diachronic and "vague" aspects of the decision problem appear stressed. We may be led to think of forest damage as a fuzzy event, where membership functions relate to forest area affected and to fuzzy tense truth values. In this frame, it would be perhaps feasible to capture the temporal dimension of the events in a time contraction (Gardies, 1975). Possible damage is measured measured at present time or perceived in the future. Conversely, necessary damage is measured at present time and assumed to remain, in absence of corrective actions. The possibility and necessity attributes would imply two different distributions which may set upper and lower bounds in the evaluation of "acid rain" control policies.

Acknowledgment. Work partially supported under a grant awarded by Italian Ministry of Public Education concerning analysis of uncertainty of accidental transients in nuclear power plants.

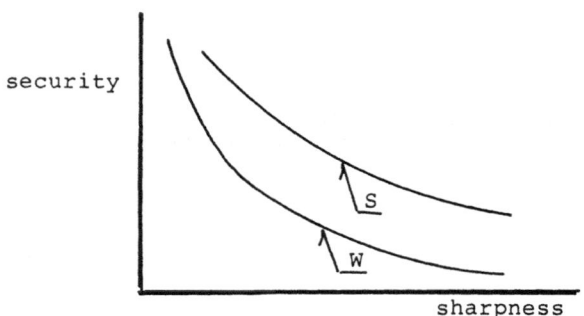

Fig. 1 - The degradation of security of a decision model with
increasing sharpeness for the cases of strong (S) and
weak (W) information structure. Arbitrary scales and
units.

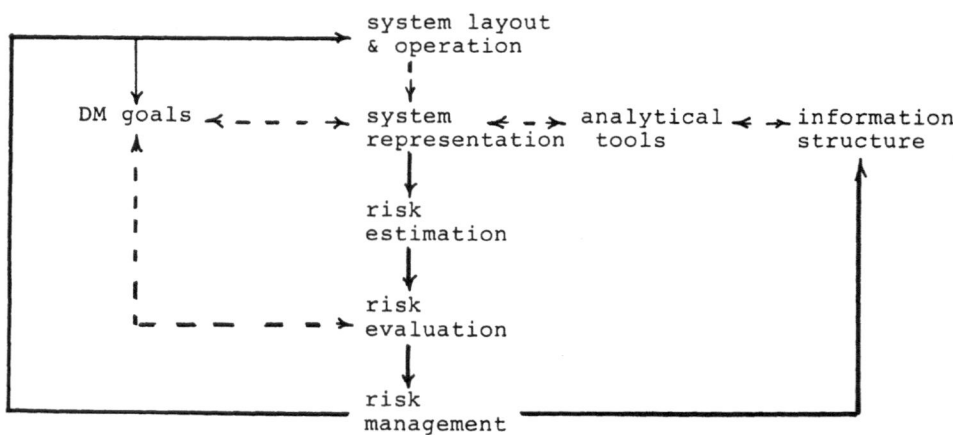

Fig. 2 - The main decision-making (DM) steps in risk assessment for a generic industrial plant. Continuous and
dashed lines show decision and information flows,
respectively.

TIME AND DECISION

Giuseppe Volta
Commission of the European Communities
Joint Research Centre — Ispra Establishment
Ispra (Varese), Italy

Summary

The lecture deals with the problem of modelling time in decision processes. It is divided into three parts. The first part gives a survey of the concept of time in the history of thought and three concepts are identified : physical, perceived, structure of possibility. Then a short survey of the consideration given to time in ancient and modern logic. It is shown how modern model logic has produced a number of formalized languages which take into account temporal evolution and is providing a substantial back-up to the simulative languages of A.I. The second part analyses the time structure of decision making. With reference to a model "decision maker - system - ambient", the times of the three terms of the model are discussed. The time of d.m. is shown to present all three aspects identified in the history of thought. The time of the system, or time horizon of the decision, is discussed under the aspects of perceived versus objective time, values in time, advantages and disadvantages of long-term vs. short-term decisions. The time of the ambient is considered in relation to the temporalized complexity of the world. The third part deals with the connections of time with uncertainty and with rationality. Uncertainty is discussed from the logical and informational points of view. A link is suggested between time and the fuzzy presentation of future events. Strategies for splitting the time horizon to better model uncertainties are also examined. Then the connection between the consideration given to time and the type of rationality adopted, sequential or synoptic, are discussed.

NATO ASI Series, Vol. F21
Intelligent Decision Support in Process Environments
Edited by E. Hollnagel et al.
© Springer-Verlag Berlin Heidelberg 1986

Introduction

1.   This Institute has been organized to study the decision process of a man placed in a technological environment and to discuss tools, methods, ideas that could help a man to make "good decisions". But what is a good decision? Before a decision has been made, can we know if it is good or not? If we know it, a decision is no more a decision: it has just the status of an automatic response. But if we cannot say, "before", if a decision is good, what can we do to avoid that, "after", we regret the consequences of the decision we have made? How can we evaluate correctly the value of waiting before making a decision? How can we consider the information that we do not yet have, but could have later, that could change the state of the world in an unexpected way ?

This type of simple but not trivial problems (Shackle, 1961) stems from the fact that decisions connect different things in time or, we could say, are embedded in time. A decision, in fact, is a separation (de-caedere), a cut in time : before and after are different "because of" that decision.

2.   Humans live in two phenomenological dimensions: space and time. But these dimensions are intrinsically different. Our body of knowledge which is the background for any decision, is certainly imbedded in time more deeply than in space. The philosophy of science has studied the problems of time much less than the problems of space. Currently, time has been considered and studied with reference to space, loosing in this way its specific characteristics.   While   the   conception   of   space   and   time   as   a quadridimensional structure was very fertile for the physico-mathematical sciences, this conception had negative effects (Reichenbach, 1958) on the comprehension of the nature and of the role of time in human knowledge and behaviour.

The experience of time is completely different from the experience of space. No specific sensory organ for time exists. Time, in contrast with space, has no symmetry, cannot be reversed, cannot have a geometry. The perception of time is developed in children later than the perception of space (Piaget, 1946). Moreover, the perception and consideration of time involves some special mental effort. The construction of a time structure is considered by some psychologists as an effort of adaptation to a changing world (Fraisse, 1957). This could explain why the world of dreams is timeless.

If decision is intimately linked with time, some elements of the complex and unique nature of time should be found in the structure of decision processes.

Time in the history of thought

3.   In the history of thought we can identify three main concepts of time: a measure of the order in which things move or change;   an experience of change;   a structure of the possibilities of existence.   Without time we cannot talk of possibility.

The first concept was first ennunciated clearly by Aristoteles: "time is number of motion in respect of "before" and "after" (Fis.,   IV, 11; 219 b1). It is the concept developed by the philosophers of nature: Newton, Kant and, more recently,   Reichenbach.   This concept establishes a correspondence between the order of time and the order of causality.   The time of physics, the so-called scientific time or objective time, the time of clocks, has its roots in this concept.

The second concept reduces time to an internal experience of the human conscience,   it is the flowing stream of the conscience itself. According to this concept,   time is a continuum "solitary present" (Shackle,   1961).   The red line linking together the people who have developed this concept passes through Plotinus,   Agustinus up to Bergson and Husserl.   Bergson has particularly stressed the difference between lived time and scientific time. Scientific time according to its criticisms,   "spatializes" time,   reducing his characteristics to the characteristics of space and loosing the sense of its true nature. This has been the concept mostly analysed by psychologists. "Phenomenological time" has been opposed to the "physical time",   and it has been found that the former has its own internal "temporal structure" corresponding to the perception of the past (short-term and long-term memory), of the present and of the future (Ornstein, 1969).

The third concept has been developed recently by the existentialist philosophers, mainly by Heidegger. One can find the root of this concept in the old Stoic and Megaric philosophy, that first introduced, with Diodorus Cronos, a temporal logic as an alternative to modal logic. The basic idea of the "master argument" of Diodorus is that "is possible what is true now or in future", assuming in this way the equivalence of time and possibility.

It is interesting to note that three fundamental works, sustaining the three different concepts of time above outlined, were published in the same year, 1928, in Germany: Reichenbach, "Philosophie der Raum-Zeit-Lehre"; Husserls, "Zur Phänomenologie des Inneren Zeitbewußtseins"; Heidegger, "Sein und Zeit". One can deduce that the three concepts correspond to three aspects of the same human experience of time.

4. A particular mention merits the attention to time in the history of logic.

Time in logic had a happy life in antiquity and a troubled life in recent history.

The greek and medioeval philosophers, together with assertive, true-false, logic also developed modal logic in which the concepts of "possible" and "necessary" were introduced. Stoics and megaric philosophers have in particular interpreted the concepts of necessary and possible in terms of existence conditional on time. For them: "p is (now) possible if and only if p is true now, or will be true in a future time"; "p is (now) necessary if and only if p is and will be always true" (Bochewski, 1956).

Modal logic, in its temporal version, allowed one to state that something true in a given moment can be false in another moment, and that the sense of a verb has a logic significance. The full acceptance of time in formal reasoning remained a characteristic of medioeval culture. But after the Renaissance modal logic and its temporal interpretation was progressively set aside. The XIX century and the first half of the XX century have been dominated by a rigorously atemporal logic, that, for a strange deformation of the historical perspective, is called "classical". The greater representative of this classic logic is Russel. An explanation of this turning point can be found in the tendency to extend to all the sciences, including the human sciences, the methodological approach that has proved so successful in physics. This approach assumes that all the laws of nature are functions, in mathematical sense, in which time enters with the same statute and characteristic of space (spatialization of time) (Delattre, 1982).

Under the pressure of the need for adequate representations of human behaviour, starting in the 1950s, a true revolution in logic is observed, with the revival of modal logic, and of its specialization in the direction of temporal logic, deontic logic, epistemic logic, etc. This revolution is associated with the names of C.F. Lewis and S. Kripke (modal logic), A.N. Prior (time logic), G.H. Von Wright (deontic and epistemic logic). This revolution is against a paradigm of classical logic that Prior calls the principle of the "comprehensive objectivity": all what exists in a moment has in every other moment a mysterious existence which is sufficient to qualify it as an object of reference, or in other words, the truth value of a proposition cannot be affected by the circumstances.

5. It is impossible, in this short review, to enter into the details of the problems that this revolution has raised. We can say only a few words about the impact of these logics on the formal representation of human reasoning.

The first product of these logics is a number of formal languages, with connectives that establish relations in time. For instance, Prior introduced in addition to the standard connectives of the predicate calculus (conjunction, disjunction, negation, implication, equivalence), the connectives: P (it has been true that), H (it has been always true that), F (it will be true that), G (it will be always true that). The proliferation of these languages represent a substantial back-up to the knowledge representations in Artificial Intelligence, as the literature in this area is showing (Allen, 1984), (Quinlan, 1985).

The second product of these logics is the recognition that coherent reasoning should not necessarily rely on a unique and rigid axiomatic support. Some axioms like the "excluded middle principle" axiom (in the form $A \lor \sim A$) can be left out, obtaining, in this way, a better adherence to human behaviour. In fact, the separation of formal classical logic from real reasoning has been pointed out by various authors (Blanché, 1967).

Time in decision making

6.   The term decision means different things and has, at this point, to be clarified.

In logic, decision means a procedure or an algorithm by which one can determine if a formula is a theorem or not. It is the explicitation of the deductive process that leads from some initial premises to some necessary conclusion.

But when dealing with human behaviour in the most general sense, for instance in economy, in policits, in day-to-day life, etc., decision means a judgement that has to be followed by some action, a design of an action which intends to predetermine the future. This design can be a simple deduction from some premises, and in this case corresponds to the decision of logic. But it can also be something more creative. The design of an action in a "non empty" decision (following Shackle's wording), is not just "pushed" by premises but it is also "pulled" by the goal that the decision maker has intentionally formed.

At the interface with machines, man is asked to take various types of decisions, varying from the simple educated response to a set of external stimuli, to some complex deductive reasoning, up to some true decisions. In the following we will call "mode 1" decisions the decisions which are just the result of deductive reasoning and "mode 2" decisions, the true decisions, which have some creative content.

7.   The representation of the decision process, to which we will refer in order to analyse the time structure, is given in the figure :

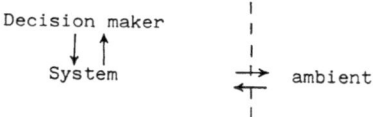

This simple model supposes that a decision maker is in front of the world and cuts the world in a part which is concerned with his decision, "the system", and a part which is not, "the ambient". The arrows represent essentially communications that guide the behaviour of the interacting parts, i.e. are manifestation of power (Luhmann, 1975). The arrows, that link the system with the ambient, do not directly affect the decision maker who, to be able to decide, being finite, has to sever the system from the ambient. The d.m. can respond questions coming from the system and in this case he decides in mode 1; or he can respond to internal demands (goals, intentions) and from the system, and in this case he decides in mode 2.

With reference to this model we can identify three times: the time of the d.m., the time of the system und the time of the ambient.

8. The time of the d.m. has a complex structure and presents all the three aspects we have summarized in paragraph 3.

First there is physical time that the mental process, supported or not by artefacts like computers, requires to retrieve information and elaborate judgements. The limitations of the human mind in handling information are well-known. A general principle of economy of thinking that governs our mental processes has been also proposed since the time of Avenarius (1876).

Intrinsic limitations, and the principle of "economy", imply together a finite time for any kind of judgement and a trade-off between the time which is considered acceptable, and the boundaries of the system (cut between ambient and system).

Second, there is the internal experience of the duration of the decision process which is independent of clock time. In order to accommodate or to adapt personal reality to the external changing world, man strives to structure his internal experience or subjective time to the objective time. But this, as we have already mentioned, implies an effort. And the decoupling of the two times can be activated for instance by stress or mental loading conditions.

In true decisions, or mode 2 decisions, when man exploits his more profound metal resources and his intentional capacity, the third aspect of time, as structure of possibility, is also emerging. This structure allows arousal of those goals that pull, as a target placed in the future, the decision process.

9. The time of the system is a "represented" time. It corresponds to the imagined system evolution that the d.m. assumes in his decision. It is

bounded by the time horizon.

Psychologists (Fraisse, 1957) have found that the individually perceived time horizons are constructed on the basis of personal recollections, so that they change with age and personality. For instance with the age the importance given to the past is increasing and that given to the future is diminishing. The congruity between the time horizon represented in decision and perceived time horizon, can be a problem. It happens that in some formal process of decision, mention is made of times of hundred or more thousand years. The congruity of what is represented and of what is perceived, in that case, is dubious. Moreover, time horizon, like memory, has an internal structure: the first hour after the decision has not the same value of the thousandth hour. In this respect, it has been suggested that time is intensional in its nature more than extensional.

The choice of the time horizon can be, to a certain extent, arbitrary, provided that it is perceivable. However, there are at least two constraints: values and trade-offs between advantages and disadvantages of long-term and short-term decisions.

About values one can assume a stable set of reference values in time or one can assume a "discount rate" or some other rule for changing values in time. The constance of values in time is sustained for instance in the name of justice by Rawls (Rawls, 1971), but it is not assumed in economics. However, in economics the time horizon is short and the meaning of the discount rate also questioned (Lind, 1980).

About the long-term and short-term decision one has to consider the possibility of shortening the time horizon, choosing a multistep decision process which has the advantage of higher flexibility. But this flexibility is real only if: a) the consequences of the decision on the system are reversible; b) the feedback channel of communication from the system to the d.m. introduces a delay much shorter than the single step. Therefore the choice of the time horizon, the type of decision, and the type of communication channel that exists between system and d.m., are not independent. Modern developments in computer science and telecommunications have introduced dramatic changes and improvement in the capabilities of information channels. "Real time" information is a possibility in many practical situations. This fact could help to shorten the time horizon for decision and to support multistep decisions rather than "one off" decisions.

10. The time of the ambient in a first approximation is neglected as is the existence of the ambient itself. This assumption is acceptable when the time horizon is short. But when it becomes larger some attention should be paid also to the ambient.

The cut between a system, a quite small portion of the world, and the ambient has been introduced in the reference model, in order to reduce the complexity of the world which is presented to the d.m. But this complexity is a temporalized complexity in the sense that time allows a continuous evolution of the selective interconnections between elements and a continuous production of new overall configurations (Luhmann, 1980).

A cut, which is a reduction of complexity and, as such, gives sense to a decision at a given time, will be in general much less meaningful at a later time. It is as if time extends the boundary of the system. To be more realistic one could consider a near field ambient that will be progressively included in the system as long as the time horizon is increasing.

Decision models, uncertainty and rationality

11. In the prevailing models and theories of decision making, only marginal attention has been given to most of the aspects outlined in the previous chapter. Decision models are usually divided in two broad categories: normative (or prescriptive) and behavioural (or descriptive).

Normative models have their prototypical example in expected utility (E.U.) or bayesian decision theory. E.U. is a theory of individual decisions, built up progressively, starting in the 17th century, and formalized in this century through the work of Ramsey, De Finetti, Savage and von Neumann. This theory assumes the existence of a set of possibilities complete and perfectly known (a crisp set), and assumes that the decision maker follows some "rationality" axioms (von Neumann and Morgenstern, 1953), the most important of them being: the decidability axiom, the transitivity axiom, the continuity axiom, and the sure thing axiom.

The model built around these axioms ignores time in the sense that it ignores learning about the boundary of the set of possibilities, about new values and preferences, etc. (Volta, 1983).

Behavioural models take into account time but with few exceptions (Shackle, 1961; Simon, 1957 and 1982), only implicitely. These models focus mainly on individual and cultural differences in decision-making (Wright, 1984) and on heuristics and bias in assigning a measure to uncertainty (Kahnemann, 1982). In between the two classes of models one can find in the literature various empirical approaches to relax the rigidity of the axioms of normative theory with the main aim of introducing some flexibility to take in account the a priori unknown flow of information.

The measure of uncertainty and the choice of a rationality principle, are common problems to all these models. The ways in which these problems are approached is tightly connected to the relevance given to the time dimension.

12. The connection between time and uncertainty can be considered from various poins of view. Let us take first the point of view of logic.

Uncertainty can concern past events and future events, but in the two cases the uncertainty is intrinsically different. About the past we are uncertain because of our knowledge limitations, but past events are unique and well defined. About the future we are uncertain because of our limitation, but also because future events belong to alternative possible worlds, only one of which will become real, when the arrow of the present will cross it. So future events, seen by an observer, which is the decision maker, present an uncertainty which is true ignorance. Moreover, to the extent that we label events with their time of occurrence, we have an ordered and a nested structure linking them. These remarks support the choice for the representation of future events of fuzzy sets, instead of ordinary crisp sets, and the use of fuzzy measures for uncertainty instead of probabilities (Dubois, 1985). Incidently we point out the tight links existing between the theory of fuzzy measures for uncertainty and the modal logic, from which the concepts of possible, necessary, plausible, etc. are originally derived.

13.   From an informational point of view, it is trivial to say that time changes information and in general reduces uncertainty. This fact raises, within the boundary of a given decision time horizon, the possibility and the interest to divide this interval and to consider various strategies for exploiting the information potential of time during it.

The simplest strategy is delaying the decision. However, this strategy, as modelled in the classical preposterior bayesian analysis, makes only apparent use of the informative value of time. In fact, acts subsequent to the first choice, after gathering information, are treated as perfectly predictable, conditional on explicitely modelled uncertainties (Brown, 1974).

Another strategy is to divide the time horizon 0-T into a short interval 0-τ and in a long interval τ-T (Parrinello, 1981). The decision is made on the short interval. But the alternative acts, amongst which the choice is made, are qualified by the set of options that remain still open at time τ for the time τ-T; i.e. the choice is made amongst "acts-with-ambient". This strategy implies the choice of a metrics for the "ambient" (values and uncertainties of the options remaining in τ-T) metrics that is decoupled by the model adopted for the short term alternatives.

The splitting of time to reduce or to better structure uncertainty is tightly linked to the type of rationality one adopts.

14.   "Rationality denotes a style of behaviour that is appropriate to the achievement of given goals, within the limits imposed by given conditions and constraints" (Simon, 1982, p. 405). According to this definition, rationality can be considered a guidance to decision, a metadecision.

Rationality selects the "appropriate" models that allow, given some premises, finding the way to reach the goals. Goals are "given", as also the conditions and constraints are "given". The former are given by a "metarationality" that sets values. The latter are imposed by the nature, by the environment.

The relevance given to time qualifies the type of rationality. The decision process can be, in principle, subdivided into finite segments corresponding to the minimum span of time needed to modify significantly the state of information of the decision maker. This possible subdivision offers the choice between a segmented and "sequential" decision process or a "synoptic" or "monolithic" decision process. These two alternatives imply a different allocation of resources. A sequential process emphasises the continuous search for information and implies also the cost of this search. A synoptic process concentrates all the effort to find and select information at the beginning. In the first type of rationality one postpones all the choices that are not strictly relevant to the present partial decision step. In the second type of rationality one adopts a rigid framework. The first type of rationality focuses on the loss of options that irreversible decisions entail, while the second type emphasizes the benefits of an irreversible decision. This second type of rationality is well examplified by Descartes (Discours de la méthode, troisième partie): if you are in a forest, maintain the straight direction you have chosen at the beginning.

15. To have some practical references we can connect these two types of rationality to two models developed for multiobjective decision making: the first is the interactive technique for satisfying decisions via multiobjective optimization, developed recently by Wierzbicki et al. (Wierzbicki, 1983) as a development of the goal programming approach (Lee, 1972); the second is the classical (in the sense of the classical logic) multiattribute utility theory (Keeney, 1976).
The first model is centered on the decision maker. He fixes "a priori" the actual aspiration levels which satisfy him, and searches for decisions which are guided (pulled) by the aspiration levels. The aspiration levels (reference point) are used to explore a limited region of the potential choices, to improve the use of resources but not to find a global optimum.
The second model is an extension of expected utility theory. It is centered on the system and it is directed to search an optimal decision.
In the first model the time horizon can be interactively modified step by step. In the second model the aim of a global optimality implies the assumption of the largest time, compatible with the analysis capability.

Conclusions

16.   Time is a fundamental dimension of human experience.   As such it is part of the structure of decision processes.   The consideration of this dimension implies logical and practical difficulties.   This explains why historically preference has been given to static,   synoptic models of decision.   This preference is connected to the preference given,   in the historical period corresponding to the scientific revolution,   to the so called classical (non-modal) logic.   The link between this historical trend with the paradigm of "scientific rationality",   typical of the physical sciences, is a matter for meditation.

In more recent years,   the rediscovery of the human dimension in technological society,   meant also the rediscovery of the time dimension, of modal logics, of broader types of measures of uncertainty and of other types of rationality centered on man more than on the system.

This new cultural perspective supports decision approaches that allow the highest flexibility and that pay weighted attention to the various steps of the time horizon. It favours also the representation of data, in general, in term of time histories. However, the dichotomy: subjective or perceived and objective or physical time,   jointly present in the process of decision, remains a challenge to our modelling capability. The congruency of these two times is essential for consistent modelling.   The strictly simulative and qualitative approach characteristic of A.I. could be a way out.

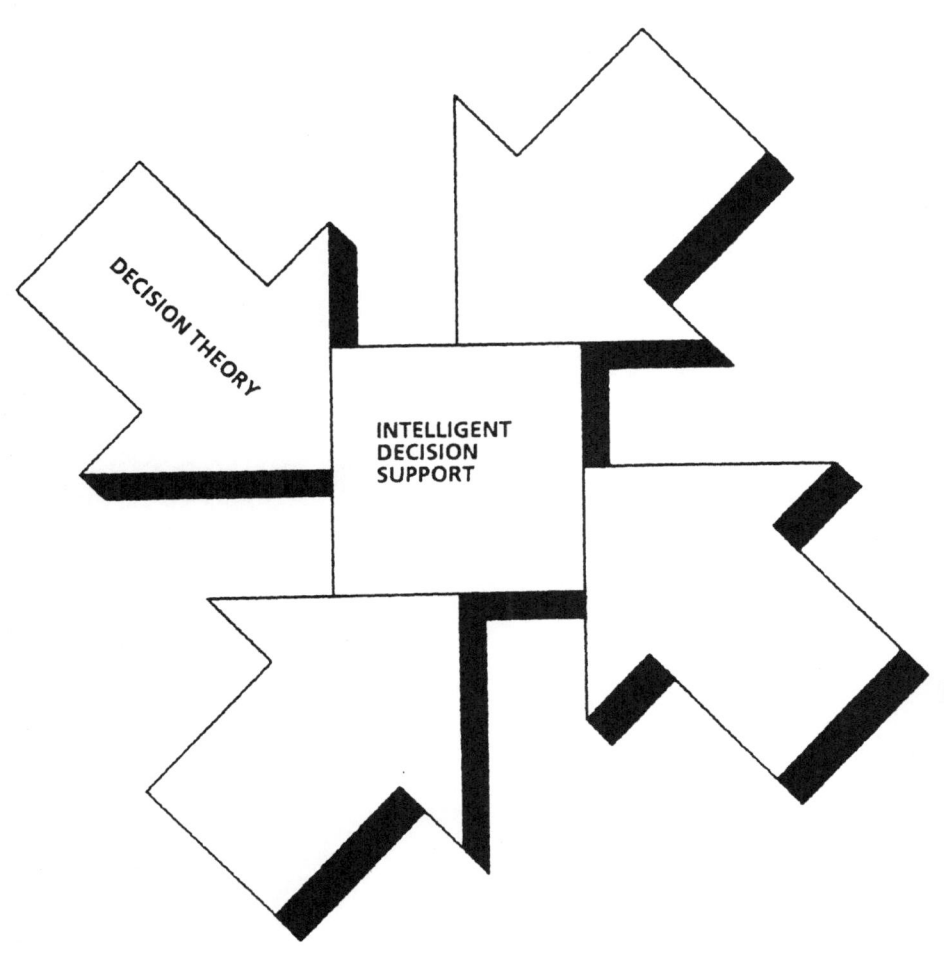

DECISION THEORY

INTELLIGENT
DECISION
SUPPORT

PSYCHOLOGY OF DECISION MAKING

DECISION MAKING IN COMPLEX SYSTEMS

Baruch Fischhoff
Decision Research
Eugene, Oregon (USA)

A Short History of Decision Aiding

It is common knowledge that decision making is problematic. One of
the clearest indications is the proliferation of decision aids, be they
consultants, analyses, or computerized support systems (Humphreys, Svenson
& Vari, 1983; Stokey & Zeckhauser, 1978; Wheeler & Janis, 1980;
vonWinterfeldt & Edwards, in press). Equally clear, but perhaps more
subtle evidence is the variety of devices used by people to avoid analytic
decision making. These include procrastination, endless pursuit of better
information, reliance on habit or tradition, and even the deferral to aids
when there is no particular reason to think that they can do better. A
common symptom of this reluctance to make decisions is the attempt to
convert decision making, which reduces to a gamble surrounded by uncertainty regarding what one will get and how one will like it, to problem
solving or pattern matching (Norman, in press), which holds out the hope of
finding the one right solution.

Somewhat less clear is just why decision making is so troublesome.
The diversity of coping mechanisms suggests a diversity of diagnoses. The
disappointing quality of the help that is obtained suggests that these
diagnoses are off target (or at least unable to engender useful solutions).
The battlefield of decision aiding is strewn with good ideas that did not
quite pan out, after raising hopes and attracting attention. Among the
aids that remain, some persist on the strength of the confidence inspired
by their proponents and some persist on the strength of the need for help,
even if the efficacy of that help cannot be established.

In retrospect, it seems as though most of the techniques that have
fallen by the wayside never really had a chance. There was seldom anything
sustaining them beyond their proponents' enthusiasm and sporadic ability to
give good advice in specific cases. The techniques drew on no systematic
theoretical base and subjected themselves to no rigorous testing.

For the past 20-30 years, behavioral decision theory has attempted to
develop decision aids with a somewhat better chance of survival (Edwards,
1961; Einhorn & Hogarth, 1981; Pitz & Sachs, 1984). Its hopes are pinned
on a mixture of prescriptive and descriptive research. The former asks how
people should make decisions, while the latter asks how they actually do
make decisions. In combination, these two research programs attempt to
build from people's strengths while compensating for their weaknesses. The
premise of the field is that significant decisions should seldom be entrusted entirely either to unaided intuition or to automated procedures.

NATO ASI Series, Vol. F21
Intelligent Decision Support in Process Environments
Edited by E. Hollnagel et al.
© Springer-Verlag Berlin Heidelberg 1986

Finding the optimum division of labor requires an understanding of where people are and where they should be. The quest for that understanding has produced enough surprises to establish that it requires an integrated program of theoretical and empirical research. Common sense is not a good guide to knowing what makes a good decision or why it is hard to identify one.

Initially, behavioral decision theory took its marching orders from standard American economics, which assumes that people always know what they want and choose the optimal course of action for getting it (Edwards 1954; Fischhoff, Goitein & Shapira, 1982; Schoemaker, 1983). Taken literally, these strong assumptions leave a narrow role for descriptive research: finding out what it is that people want by observing their decisions and working backward to identify the objectives that were optimized. These assumptions leave no role at all for prescriptive research, because people can already fend quite well for themselves. As a result, the economic perspective is not very helpful for the erstwhile decision aider—if its assumptions are true.

However, the perceived need for decision aiding indicates that the assumptions are not true—people do need help. The first, somewhat timorous, response of researchers to this discrepancy between the ideal and the reality was to document it. It proved not hard to show that people's actual performance is suboptimal (Fischhoff et al., 1982; Kahneman, Slovic & Tversky, 1982). Just knowing the size of the problem can be helpful in several ways: It can show how much to worry, where to be ready for surprises, where help is most needed, and how much to invest in that help. However, size estimates are not very informative about how to make matters better.

Realizing this limitation, researchers turned their attention from what people are not doing (making optimal decisions) to what they are doing and why it is not working. Aside from their theoretical interest, such psychological perspectives offer several points of leverage for erstwhile decision aiders. One is that they allow one to predict where the problems will be greatest by describing how people respond to different situations. A second is that they help decision aiders talk to decision makers by showing how the latter think about their tasks. A third is that they show the thought processes that must be changed if people are to perform more effectively. Although it would be nice to make people over as model decision makers, the reality is that they have to be moved in gradual steps from where they are now.

As behavioral decision theory grew, it generated an adjunct group of practitioners, hoping to make a living through behaviorally informed decision aiding. "Decision Analysis" (Behn & Vaupel, 1982; Brown, Kahr & Peterson, 1974; Raiffa, 1968) is the generic name for these methods which create models of the problems faced by individual decision makers and, then, rely on the formal procedures of decision theory to identify the best course of action in each. These methods were descriptive in the sense of trying to capture the subjective reality faced by the decision maker and prescriptive in the sense of providing advice on what to do. The behavioral research affects the procedures used to elicit information from decision makers, the credence given the information that they provide, and the division of labor among the decision maker, decision analyst and formal model.

Like other good-looking products, decision analysis has taken on a life of its own, with college courses, computer programs, and consulting firms. Its relative success and longevity may owe something to the initial attention paid to its behavioral foundations. That research probably helped both by sharpening the technique and by giving it an academic patina that enhanced its marketability. Moreover, there is still a flow of basic research looking at questions such as, Can people assess the extent of their own knowledge? Can people tell when something important is missing from the description of a decision problem? and, Can people describe quantitatively the relative importance of different objectives (e.g., speed versus accuracy)?[1]

The better work in the field, both basic and applied, carries strong caveats regarding the quality of the help that it is capable of providing and the degree of residual uncertainty surrounding even the most heavily aided decisions. Such warnings are essential, because it is hard for the buyer to beware. People have enough experience to evaluate quality in toothpaste and politicians. However, it is hard to evaluate advice, especially when the source is unfamiliar and the nature of the difficulty is unclear. Without a sharp conception of why decision making is hard, one is hard put to evaluate attempts to make it better (Fischhoff, 1980).

Why Is Individual Decision Making So Hard?

According to most prescriptive schemes, good decision making involves the following steps:

  a. Identify all possible courses of action (including, perhaps, inaction).

  b. Evaluate the attractiveness (or aversiveness) of the consequences that may arise if each course of action is adopted.

  c. Assess the likelihood of each consequence actually happening (should each action be taken).

  d. Integrate all these considerations, using a defensible (i.e., rational) decision rule to select the best (i.e., optimal) action.

The empirical research has shown difficulties at every step of this way:

Option generation. When they think of action options, people often neglect seemingly obvious candidates. Moreover, they seem relatively insensitive to the number or importance of the omitted alternatives

(Fischhoff, Slovic & Lichtenstein, 1978; Pitz, Sachs & Heerboth, 1980).
Options that would otherwise command attention are out of mind when they
are out of sight, leaving people with the impression that they have analyzed problems more thoroughly than is actually the case.

Those options that are noted are often defined quite vaguely, making
it difficult to evaluate them precisely, communicate them to others, follow
them if they are adopted, or tell when circumstances have changed enough to
justify rethinking the decision.[2] Imprecision also makes it difficult to
evaluate decisions in the light of subsequent experience, insofar as it is
hard to reconstruct exactly what one was trying to do and why. That
reconstruction is further complicated by hindsight bias, the tendency to
exaggerate in hindsight what one knew in foresight (Fischhoff, 1975). The
feeling that one knew all along what was going to happen leads one to be
unduly harsh on past decisions (if it was obvious what was going to happen,
then failure to select the best option must mean incompetence) and to be
unduly optimistic about future decisions (by encouraging the feeling that
things are generally well understood, even if they are not working out so
well).

Value assessment. Evaluating the potential consequences might seem to
be the easy part of decision making, insofar as people should know what
they want and like. Although this is doubtless true for familiar and
simple consequences, many "interesting" decisions present novel outcomes in
unusual juxtapositions. For example, two potential consequences that may
arise when deciding whether to dye one's graying hair are reconciling
oneself to aging and increasing the risk of cancer 10-20 years hence. Who
knows what either event is really like, particularly with the precision
needed to make tradeoffs between the two? In such cases, one must go back
to some set of basic values (e.g., those concerned with pain, prestige,
vanity), decide which are pertinent, and determine what role to assign
them. As a result, evaluation becomes an inferential problem (Rokeach,
1973).

The evidence suggests that people have trouble making such inferences
(Fischhoff, Slovic & Lichtenstein, 1980; Hogarth, 1982; National Research
Council, 1982; Tversky & Kahneman, 1981). They may fail to identify all
relevant values, to recognize the conflicts among them, or to reconcile
those conflicts that they do recognize. As a result, the values that they
express are often highly (and unwittingly) sensitive to the exact way in
which evaluation questions are posed, whether by survey researchers, decision aids, politicians, merchants, or themselves. Formally equivalent
versions of the same question can evoke quite different considerations and,
hence, lead to quite different decisions. To take just three examples, (a)
the relative attractiveness of two gambles may depend upon whether people
are asked how attractive each is or how much they would pay to play it
(Grether & Plott, 1979; Slovic & Lichtenstein, 1983); (b) an insurance
policy may become much less attractive by calling its premium as a "sure

loss" (Fischhoff et al., 1980; Hershey, Kunreuther & Schoemaker, 1982); (c) a risky venture may seem much more attractive when described in terms of the lives that will be saved by it, than in terms of the lives that will still be lost (Kahneman & Tversky, 1979).

People can view most consequences in a number of different lights. How richly they do view them depends upon how sensitive the evaluation process is. Questions have to be asked some way and how they are may induce random error (by confusing people), systematic errors (by emphasizing some perspectives and neglecting others) or unduly extreme judgments (by failing to evoke underlying conflicts). People appear to be ill-equipped to recognize the ways in which they are manipulated by evaluation questions, in part because the idea of uncertain values is counter-intuitive, in part because the manipulations "prey" (perhaps unwittingly) on their own lack of insight. Even consideration of their own past decisions does not provide a stable point of reference, because people have difficulty introspecting about the factors that motivated their actions (i.e., why they did things) (Ericsson & Simon, 1980; Goldberg, 1968; Nisbett & Wilson, 1977). Thus, uncertainty about values can be as serious a problem as uncertainty about facts (March, 1978).

Uncertainty assessment. Although people are typically ready to recognize uncertainty about what will happen, they are not always well prepared to deal with that uncertainty (by assessing the likelihood of future events). How people do (and do not) make judgments under conditions of uncertainty has been a major topic of research for the past 15 years (Kahneman et al., 1982). A rough summary of its conclusions would be that people are quite good at tracking repetitive aspects of their environment, but not very good at combining those observations into inferences about what they have not seen (Kahneman et al., 1982; Hasher & Zacks, 1984; Peterson & Beach, 1967). Thus, they might be able to tell how frequently they have seen or heard about a particular cause of death, but not how unrepresentative their experience has been--leading them to overestimate risks to which they have been overexposed (Tversky & Kahneman, 1973). They can tell what usually happens in a particular situation and recognize how a specific instance is special, yet not be able to integrate those two (uncertain) facts--most often focusing on the specific information and ignoring experience (Bar Hillel, 1980). They can tell how similar a specific instance is to a prototypical case, yet not how important similarity is for making predictions--usually relying on it too much (Bar Hillel, 1984; Kahneman & Tversky, 1972). They can tell how many times they have seen an effect follow a potential cause, yet not infer what that says about causality--often perceiving correlations when none really exist (Beyth-Marom, 1982b; Einhorn & Hogarth, 1978; Shaklee & Mimms, 1982.

In addition to these difficulties in integrating information, people's intuitive predictions are also afflicted by a number of systematic biases in how they gather and interpret information. These include overconfidence in the extent of their own knowledge (Fischhoff, 1982; Lichtenstein, Fischhoff & Phillips, 1982; Wallsten & Budescu, 1982), underestimation of the time needed to complete projects (Armstrong, in press; Kidd, 1970; Tihansky, 1976), unfair dismissal of information that threatens favored beliefs (Nisbett & Ross, 1980), exaggeration of personal immunity to various threats (Weinstein, 1980), insensitivity to the speed with which exponential processes accelerate (Wagenaar & Saginaw, 1976), and oversimplification of others' behavior (Mischel, 1968; Ross, 1977).

Option choice. Decision theory is quite uncompromising as far as the rule that people should use to integrate all of these values and probabilities in the quest of a best alternative. It should be an expectation rule, whereby an option is evaluated according to the attractiveness of its consequences, weighted by their likelihood of being obtained (Schoemaker, 1983). Since it has become acceptable to question the descriptive validity of this rule, voluminous research has looked at how well it predicts behavior (Feather, 1982).

A rough summary of this work would be that: (a) it often predicts behavior quite well—if one knows how people evaluate the likelihood and attractiveness of consequences; (b) with enough ingenuity, one can usually find some set of beliefs (regarding the consequences) for which the rule would dictate choosing the option that was selected—meaning that it is hard to prove that the rule was not used; (c) expectation rules can often predict the outcome of decision-making processes even when they do not at all reflect the thought processes involved—so that predicting behavior is not sufficient for understanding or aiding it (Fischhoff et al., 1982).

More process-oriented methods revealed a more complicated situation. People seldom acknowledge using anything as computationally demanding as an expectation rule or feel comfortable using it when it is proposed to them (Lichtenstein, Slovic & Zink, 1969). To the extent that they do compute, they often seem to use quite different rules (Kahneman & Tversky, 1979; Beach & Mitchell, 1978; Payne, 1982). Indeed, they even seem unimpressed by the assumptions used to justify the expectation rule (Slovic & Tversky, 1974). To the extent that they do not compute, they use a variety of simple rules whose dictates may be roughly similar to those of the expectation rule or may be very different (Beach & Mitchell, 1978; Payne, 1982; Janis & Mann, 1977; Tversky, 1969). Many of these can be summarized as an attempt to avoid making hard choices by finding some way to view the decision as an easy choice (e.g., by eliminating consequences on which the seemingly best option rates poorly) (Montgomery, 1983).

Cognitive assets and biases. This (partial) litany of the problems described by empirical researchers paints quite a dismal picture of people's ability to make novel (or analytical) decisions, so much so that the investigators doing this work have been accused of being problem mongers (Berkeley & Humphreys, 1982; Jungermann, 1984). Of course, if one hopes to help people (in any arena), then the problems are what matter, for they provide a point of entry. In addition to meaning well, investigators in this area have also had a basically respectful attitude toward the objects of their studies. It is not people, but their performance that is shown in a negative light. Indeed, in the history of the social sciences, the interest in judgmental biases came as part of a "cognitive" backlash to psychoanalysis with its dark interpretation of human foibles. The cognitive perspective showed how biases could emerge from honest, unemotional thought processes.

Typically, these mini-theories show people processing information in reasonable ways which often work well, but can lead to predictable trouble. A simple example would be relying on habit or tradition as a guide to decision making. That might be an efficient way of making relatively good decisions, but would lead one astray if conditions had changed or if those past decisions reflected values that were no longer applicable. A slightly more sophisticated example is reliance on the "availability heuristic" for estimating the likelihood of events for which adequate statistical information is missing. This is a rule of thumb by which events are judged likely if it is easy to imagine them happening or remember them having occurred in the past. Although it is generally true that more likely events are more available, use of the rule might lead to exaggerating the likelihood of events that have been overreported in the media or are the topic of personal worry (Tversky & Kahneman, 1973).

Reliance on these simple rules seems to come from two sources. One is people's limited mental computation capacity; they have to simplify things in order to get on with life (Miller, 1956; Simon, 1957). The second is their lack of training in decision making, leading them to generate rules that make sense, but have not benefited from rigorous scrutiny (Beyth-Marom et al., 1985). Moreover, people's day-to-day experience does not provide them with the conditions (e.g., prompt unambiguous feedback) needed to acquire judgment and decision making as learned skills. Experience often allows people to learn the solutions to specific repeated problems through trial-and-error. However, things get difficult when one has to get it right the first time.

What Can Be Done about It?

The down side of this information-processing approach is the belief that many problems are inherent in the way that people think about making decisions. The up side is that it shows specific things that might be done to get people to think more effectively.

Just looking at the list of problems suggests some procedures that might be readily incorporated in automated (on-line) decision aids (as well as their low-tech human counterparts). To counter the tendency to neglect significant options or consequences, an aid might provide checklists with generic possibilities (Beach, Townes, Campbell & Keating, 1976; Hammer,

1980; Janis, 1982). To reduce the tendency for overconfidence, an aid might force users to list reasons why they might be wrong before assessing the likelihood that they are right (Koriat, Lichtenstein & Fischhoff, 1980). To discourage hindsight bias, an aid might preserve the decision makers' history and rationale (showing how things once looked) (Slovic & Fischhoff, 1977). To avoid incomplete value elicitation, an aid might force users to consider alternative perspectives and reconcile the differences among them. Or, at least these seem like plausible procedures. Whether they work is an empirical question. For each intervention, one can think of reasons why it might not work, at least if done crudely (e.g., long checklists might reduce the attention paid to individual options, leading to broad but superficial analysis).

Modeling languages. One, or the, obvious advantage of computerized aids is their ability to handle large amounts of information rapidly. The price paid for rapid information handling is the need to specify a model for the computer's work. This model could be as simple as a list of key words for categorizing and retrieving information or as complex as a full-blown decision analysis (Behn & Vaupel, 1982; Raiffa, 1968) or risk analysis (McCormick, 1981; US Nuclear Regulatory Commission, 1983) within which all information is incorporated. However "user friendly" an aid might be, using a model means achieving a degree of abstraction that is uncommon for many people. For example, even at the simplest level, it may be hard to reduce a substantive domain to a set of key words. Moreover, any model is written in something like a foreign language, with a somewhat strange syntax and vocabulary. Successful usage means being able to translate what one knows into terms that the modeling language (and the aid) can accommodate. Any lack of fluency on the part of the user or any restrictions on the language's ability to capture certain realities reflects a communication disorder limiting the aid's usefulness.

For example, probabilistic risk analyses provide a valuable tool for figuring out how complex technical systems, such as nuclear power or chemical plants, operate and how they will respond to modifications (McCormick, 1981; US Nuclear Regulatory Commission, 1983). They do this by representing the system by the formal connections among its parts (e.g., showing how failures in one sector will affect performance in others). Both judgment and statistics are used to estimate the model's parameters. In this way, it is possible to pool the knowledge of many experts, expose that knowledge to external review, compute the overall performance of the system, and see how sensitive that performance is to variations (or uncertainties) in those parameters. Yet, current modeling languages require the experts to summarize their knowledge in quantitative and sometimes unfamiliar terms; and they are ill-suited to represent human behavior (such as that of the system's operators) (Fischhoff, in press). As a result, the model is not reality. Moreover, it may differ in ways that the user understands poorly,

just as the speaker of a foreign language may be insensitive to its nuances. At some point, the user may lose touch with the model without
realizing it. The seriousness of this threat with particular aids is an
empirical question that is just beginning to receive attention (National
Research Council, 1983).

Skilled judgment. Whether or not one relies on an aid, a strong
element of judgment is essential to all decision making. With unaided
decision making, judgment is all. With an aid, it is the basis for creating the model, estimating its parameters, and interpreting its results.
Improving the judgments needed for analysis has been the topic of intensive
research, with moderately consistent (although incomplete) results, some of
them perhaps surprising (Fischhoff, 1982). A number of simple solutions
have proven rather ineffective. It does not seem to help very much to
exhort people to work harder, to raise the stakes hinging upon their performance, to tell them about the problems that other people (like them)
have with such tasks, or to provide theoretical knowledge of statistics or
decision theory. Similarly, it does not seem reasonable to hope that the
problems will go away with time or when the decisions are really important.
Judgment is a skill that must be learned. Those who do not get training or
who do not enjoy a naturally instructive environment (e.g., one that rewards people for wisdom, rather than, say, for exuding confidence) will
have difficulty going beyond the "hard" data at their disposal.

Although training courses in judgment per se are rare, many organized
professions hope to inculcate good judgment as part of their apprenticeship
program. This learning is expected be a byproduct of having one's behavior
shaped by masters of the craft (be they architects, coaches, officers, or
graduate advisors). What is learned is often hard to express in words,
hence, must be attributed to judgment (Polanyi, 1962). What is unclear is
whether that learning extends to new decisions, for which the profession
has not acquired trial-and-error experience to shape its practices.

Where attempts have been made to improve judgment, a number of approaches have proven promising (Fischhoff, 1982). One is to provide the
conditions that learning theory holds to be essential for skill acquisition; for example, weather forecasters show great skill in assessing the
confidence to be placed in their precipitation forecasts—for which they
receive prompt, pertinent, and unambiguous feedback that they are required
to consider (Murphy & Winkler, in press). If these conditions do not exist
in life, then they might be simulated in the laboratory; for example,
Lichtenstein and Fischhoff (1980) have been able to improve confidence
assessment with moderate generalization to untrained tasks. A second
apparently effective approach is to restructure how people perform judgment
tasks, so as to enable them to use their own minds more effectively. For
example, hindsight bias has been reduced by forcing people to imagine how
events that did happen might not have happened (Slovic & Fischhoff, 1977);
availability bias may be reduced by encouraging people to search their

minds in a variety of ways so as to get a more diverse set of examples
(Behn & Vaupel, 1982; Raiffa, 1968); new evidence may be interpreted more
appropriately by having people consider how it might be consistent with
hypotheses that they doubt (Fischhoff & Beyth-Marom, 1983; Kahneman &
Tversky, 1979). Developing such procedures requires an understanding of
how people do think as well as of how they should think. Finally, there is
the obvious suggestion to train people in the principles of decision making, along with exercises in applying those principles to real problems
(Beyth-Marom et al., 1985). Researchers working in this area typically
feel that they have learned something from observing everyone else's problems. Whether this is an accurate perception and whether similar understanding can be conferred on others is an empirical question.

How Is Distributed Decision Making Different?

If life is hard for single individuals wrestling with their fate, then
what happens in complex systems, with interdependent decision makers responsible for incompletely overlapping portions of dynamic problems? Addressing these situations is a logical next step for behavioral decision
theory, although not one that it can take alone. The existential problem
in complex systems is still individuals pondering the unknown. However,
there are now rigid machines, rules, and doctrines in the picture, along
with more fluid social relations. These require the skills of computer
scientists, human factors specialists, substantive experts, and organizational theorists.

What follows is our attempt to pull together these perspectives as
part of a theory of decision making in complex systems. In doing so, the
general problem is characterized as distributed decision making, defined as
any situation in which decision-making information is not shared completely
by those with a role in shaping the decision. The set of systems having
this property includes high tech examples, such as air traffic control and
satellite management of a multinational corporation, mid tech examples,
such as forest-fire fighting and police dispatch, and low tech examples,
such as a volunteer organization's coordination of its branches' activities
or a couple's integration of their child-rearing practices (or their use of
a common checking account).

Whatever one's focus, looking far afield for examples provides the
opportunity to understand one's own situation better by contrasting it with
circumstances of others. They may do things so well or so poorly as to
cast the viability of different strategies into sharp relief, as was the
goal of In Search of Excellence (or might be the goal of its complement, In
Search of Dereliction?). Synthesizing the experience of diverse systems
may highlight the significant dimensions in characterizing and designing
other systems.

Although geographical separation is often considered a distinguishing
characteristic of distributed decision making, there can be substantial

difficulties in coordinating the (current and past) information of individuals in the same room or tent. (In the 60s, some of these problems were
called "failures to communicate.") As a result, the importance of different kinds of separation is left as a matter for further investigation. The
distribution of decision-making authority might seem to be another distinguishing characteristic, however it seems to be trivially achieved in
almost all human organizations. Few are able, even if they try, to cen-
tralize all authority to make decisions. For example, even when there are
focal decision makers who choose courses of action at clearly marked points
in time, their choice is often refined through interactions with their
subordinates, shaped by the information (and interpretations) reported by
others, and constrained by the pre-decisions of their predecessors and
superiors (e.g., avoid civilian casualties, avoid obviously reversing directions). If some distribution of decision-making authority is taken as a
given, then the form of distribution becomes a topic for future elaboration.

From this perspective, a useful unifying concept seems to be that of a
shared model. Those living in distributed decision-making systems have to
keep in mind some picture of many parts of that system, including how
external forces affect it, what communications links exist within it, what
the different actors in the system believe about its internal and external
situation, and what decisions the actors face (in terms of the options,
values, constraints, and uncertainties). These beliefs about the system
are some times dignified by terms like "mental representation." It seems
unlikely than anyone (in our lifetimes, at least) will ever actually observe what goes on in people's minds with sufficient clarity to be able to
outline their contents. What investigators can see is a refined and disci-
plined version including those aspects of people's beliefs that they are
able to communicate or act upon consistently. That communication might be
in terms of unrestricted natural language, in terms of the restricted
vocabulary of a formal organization, or in terms of a structured modeling
language.

In all cases, though, people need to translate their thoughts into
some language before those can be shared with others. Their ability to use
the language sets an upper limit on the system's coordination of decision
making, as do the system's procedures for information sharing. Looking at
the language and procedures provides a way of characterizing a system's
potential (and anticipating its problems). Looking at the knowledge that
has been shared provides a way to characterize its current state of affairs
(and anticipate its problems).

How Do Distributed Decision Making Systems Differ?

At the core of distributed decision-making systems are the people who
have to get the work done. As a result, a natural way to begin an analysis
of such systems is with the reality faced by those individuals, wherever

they find themselves within them. Sensible complications are to look, then, at the issues that arise when decision making is distributed over two individuals and, finally, when multiple individuals are involved.

What follows is a generalized task analysis, which attempts to characterize such systems in terms of the behaviorally significant dimensions that must be considered when designing systems and adapting people to them. It asks how people understand and manipulate their environment under reasonably unemotional conditions. Insofar as pressure and emotion degrade performance, problems that are unresolved at this level constitute a performance ceiling. For example, people need to stretch themselves to communicate at all. The risk that they may not stretch enough or in the right direction to be understood is part of the human condition. These risks can, however, be exacerbated (in predictable ways) by designers who do things such as compose multi-profession working groups without making basic human communication and understanding a fundamental concern.

Single-person systems. The simplest situation faced by an individual decision maker involves a static world about which everything can be known and no formal representation of knowledge is required. The threats to performance in this "basic" situation are those identified in the research on individual decision making (described above). They include the difficulties that arise in identifying relevant options, assembling and reviewing the knowledge that should be available, determining the values that are pertinent and the tradeoffs among them, and integrating these pieces in an effective way. The aids to performance should also be those identified in the existing literature, such as checklists of options, multi-method value elicitation procedures, and integration help.[3]

A first complication for individual decision making is the addition of uncertainty. With it, come all the difficulties of intuitive judgment under uncertainty, such as the misperception of causality, overconfidence in one's own knowledge, and heuristic-induced prediction biases. The potential solutions include training in judgmental skills, restructuring tasks so as to overcome bad habits, and keeping a statistical record of experience so as to reduce reliance on memory. A second complication is going from a static to a dynamic external world. With it, come new difficulties, such as undue adherence to currently favored hypotheses, as well as the accompanying potential solutions, such as reporting forms that require stating how new evidence might be consistent with currently unfavored hypotheses. A third complication is use of a formal modeling language for organizing knowledge and decision making. One associated problem is the users' inability to speak the modeling language; it might be addressed by using linguists or anthropologists to develop the language and train people to it. Another associated problem is the language's inability to describe certain situations (such as those including human factors or unclear intentions); it might be addressed by providing guidelines for when and how to override the conclusions produced by models using the language.

Two-person systems. Adding a second person to the system raises additional issues. However, before addressing them, it is important to ask what happens to the old issues. That is, are they eliminated, exacerbated, or left unchanged by the complications engendered by each kind of two-person system?

In behavioral terms, the simplest two-person system involves individuals with common goals, common experience, and a hardened communications link. Thus, they would have highly shared models and the opportunity to keep them consistent. Having a colleague can reduce some difficulties experienced by individuals. For example, information overload can be reduced by dividing information-processing responsibilities; some mistakes can be avoided by having someone else to check one's work. On the other hand, having someone who thinks similarly in the system may just mean having two people prone to the same judgmental difficulties. It might even make matters worse if they drew confidence from the convergence of their (similarly flawed) judgmental processes.

More generally, agreement on any erroneous belief may increase confidence without a corresponding increase in accuracy, perhaps encouraging more drastic (and more disastrous) actions. "Risky shift" is a term for groups' tendency to adopt more extreme positions than do their individual members (Davis, 1982; Myers & Lamm, 1976); "groupthink" is a term for the social processes that promote continued adherence to shared beliefs (Janis, 1972). Restricting communication would be one way to blunt these tendencies, however, at the price of allowing the models to drift apart, perhaps without the parties realizing it. Even with unrestricted communication, discrepant views can go a long while without being recognized. "False consensus" refers to the erroneous belief that others share one's views (Nisbett & Roths, 1980); "pluralistic ignorance" refers to the erroneous belief that one is the odd person out (Fiske & Taylor, 1984). Both have been repeatedly documented; both can be treated if the threat is recognized and facing the discrepancy is not too painful.

Such problems arise because frequent interaction can create a perception of completely shared models, when sharing is inevitably incomplete. An obvious complication in two-person distributed decision-making systems is for experience to be obviously incomplete. Although this situation reduces the threat of unrecognized disagreement and increases the chances of erroneous beliefs being challenged, it raises additional problems. One is the possibility that terms will unwittingly be used differently by the two, without the recognition that, say, "risk" or "threat" or "likely" or "destructive power" have different meanings (Beyth-Marom, 1982a; Bunn & Tsipis, 1983; Fischhoff, Watson & Hope, 1984). If such discrepancies go undetected, then the parties' perceptions will drift apart until some dramatic and unpredictable act occurs. To avoid having that happen at some inopportune time, inconsistencies must be actively sought and resolved

(National Interagency Incident Management System, 1982). A contrasting problem is the inability to reconcile or even to face differences between models. Where the differences are great enough to affect decisions (vonWinterfeldt & Edwards, 1982), the skills of a mediator are needed to bring them together.

A common complication in two- (or multi-) person systems is unreliable communication links. Often, this situation is imposed by external pressures, either directly (as when they attempt to disrupt communications) or indirectly (as when a hardened communication link is foregone to facilitate freedom of action within the team or to enhance the security of those communications that are made). Here, as elsewhere, the shared-models perspective suggests that this complication reduces some problems, raises others, and requires different systems design. For example, interrupting the communications links between individuals (or units) with deeply shared common experience allows them to acquire different information and formulate somewhat independent perspectives, thereby reducing the risk of groupthink. However, it may lead to unrecognized drift in their beliefs and unpredictable behavior when they are called to action. One possible protective device is to provide efficient checking procedures, enabling system members to detect and diagnose discrepancies in their models. The mixture of theoretical and empirical research needed to produce such procedures should also be able to produce general estimates of systems reliability, showing the level of vigilance appropriate for different circumstances.

One complication of concern for efficiency in communication is the imposition of institutional structures. These could include restricted vocabularies, time- or event-related reporting requirements, interaction protocols, or confirmation procedures ranging from repeating a message to deriving its implications. At the extreme, communication might be through an analytical language, designed to create the system's formal model of its world. The problems and possibilities of these languages for individual decision makers remain with multiple decision makers. An additional advantage that accrues at this level is the ability to pool the knowledge of diverse individuals (e.g., having different specialties, observing different fronts) in a single place. An additional disadvantage is that the language may suppress the nuances of normal communication that people depend upon to understand and make themselves understood. It is unclear what substitutes people will find (or even if they will recognize the need to find them), when deprived of facial expression, body language, intonation, and similar cues. These problems may be exacerbated further when knowledge comes to reside in a model without indication of its source, so that model users do not know who said it, much less how it was said. Finally, models that cannot express confidence in their subordinate's reports probably cannot generate the confidence needed to follow a superior's choice of action--making it hard to lead through electronic mail.

The great memory capacity of automated aids makes it possible, in principle, to store such information. However, there are "human" problems both in getting and in presenting those additional cues. On the input side, one worries about people's inability to characterize the extent of their own knowledge, to translate it into the precise terms demanded by a language, or to see how they themselves relied on non-verbal cues to be understood.[4] On the output side, one worries about creating meaningful displays of such qualifications. If shown routinely, they may clutter the picture with "soft" information which, in any case, gets lost when users attempt to generate best guesses at what is happening (Peterson, 1973). If available upon request, qualifications may slip the mind of decision makers who want clear-cut answers to their questions. Because it is so difficult to tell when qualifications are not in order, such systems require careful design and their operators require careful training. Unless users have demonstrated mastery of the system, it may be appropriate to sacrifice sophistication for fluency.[5]

A final, behaviorally significant complication that can arise with two-person distributed decision-making systems is inconsistencies in the goals of the parties. They have similar values, but differ over the goals relevant to a particular case; they may have a common opponent, yet stand to share differently from the spoils of victory; they may strive for power within the system, while still being concerned about its ability to meet external challenges.[6] Like other complications, these can be useful. For example, disagreement over the application of general values can uncover labile values that might not be detected by an individual; competition might sharpen the wits of the competitors; by some accounts, conflict itself is part of what binds social units together (Coser, 1954). For the effects of conflict to be managed, they must be recognized by those who design and operate systems. One common, but difficult response is introducing sharply defined reward systems to create the correct mix of incentives.[7]

Multiple-person systems. Most of the issues arising in the design and diagnosis of two-person systems remain with multiple decision-maker systems, although with somewhat new wrinkles. The simplest level involves common-goaled individuals, with shared experience and hardened communication links. As before, having more people around means having the opportunity for more views to evolve and be heard. Yet, this advantage may backfire if the shared (past and present) experience leads them to think similarly while taking confidence in numbers (Lanir, 1982). As the number of parties multiplies, so does the volume of messages (and perhaps information).

If hardened communications means that everyone hears everything, then there may be too much going on to ensure that everyone hears anything. It may also be hard to keep track of who knows what. With an automated aid, it may be possible to track (or reconstruct, if needed) who heard what.

Modeling the decison-making situations faced by different individuals may help discern who needs to know what.

As organizational size increases, the possibility of completely shared experiences decreases. The maximum might be found in a hierarchical organization whose leaders had progressed through the ranks from the very bottom, so that they have a deep understanding of the reality of their subordinates' worlds, allowing them to imagine what others might be thinking and how they might respond in particular circumstances. In such situations, less needs to be said and more can be predicted, making the organization more intimate than it seems.

However, size also makes the liabilities of commonality more extreme. Not only is shared misunderstanding more likely, but it is also more difficult to treat because it is so broadly entrenched and the organizational climate is likely to be very rough for those who think differently. Indeed, the heterogeneity of an organization's selection and retention policies may be a good indicator of its resilience within a complex and changing reality. If there are any common biases in communications between individuals (e.g., underestimation of costs, exaggeration of expectations from subordinates), then the cumulative bias may be out of hand by the time communications have cascaded up or down the organizational chart. When the world is changing rapidly, then the experience of having once been at every level in the organization may give an illusory feeling of understanding its reality. For example, the education, equipment, and challenges of foot soldiers or (sales representatives) may be quite different now than when their senior officers were in the trenches. An indicator of these threats might be the degree of technological change (or instability) in the organization and its environment. One treatment might be periodic rotation through the ranks, another might be the opportunity to cut through the normal lines of communication in order to find out what is really happening at diverse places; either might reveal discrepancies in the models held by different parties. These problems might be avoided somewhat by resisting opportunities to change the organization, unless the promised improvements (in coping with the external world) will clearly compensate for the likely decrements in internal understanding.

Both the problems and promises of unshared experience increase as one goes from two- to multi-person systems. More people do bring more perspectives to a problem and with them increase the chances of challenging misconceptions. However, the intricacies of sharing and coordinating that information may become unmanageable. Even more seriously, with so many communication links, it may become nearly impossible even to discover the existence of misunderstandings, such as differences in unspoken assumptions or in the usage of seemingly straightforward terms. If communications are decentralized, then various subunits may learn to speak to one another, solving their local problems but leaving the system as a whole unstable.

If communications are centralized, then the occupants of controlling nodes have an opportunity to create a common picture, but doing so requires extraordinary attention to detail, regarding who believes what when and how they express those beliefs. One aid to tracking these complex realities is to maintain formal models of the decision-making problems faced at different places. Even if these models could only capture a portion of those situations, comparing the models held at headquarters and in the field might provide a structured way of focusing on discrepancies. When theory or data suggest that those discrepancies are large and persistent, then it may be command, rather than communications, strategies that require alteration. When superiors cannot understand their subordinates' world, one suggestion is to concentrate on telling them what to do, rather than how to do it, so as to avoid micromanagement that is unlikely to be realistic. A second is to recognize (and even solicit) signals that they, the leaders, are badly out of touch with their subordinates' perceptions (suggesting that one or both sets of beliefs need adjustment).

Reliability problems in multi-person systems begin with those of two-person systems. As before, their source may be external (e.g., disruptions, equipment failure) or internal (e.g., the desire for flexibility or autonomy). As before, the task of those in them is to discern when communications have failed, how they have failed (i.e., what messages have been interrupted or garbled), and how the system can be kept together. The multiplicity of communications means a greater need for a prestructured response—if the threat of unreliability is recognized. Depending upon the organization's capabilities, one potential coping mechanism might be a communications protocol that emphasized staying in touch, even when there was nothing to say, in order to monitor reliability continually; another might be analyses of the "backlash" effect of actions or messages, considering how they discourage or restrict future communications (e.g., by suggesting the need for secrecy or revealing others' positions); another might be reporting intentions along with current status, to facilitate projecting what incommunicant others might be doing now; another might be creating a "black box" from which one could reconstruct what had happened before communications went down.

A complicating factor in reliability problems, which emerges here but could be treated with two-person systems, is that lost communications may reflect loss of the link or loss of the communicator at the other end of the link. That loss could reflect defection, disinterest, or destruction. Such losses simplify communications (by the number of links involving that individual) and can provide diagnostic information (about possible threats to the rest of the system). However, they require reformulation of all models within the system involving the lost individual. Where that reformulation cannot be confidently done or disseminated, then contingency plans are needed, expressing a best guess at how to act when the system may be shrinking. Whether drawn for vanishing links or individuals, those plans should create realistic degrees of autonomy for making new decisions and for deviating from old ones (e.g., provide answers to: Are my orders still valid? Will I be punished for deviating from them?).

A final complicating factor with multiple-person systems, for which the two-person version exists but is relatively uninteresting, concerns the heterogeneity of its parts. At one extreme, lies a homogeneous organization whose parts interact in an "additive" fashion, with each performing roughly the same functions and the system's strength depending on the sum of such parts. At the other extreme, lies a heterogenous organization whose specialized parts depend on one another for vital services, with its strength coming from the sophistication of its design and the effectiveness of its dedicated components. Crudely speaking, a large undifferentiated infantry group might anchor one end of this continuum and an integrated carrier force the other.

The operational benefits of a homogeneous system are its ability to use individuals and materials interchangeably, as well as its relative insensitivity to the loss of particular units (insofar as their effect is additive). Common benefits of homogenous systems for distributed decision-making are the existence of a shared organizational culture (allowing components to interpret one another's actions), the relative simplicity of organizational structure, and the opportunity to create widely applicable organizational policies. Inherent limitations may include homogeneity of perspectives and skills, leaving the system relatively vulnerable to deeply shared misconceptions (what might be called "intellectual common-mode failure") and relatively devoid of the personnel resources needed to initiate significant changes (or even detect the need for them without very strong, and perhaps painful, messages from the environment).

The operational benefits of a heterogenous system lie in its ability to provide a precise response to any of the anticipated challenges posed by a complex environment. Its advantages as a distributed decision-making system lie in its ability to develop task-specific procedures, policies, and communications. One inherent disadvantage in this respect may be the difficulty of bearing in mind or modeling the operations of a complex interactive system, so it is hard to know who is doing what when and how their actions affect one another. For example, backlash and friendly fire may be more likely across diverse units than across similar ones. Even if they do have a clear picture of the whole, the managers of such a system may find it difficult to formulate an organizational philosophy with equivalent meanings in all the diverse contexts it faces. The diversity of parts may also create interoperability problems, hampering the parts' ability to communicate and cooperate amongst themselves.

Both kinds of systems may be most vulnerable to the kinds of threats against which the other is most strongly defended. The additive character of homogeneous systems means that it is numbers that count. A command system adapted to this reality may be relatively inattentive to those few ways in which individual units are indispensible, such as their ability to reveal vital organizational intelligence or to embarrass the organization as a whole. Conversely, the command structure that has evolved to orchestrate the pieces of a heterogeneous system may be severely challenged by situations in which mainly numbers matter. An inevitable byproduct of specialization is having fewer of every specialty and less ability to transcend specialty boundaries. There may, therefore, be less staying power in protracted crises.

Perhaps the best treatment for these limitations is incorporating some properties of each kind of system in the other. Thus, for example, homogeneous organizations could actively recruit individuals with diverse prior experience in order to ensure some heterogeneity of views; they might also develop specialist positions for dealing with non-additive issues wherever those appear in the organization (e.g., intelligence officers, publishers' libel watchdogs). Heterogeneous organizations might promote generalists with the aim of mediating and attenuating the differences among their parts; they might also transfer specialists across branches so as to encourage the sharing of perspectives (at the price of their being less well equipped to do the particular job); they might create roles for a few freefloating individuals, entrusted with wandering through the organization to identify and tie up loose ends. Whether such steps are possible, given how antithetical they are to the ambient organizational philosophy, would be a critical design question.

Principles in Designing Distributed Decision Making Systems

Goals of the analysis. The preceding task analysis began with the problems faced in designing the simplest of decision-making systems, those involving single individuals grappling with their fate under conditions of certainty, with no attempt at formalization. It proceeded, first, to complicate the lives of those single individuals and, then, to consider several levels of complication within both two-person and multi-person organizations. A full-blown version of this analysis[8] would consider, at each stage, first, how the problems that arose in simpler systems were complicated or ameliorated and, second, what new problems arose. For each set of problems, it would try to develop a set of solutions based, as far as possible, on the available research literature in behavioral decision theory, cognitive psychology, human factors, communications research, or organizational theory. The recommendations offered here are, therefore, but speculations, suggestive of what would emerge from a fuller analysis.

That analysis would proceed on two levels. One is to investigate solutions to highly specific problems, such as the optimal communications protocol or visual display for a particular heterogeneous system. The second is to develop general design principles, suggesting what to do in lieu of detailed specific studies. In reality, these two efforts are highly intertwined, with the general principles suggesting the behavioral dimensions that merit detailed investigation and the empirical studies substantiating (or altering) those beliefs. Were a more comprehensive analysis in place, a logical extension would be to consider the interaction between two distributed decision-making systems, each characterized in the same general terms. Such an analysis might show how the imperfections of each might be exploited by the other as well as how they might lead to mutually undesirable circumstances. For example, an analysis of the National Command Authorities of the US and USSR might show the kinds of challenges that each is least likely to handle effectively. That diagnosis might, in turn, suggest unilateral recommendations (or bilateral agreements) to the effect, "Don't test us in this way unless you really mean it. We're not equipped to respond flexibly."[9]

Design guidelines. Although still in its formative stages, the above analysis suggests a number of general conclusions that might emerge from a more comprehensive analysis of distributed decision-making systems. One is that the design of the system needs to bear in mind the reality of the individuals at each node in it. If there is a tendency to let the design process be dominated by issues associated with the most recent (or highest order) complication, then it must be resisted. If the designers are unfamiliar with the world of the operators, then they must learn about it. For example, one should not become obsessed with the intricacies of displaying vast quantities of information when the real problem is not knowing what policy to apply. Given the difficulty of individual decision making, one must resist the temptation to move on to other, seemingly more tractable problems.

A second general conclusion is that many group problems may be seen as variants of (perhaps unsolved) individual problems. For example, a common crisis in the simplest individual decision-making situations is determining what the individual wants from them. The group analog is determining what general policies are relevant and how they are to be interpreted in specific circumstances. As another example, individuals' inability to deal coherently with uncertainty may underlie their (unrealistic) demands for certainty in communications from others.

A third general conclusion is that many problems that are attributed to the imposition of novel technologies can be found in quite low-tech situations. Two people living in the same household can have difficulty communicating; allowing them to use only phone or telex may make matters better or worse. The speed of modern systems can induce enormous time pressures, yet many decisions cannot be made comfortably even with unlimited time. Telecommunications systems can generate information overload, yet the fundamental management problem remains the simple one of determining what is relevant. In such cases, the technology is best seen is giving the final form to problems that would have existed in any case and as providing a possible vehicle for either creating solutions or putting solutions out of reach.

A fourth conclusion is that it pays to accentuate the negative when evaluating the designs of distributed decision-making systems, and to accentuate the positive when adapting people to those systems. That is, the design of systems is typically a top-down process beginning with a set of objectives and constraints. The idealization that emerges is something for people to strive for but not necessarily something that they can achieve. Looking at how the system keeps people from doing their jobs provides more realistic expectations of overall system performance as well as focuses attention on where people need help. The point of departure for that help must be their current thought processes and capabilities, so that they can be brought along from where they are toward where one would like them to be. People can change, but only under carefully structured conditions and not that fast. When they are pushed too hard, then they risk losing touch with their own reality.

Design ideologies. A fifth conclusion is that the design of distributed decision-making systems requires detailed empirical work. A condition for doing that work is resisting simplistic design philosophies. Each design arena spawns a number of such slogans, each having the superficial appeal capable of generating strong organizational momentum, while frustrating efforts at more sensitive design. One such family of simple principles concentrates on dealing with a system's mistakes, by claiming either to avoid them entirely in prospect (as expressed in "zero defects" or "quality is free"), to adapt to them promptly in process (as expressed in "muddling through"), or to respond to them in hindsight ("learning from experience"). A second family of slogans concentrates on being ready for all contingencies, by instituting either rigid flexibility or rigid inflexibility, either leaving all options open or planning for all contingencies. A third family emphasizes controlling the human element in systems, either by selecting the right people or by creating the right people (through proper training and incentives). A fourth family of principles proposes avoiding the human element either where it is convenient (because viable alternatives exist), where it is desirable (because humans have known flaws), or in all possible circumstances whether or not human fallibility has been demonstrated (in hopes or increasing system predictability).

Rigid subscription to any of these principles gives the designers (and operators) of a system an impossible task. For example, the instruction "to avoid all errors" implies that time and price are unimportant. Where this is not the case, the designers are left adrift, forced to make tradeoffs without explicit guidance. Where fault-free design is impossible, then the principle discourages treatment of those faults that do remain. Many failsafe systems only "work" because the people in them have learned, by trial and error, to diagnose and respond to problems that are not supposed to happen. Because the existence of such unofficial intelligence has no place in the official design of the system, it may have to be hidden, it may be unable to get needed resources (e.g., for record keeping or realistic training exercises), and it may be destroyed by any uncontrollable change in the system (which invalidates operators' understanding of those intricacies of its operation that do not appear in any plans or training manuals). From this perspective, where perfection is impossible, it may be advisable to abandon near-perfection as a goal as well, so as to ensure that there are enough problems for people to learn to cope with them. In addition, where perfection is still (but) an aspiration, steps toward it should be very large before they justify disrupting accustomed (unwritten) relationships. That is, technological instability threatens system operation. Additional threats from this philosophy include unwillingness to face those intractable problems that do remain and setting the operators up to "take the rap" when their use of the system proves impossible.

Similar analyses exist for the limitations of each of the other simple rules. In response, proponents might say that the rules are not meant to be taken literally and that compromises are a necessary part of all design. Yet, the categorical nature of such principles is an important part of their appeal and, as stated, they provide no guidance or legitimation for compromises. Moreover, they often tend to embody a deep misunderstanding of the role of people in person-machine systems, reflecting, in one way or another, a belief in the possibility of engineering the human side of the operation as completely as one might hope to engineer the mechnical or electronics side.

Human factors. As the long list of apparent operator failures in technical systems suggests, the attempts to implement this belief in engineering people are often needlessly clumsy (National Research Council, 1983; Perrow, 1984; Rasmussen & Rouse, 1981). The extensive body of human-factors research is either unknown or invoked at such a late stage in the design process that it can amount to little more than the development of warning labels and training programs for coping with inhuman systems. It is so easy to speculate about human behavior (and provide supporting anecdotal evidence) that systematic empirical research hardly seems needed. Common commitants of insensitive design are situations in which the designers (or those who manage them) have radically different personal experiences than the operators, work in organizations that do not function very well interpersonally, or are frustrated in trying to understand why some group of others (e.g., the public) does not like them.

However, even when the engineering of people is sensitive, its ambitions are often misconceived. The complexity of systems places some limits on their perfectability, making it hard to understand the intricacies of a design. As a result, one can neither anticipate all problems or confidently treat those one can anticipate, without the fear that corrections made in one domain will create new problems in another. Part of the "genius" of people is their ability to see (and, hence, respond to) situations in unique (and, hence, unpredictable) ways. Although this creativity can be seen in even the most structured psychomotor tasks, it is central and inescapable in any interesting distributed decision-making system. Once people have to do any real thinking, the system becomes complex (and, hence, unperfectable). In such cases, the task of engineering is to help the operators understand the system, rather than to manage them as part of it. A common sign of insensitivity in this regard is use of the term "operator error" to describe problems arising from the interaction of operator and system. A sign of sensitivity is incorporating operators in the design process. A rule of thumb is that human problems seldom have purely technical solutions, while technical solutions typically create human problems.

The Possibility of Distributed Decision Making

Pursuing this line of inquiry can point to specific problems arising in distributed decision-making systems and focus technical efforts on solving them. Those solutions might include displays for uncertain information, protocols for communication in complex systems, training programs for making do with "unfriendly" systems, contingency plans for coping with predictable system failures, and terminology for coordinating diverse units. Deriving such solutions is technically difficult, but part of a known craft. Investigators know how to describe such problems, devise possible remedies, and subject those remedies to empirical test.[10] Where the opportunities to develop solutions are limited, these kinds of perspectives can help characterize existing systems and improvise balanced responses to them.

However, although these solutions might make systems better, they cannot make them whole. The pursuit of them may even pose a threat to systems design if it distracts attention from the broader question of how systems are created and conceptualized. In both design and operation, healthy systems enjoy a creative tension between various conflicting pressures. One is between a top-down perspective (working down toward reality from an idealization of how the system should operate) and a bottom-up perspective (working up from reality toward some modest improvement of the current presenting symptoms). Another is between bureaucratization and innovation (or inflexibility and flexibility). Yet others are between planning and reacting, between a stress on routine and crisis operations, between risk acceptance and risk aversion, between human and technology orientation. A common thread in these contrasts is the system's attitude toward uncertainty, does it accept that as a fact of life or does it live in the future, oriented toward the day when everything is predictable or controllable?

Achieving a balance between these perspectives requires both the insight needed to be candid about the limitations of one's system and the leadership needed to withstand whichever pressures dominate at the moment. When a (dynamic) balance is reached, the system can use its personnel most effectively and develop realistic strategies. When it is not reached, the organization is in a state of crisis, vulnerable to events or hostile actions that exploit its imbalances. The crisis is particularly great when the need for balance is not recognized or cannot be admitted (within the current organizational culture), and when an experimental gulf separates the management and operators. In this light, one can tell a great deal about how a system functions by looking at its managers' philosophy. If that is oversimplified or overconfident, then the system will be, too, despite any superficial complexity. The goal of a task analysis then becomes to expose the precise ways in which this vulnerability expresses itself.

Notes

Support for this research was provided by the Office of Naval Research, under Contract No. N00014-85-C-0041 to Perceptronics, Inc., "Behavioral Aspects of Distributed Decision Making." Special thanks to Stephen Johnson for help on a previous draft.

[1] All three of these questions refer to essential skills for effective use of decision analysis. The empirical evidence suggests that the answer to each is, "No, not really." However, there is some chance for improving performance by properly structuring tasks (Fischhoff, Svenson & Slovic, in press; Kahneman et al., 1982; Slovic, Lichtenstein & Fischhoff, in press).

[2] For discussion of such imprecision in carefully prepared formal analyses of government actions, see Fischhoff (1983, 1984) and Fischhoff & Cox (in press).

[3] In the absence of a formal model, computational help is impossible. However, there are integration rules following other logics, such as flow charts, hierarchical lists of rules, or policy-capturing methods for determining what consistency with past decisions would dictate (Dawes, 1979; Meehl, 1954).

[4] For example, Samet (1975) showed that a commonly used military system required information to be characterized in terms of "reliability" and "validity" even though these concepts were not distinguished in the minds of users.

[5] A case to consider in this regard is the "hot line" between the US and USSR. Although it might seem like a technical advance to upgrade the quality of the line so that the leaders could talk to one another directly (e.g., through videophone), perhaps the quality of the communication is better with the current telex systems having human operators who spend eight hours a day "talking" to one another. By contrast, given the differences between the cultures, who knows what unintentional cues would be sent by the leaders through their voices, postures, intonation, etc.

[6] A common variant within larger organizations is that they reward individuals within them for growth (i.e., making their own subunits larger), while striving as a whole for profit (Baumol, 1959).

[7] One example of the difficulty of diagnosing and designing such systems may be seen in the current debate over whether competition among the armed services improves national defense (by ensuring that there are technically qualified critics of each service's new weapons proposals) or degrades it (by fostering waste, duplication, and interoperability problems).

[8] This is one goal of the project sponsoring this work.

[9] This is another component of the project sponsoring this work.

[10] We are even exploring some of these ourselves. Current projects "in the lab" include the development of methods to (a) make contingency plans more realistic, (b) generate options for novel decision-making situations, (c) improve the accessibility of information in computerized databases, (d) structure judgmental tasks to make better use of people's mental capabilities, (e) formulate policies that will be meaningful in varied circumstances, and (f) screen options rapidly in crisis situations.

DOES THE EXPERT KNOW? THE RELIABILITY OF PREDICTIONS AND CONFIDENCE RATINGS OF EXPERTS

Willem A. Wagenaar
University of Leiden, The Netherlands

Gideon B. Keren
TNO Institute for Perception, The Netherlands

Experts are often asked to assess two different kinds of probabilities. One is THE PROBABILITY THAT SOMETHING WILL HAPPEN: rain, hitting an oil well, dying in an operation, tube fracture, a total melt-down. We will call these probabilities "predictions". Experts' predictions are widely used in all sorts of personal and public decision making. Two examples of formal usage of expert opinion are risk analyses and expert systems used for diagnostic tasks. The second kind of probability assessed by experts is THE PROBABILITY THAT THEIR ANSWERS ARE CORRECT. We will call these probabilities "confidence ratings". A little later we will demonstrate that predictions and confidence ratings have often been confused in the literature.

Let us first point out some differences between the two assessments. A prediction is always related to a future event, like rain tomorrow in the central part of The Netherlands. No one knows for sure whether it will rain tomorrow or not. The expertise of the expert consists of knowing better than a lay person which factors determine the occurrence of an event, and how the available data should be aggregated to arrive at a prediction. Absolute certainty is not obtainable because the necessary scientific insights are just not available, not even to the best expert in the world. Without going into the question whether all causes of events can be known in principle, we can state as a matter of fact that we tend to rely on expert judgment exactly in those situations in which not all causes are known.

A confidence rating, for instance on the question "Did TMI occur in 1979 or in 1980?", is usually related to a past event. TMI occurred in one of the two years only. The source of the assessor's uncertainty is not that the factors determining the outcome are unknown to mankind; the source is lack of expertise. Prediction presupposes expertise; confidence ratings are used to assess expertise.

A related distinction between prediction and confidence rating is that subjects in studies on prediction are mostly experts, like meteorologists or engineers.

NATO ASI Series. Vol. F21
Intelligent Decision Support in Process Environments
Edited by E. Hollnagel et al.
© Springer-Verlag Berlin Heidelberg 1986

Subjects in studies on confidence rating are typically non-experts like college students.

To complicate the distinction a little, we will explain that confidence ratings can also be given with respect to predictions. Consider the meteorologist assessing the probability of rain at 70%. This number represents a central tendency like a mean or a mode. The prediction could have been based on a comparison with those 100 days in the past 40 years, that showed the same type of situation as today's weather. The percentage could be based on the finding that rain followed in 70 of these 100 cases. Now it is also possible that in the record of the past 40 years only ten days were found that resembled today. If seven out of these ten were followed by rain, the expert would again arrive at an estimate of 70%. But the confidence interval would be considerably wider. Accepting some statistical assumptions the 95% confidence interval in the two cases are $\pm$ 9% and $\pm$ 28%, respectively. Hence, the expert could produce the same PREDICTIONS in both cases, and still attach different amounts of CONFIDENCE to them.

Unfortunately the two tasks have been somewhat confounded in the literature. One reason could be that the normative requirement of CALIBRATION holds for both types of assessment. Good calibration of predictions means that out of 100 cases in which the probability of an event was estimated at p%, indeed the event occurred p times. It should rain on 70% of the days for which the probability of rain was assessed at 70%. The same is true for confidence ratings: Taking together all questions in which a person expressed 70% confidence that an answer was correct, we should not find that in fact 50% of the answers was wrong. Such a result would signify OVERCONFIDENCE in the ability to answer those questions. Calibration has been studied both for predictions and confidence ratings. For both types of asessment it has been found that people tend to be overconfident (cf. Lichtenstein, Fischhoff & Phillips, 1982, for a general and very useful overview). The indiscriminate use of the term overconfidence in both cases reflects the confounding described above. Too high predictions are not the product of overconfidence, but of the application of incorrect models or databases. As in the example of predicting rain, less confidence in models and data would not change the central tendency of the prediction. The harm done by confusion of the two notions becomes clear when it comes to explanations. It is obvious that one possible explanation of too high confidence ratings is some sort of unwillingness to express lack of knowledge in an area where knowledge should exist. Doctors may not like to express a very low confidence in their diagnoses, but there is no reason why an expert meteorologist would shun a 10% probability of rain.

Another drawback of confusing predictions and confidence ratings is the habit to generalise studies on confidence ratings of lay people to predictions by experts. In this paper we will argue that the processes underlying confidence ratings and predictions are different. Generalization of one area to the other is therefore impossible.

In order to avoid the confusion described above we will reserve the terms OVERCONFIDENCE and UNDERCONFIDENCE exclusively for errors in confidence ratings. Prediction errors will be called OVERESTIMATES and UNDERESTIMATES.

The main question addressed in this contribution is not whether predictions and confidence ratings of experts are good. We know quite well that they sometimes are appalling, like in the case of medical doctors assessing the probability of a diagnosis (Christensen-Szalansky & Busyhead, 1981; Lusted, 1977; DeSmet, Frybank & Thornbury, 1979). It might be more interesting and relevant for practical aplication to know why these assessments are unreliable. What factors are involved; when can we trust experts, and when can we not?

Time has not yet come to answer these questions unequivocally. The present paper will contain only some demonstrations of factors that influence calibration of predictions and confidence ratings.

EXPERIMENT 1: THE CONFIDENCE OF EYEWITNESSES

The literature on eyewitness testimony (Loftus, 1979) suggests that confidence is not always a good indicator of accuracy, and that people are often overconfident. Quite surprisingly much of the literature focusses on accuracy, while neglecting confidence. It is evident that inaccuracy is the more devastating when it goes together with overconfidence. Juries and judges tend to rely more on eyewitness testmony when expressed with great confidence. Classic examples of inaccuracy and extreme confidence are of course found in the class of visual illusions. How many memories in everyday are of the same illusory type, in the sense that they are wrong, but revealed with great confidence?

The experiment was quite simple. A series of 22 slides, presenting a car-pedestrian accident, was shown to 200 students at a rate of one slide in three seconds. After about half an hour of other paper-and-pencil testing subjects were presented with picture pairs. One of the pictures they had seen before, the other was slightly altered. The task was to distribute 100 probability points between the two pictures, indicating the probabilities that they had

seen the one or the other picture before. Some calibration curves are presented
in Fig. 1. The convention used in such pictures is that accuracy is plotted as
a function of the probability assessment. The percentages of observations on
which the data points are based are shown along the curves. The horizontal axis
represents classes of responses, but since subjects tend to use multiples of
ten in the large majority of the cases, the best representation of the class
mean is the multiple of ten in that class.

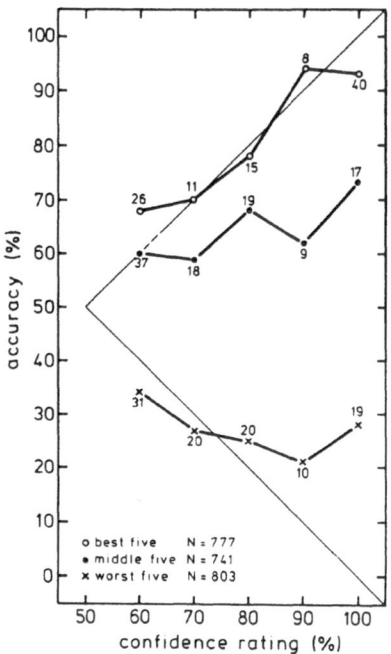

Fig. 1   Calibration curves of Experiment 1, eyewitnessing a car-
pedestrian accident.

The effect of overconfidence was quite pronounced for ten out of 15 picture
pairs. The eyewitnesses' confidence is in no way an indicator of their accuracy. What was wrong? Are confidence ratings mere guesses? One piece of
information that reveals the nature of the problem is the distribution of
accuracy scores over questions. Five out of the 15 questions scored less than
50% correct. The two worst results were 18% and 21% correct. The subjects were
not guessing; they thought that there was a good reason to pick the wrong
answer. Fig. 1 reveals that for the five best answered questions calibration
was nearly perfect. However, for the five most difficult questions (18 to 43%

correct) calibration was almost completely inverted. This finding suggests that the most difficult questions are of the illusory type: Subjects conclude from an incorrect reasoning that they saw the non-presented picture. The more plausible the reasoning, the more confident subjects are. High confidence scores do not reflect the clarity of a representation in memory, but the plausibility of the inferences process that substituted memory.

The relation between difficulty and overconfidence was also noted using general knowledge questions, e.g. by Lichtenstein & Fischhoff (1977). These authors do not discuss what is assessed by subjects: Knowledge or the quality of inference, although it is perfectly clear that the questions are in an area about which subjects do not always have knowledge. Loftus, Miller & Burns (1978) report that eyewitnesses can be more confident when incorrect than when correct, which is suggestive of a calibration curve with a negative slope. Is it true that overconfidence stems from trust in misapplied inference?

EXPERIMENT 2: CONFIDENCE WITHIN A PSYCHOPHYSICAL SETTING

The results of the first experiment suggest that overconfidence is caused by too much reliance on inference processes. This hypothesis can be further tested by looking at calibration in psychophysical studies, where conscious inferences do not play a dominating role. Our prediction is that calibration would be good under such conditions. The same prediction was put forward by Dawes (1974, 1980), who noted that evidence for poor calibration has been mainly obtained from tasks which required judgments about intellectual knowledge. He proposed that humans tend to overestimate their intellectual capacity, which would be a sufficient explanation of overconfidence in calibration studies. Testing calibration in perceptual tasks, Dawes (1980) still observed overconfidence, which lead him finally to abandon the hypothesis. The tasks employed by Dawes were naming the color of the eyes of colleagues at the Psychology Department of the University of Oregon, and comparison between areas of circles and squares. Both tasks do still involve a considerable amount of inferential reasoning. This induced Keren (1985) to study confidence in a more pure psychophysical task. Two letters were presented briefly side by side, followed by a post-stimulus position cue indicating which of the two letters was to be reported. The cued location contained always one letter from the target set [A, E]. The uncued position contained a neutral letter, such as K or N. Each of nine subjects performed in about 400 trials. In each trial subjects were asked to report the letter perceived at the cued location, and to indicate by a percentage between

50 and 100 how confident they were that the response was correct. Two levels of difficulty were created by appropriate manipulation of the exposure time. The resulting calibration curves are shown in Fig. 2.

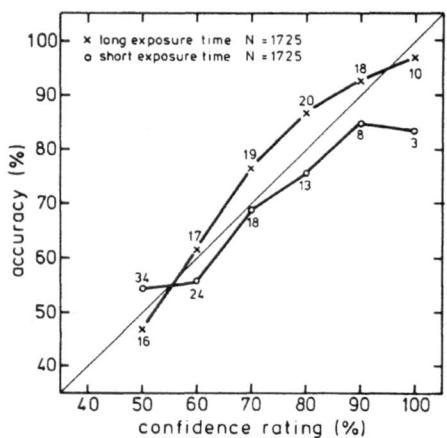

Fig. 2 Calibration curves of Experiment 2, psychophysical stimuli.

Performance with long exposure times was 77%, with short exposures 63%. There was an effect of overconfidence in the difficult condition, but the deviations from the ideal line were very small. More important is that subjects seemed to be aware of their level of performance. In the difficult condition subjects used most frequently the 50% confidence level, versus the 80% level in the easier condition. Mean confidence was 65% in the difficult condition, and 74% in the easy condition.

Without going into lengthy discussions about the psychophysical experiment, we can simply conclude that subjects are in principle able to react to differences in accuracy by adjustment of their confidence ratings. It is obvious that feedback about difficulty is a necessary condition for this to happen. Difficulty remains undetected when stimuli have an illusory effect or promote illusory inference. In the present study task difficulty was increased by manipulation of the physical energy in the stimuli. This was detected by subjects. Even without knowledge of results they noticed that accurate perception of the stimuli was more difficult, and they adjusted confidence in their responses accordingly.

EXPERIMENT 3: SPREADING CONFIDENCE

If it is true that overconfidence is due to the illusory character of the problems rather than to a basic trait in the assessors, it must be possible to elicit well-calibrated and overconfident assessments from the same subjects, just by changing the problem set. This will be demonstrated in this experiment.

The problems were, as in Experiment 1, framed within the setting of a memory experiment. The material to be remembered was a nationwide TV program on a speed skating race over 200 km in Friesland, which attracted much public attention. About one month after the event 192 students were asked to answer questions on the program; only one student had not seen the program because he had been abroad. All other students had watched it for periods ranging from 30 minutes to 16 hours. On the basis of numbers of hours watched, subjects were divided into two groups: Intensive watchers (three hours or more) and casual watchers (less than three hours). All subjects answered 20 questions about the event. Half of the questions were on details that could have been seen, like the color of the winner's suit, or the make of the car following the leading group. These are the VISUAL items. Half of the questions were on details that one could know, like how the winner's name was spelled, or at what time the finish control was closed. These are the FACTUAL items. Each item was phrased as a two-choice question. The task was again to distribute 100 probability points between the two alternatives.

Table 1 Mean accuracy and confidence in Experiment 3.

group	question type	correct answers	confidence	deviation
intensive	visual	81%	88%	8.2%
	factual	60%	87%	27.1%
casual	visual	73%	82%	9.7%
	factual	61%	83%	21.9%

The major results are presented in Table 1. In both groups visual questions were more accurately answered than factual questions. However, in neither group did subjects adjust their confidence to the difficulty level. The high confidence in their ability to answer questions was maintained for the factual questions.

The overall deviation in the last column of Table 1 is the weighted mean difference between confidence and accuracy in each class (cf. Lichtenstein & Fischhoff, 1977).
Intensive and casual watchers were about equally overconfident with visual questions; with factual questions the intensive watchers were more overconfident than the casual watchers.
The stronger effect of illusion in the intensive group implies that inferences are trusted more when they are made in a context of a larger set of other questions within one's field of expertise. One could describe this effect as SPREADING EXPERTISE.

EXPERIMENT 4: CONFIDENCE RATINGS OF WEATHER FORECASTERS

The results of Experiment 3 suggested that experts may become overconfident when they transgress the limits of their field of expertise. This conclusion would have farreaching consequences for the use of expert opinion in general. Therefore it is useful to test this idea in a different setting.
The setting chosen is that of meteorologists answering questions about population sizes (Keren, 1984). The 21 meteorologists were employed at the Royal Dutch Meteorological Institute. These experts were well-calibrated on meteorological predictions, as was established in a study executed at the same location by Daan & Murphy (1982). Still they might be the victim of illusory inference when faced with problems that are new to them.
The chosen problem area concerned the population of countries and cities. Which has the larger population, Israel or Nepal; The Hague or Rotterdam? Fifty questions of this type were administered to meteorologists and students.

The overall accuracy was 76% correct responses for both groups. The calibration curves showed that meteorologists were generally overconfident, and more so than students. The difference between the two groups was significant, and suggests that again confidence in making probabilistic statements in one area spreads to another area. The difference between population questions and weather forecasting is evident from the fact that meteorologists expressed 100% certainty in about 16% of the cases. In weather forecasting they assign hardly ever 100% certainty to their predictions, because they know that their databases and models are just not good enough.

The conclusion is this: people can be well-calibrated on predictions and still be overconfident when confronted with problems that evoke illusory inference. May be it is the experience that one is perfectly calibrated in one field of expertise that causes the greater amount of confidence in another field.

EXPERIMENT 5: PREDICTION OF MEMORY PERFORMANCE

Thus far we have only discussed confidence ratings. We will now shift to predictions. The first experiment will be on prediction of retrieval from biographical memory (see Wagenaar, 1985, for a more extensive discussion). One of the authors used himself as a subject in an experiment that stretched over a period of six years. During a four year period 1605 events were recorded. For each event six questions were formulated, and at the same time the subject estimated the probability that he could answer the questions correctly at the end of the retention period. All events were scaled on three dimensions: salience (on a 7-point scale), emotional involvement (on a 5-point scale), and pleasantness (on a 7-point scale). During recall the subject responded to six questions for each of the 1605 events, which resulted in a total of 9630 responses.

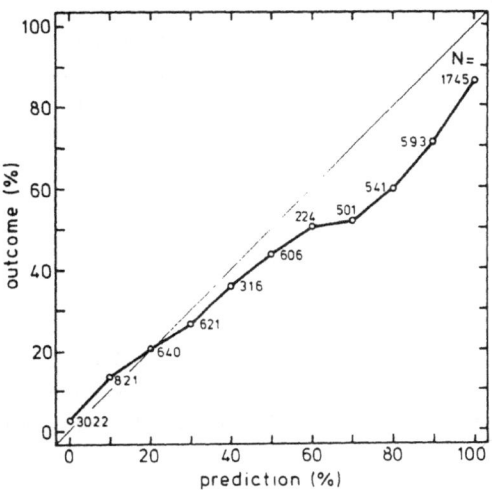

Fig. 3 Calibration curve of retrieval from biographical memory.

Predictions and accuracy scores are presented in Fig. 3. There is a clear effect of overestimation. In a way the subject could be called an expert, since he is an experimental psychologist who is familiar with memory research for at least 20 years. Hence the predictions could have been based on the knowledge that recall becomes more difficult when retention period increases. Also the ratings of salience, emotional involvement and pleasantness could have been taken into account. Since the success rate was predicted for all questions separately, the data base consists of two complete sets of results: predictions and real outcomes. The dependence on the factors mentioned above can be analysed both for predictions and outcomes. In this manner we can establish how the subject weighted the factors in the mental model that produced the predictions, and how much these factors in fact contributed to the outcome.

The probability of correct retrieval of features as a function of retention period is presented in Table 2, together with the predicted probabilities. The retention period stretches over five years, because the recall itself took another year.

Table 2   Observed and predicted probabilities of retrieval from autobiographical memory.

retention period	probability of recall	
(years)	predicted	observed
1	0.46	0.56
2	0.40	0.43
3	0.37	0.41
4	0.42	0.34
5	0.44	0.32

It is clear from this table that retention period, although fully known at the time of recording, was not weighted in the predictions. The subject lived under the illusion that longer retention periods would not affect memory.

When we look at salience, emotional involvement and pleasantness (Table 3) we can first observe that recall was positively related to salience and emotional involvement; pleasant events (category 5 and 6) were better recalled than neutral (category 4) and unpleasant events (category 2 and 3).

Table 3 Predicted and true probabilities of retrieval from autobiographical memory, as a function of salience, emotional involvement
and pleasantness.

category number	salience			emot.involv.			pleasantness		
	pred.	outc.	N	pred.	outc.	N	pred.	outc.	N
1	0.35	0.31	790	0.39	0.35	608	–	–	1
2	0.47	0.31	622	0.42	0.37	686	0.41	0.31	47
3	0.52	0.43	168	0.44	0.35	278	0.39	0.34	204
4	0.62	0.53	21	0.65	0.64	32	0.38	0.32	600
5	–	–	3	–	–	1	0.44	0.40	679
6	–	–	1	–	–	–	0.52	0.56	73
7	–	–	–	–	–	–	–	–	1

The prediction of the general direction of these relations was quite good, but
the details contained some errors. Predictions were too optimistic about the
recall of events in the second salience category (events that occur only once a
week). The same optimism was visible in category 3 and 4, but the impact was
much smaller, because there were less observations in these categories. An
effect of emotional involvement on recall was predicted to show up within the
three first categories, but it did in fact not occur. The recall of unpleasant
and neutral events was predicted too optimistically, thus underestimating the
importance of the pleasantness dimension.

Let us, for the sake of argument, assume that prediction of memory performance
is based on a weighted linear combination of factors that have a predictive
value. The present data show two kinds of shortcomings in the application of
such a model. First too small weights were attached to some factors, like
retention period and pleasantness. Omitting a factor in a linear prediction
model will in principle result in too low predictions over the full extent of
the scale. This would become visible as underestimation. The second shortcoming
is a too large weight attached to some other factors, like salience and emotional involvement. In principle this will result in too high predictions or
overestimation. Both shortcomings affect mean and standard deviation of the
distribution of predictions. It is not improbable that expert predictors would
adjust mean and standard deviations of the predictions such that these are
close to the corresponding parameters of the outcome distribution. Both types
of errors will after adjustment of mean and standard deviation reveal overestimation at the higher end of the scale. This can be easily understood when
it is realised that a 100% prediction based on faulty information can only be
too high, never too low. Therefore it is not possible to infer from the general
form of the calibration curve whether factors received too little or too much

weight. Only a detailed analysis like the one presented here will reveal what shortcomings of the prediction model caused the overestimation.

One reason for the overestimation in this experiment was that the subject was in one respect not a real expert: He lacked feedback about the quality of the assessments. There is no logical reason why prediction models should be perfect without feedback. The operation of feedback depends on two conditions: The availability of data, and the experts' attention. The second principle will be further investigated in the next experiment.

EXPERIMENT 6: PREDICTION OF HAND COMPOSITIONS IN BLACKJACK

The question posed to subjects in this experiment was to assess the probability that a particular card combination would occur in the game of Blackjack (cf. Wagenaar & Keren, 1985). Players in that game try to collect 21 points by buying cards from the dealer, without exceeding that number (busting). The dealer has the same objective, but plays according to a fixed strategy, which is buying with 16 points or less, and stopping with 17 points or more. The players win whenever they collect more points than the dealer, without busting.

The 17 questions were phrased in the form "Out of 1000 hands, how many will ....". The outcomes could have been obtained by recording how often such predictions were correct. This method would have been extremely inefficient, since the probabilities of the various outcomes can be computed analytically.

Three groups of subjects participated in the experiment. One group consisted of 22 professional Blackjack dealers, each of them with several years of experience. None of the dealers had any fundamental insight in statistics and the theory of combinations and permutations. The second group consisted of ten statistical experts who had never played Blackjack. Each of them held an academic degree in a curriculum that included a course in mathematical statistics. The third group consisted of 14 lay people, who had no experience at the Blackjack table, nor any knowledge of statistical theory.

The resulting calibration curves are presented in Fig. 4. First it should be noted that all groups are rather well calibrated. The mean deviation (computed as in Experiment 3) was 0.06 for dealers, -0.03 for statistical experts, -0.07 for lay people. The differences are statistically significant.

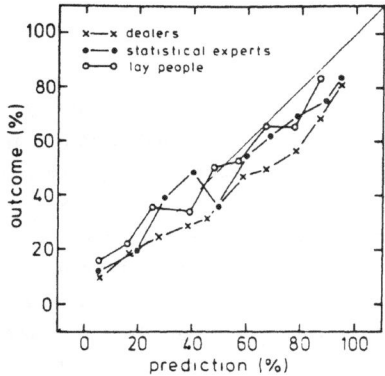

Fig. 4 Calibration in Experiment 6, predictions of hands in Blackjack.

As in the previous experiment, calibration curves do not allow us to attribute the effects of overestimation and underestimation to specific causes. But we can use the responses to perform a discriminant analysis, in order to determine on which questions the groups differed most. Three of the four questions that revealed inter-group differences, were the only ones that required subjects to assess conditional probabilities. Dealers overestimated, and lay people under-estimated these. The statistical experts seemed to do quite well at these questions, with the exception of one question which contained the additional difficulty of cumulative probability, introduced by the term "at least". This problem seemed to baffle statistical experts and lay people alike.

The most surprising aspect of the results is of course that neither of the two groups of experts was considerably better calibrated than the group of lay people. Especially the overestimation by professional dealers came as a sur-prise, since each of these people deal some two-and-a-half million cards a year, thus observing huge numbers of the events about which they were stating their estimates. Apparently being exposed to the experience is not enough, as was argued before by Brehmer (1980). Feedback information is processed, only if it has some relevance. Knowledge about the probability of card combinations is irrelevant to professional dealers. Therefore they will never become experts at the prediction of hands in Blackjack. In the next experiment we will look at a situation in which experts receive feedback that is relevant for the practice of their profession.

EXPERIMENT 7: CALIBRATION OF LAWYERS

This experiment is jointly run in the United States by Elizabeth F. Loftus of the University of Washington, and in the Netherlands by one of the authors. Since the collection of data will take a long time, only some preliminary results can be presented here.
We asked lawyers to predict the outcomes of court trials in which they represented one side, shortly before the trial started. Altogether, 136 observations have been collected, 103 in the Netherlands, 33 in the U.S. Of course these data are not sufficient to plot complete calibration curves. A rough impression of the degree of calibration is obtained by inspection of Table 4.

Table 4   Calibration of U.S. and Dutch lawyers.

predicted	mean prediction	true outcome	p(binomial)
U.S. lawyers N=33			
50% or less	32%	53%	0.06
over 50%	73%	56%	0.11
Dutch lawyers N=103			
50% or less	25%	35%	0.10
50% to 89%	72%	68%	0.37
over 89%	97%	77%	0.00

Both groups exhibited overestimation. The predictions of U.S. lawyers seemed to be extremely insensitive to factors that determined the true outcomes. Dutch lawyers were somewhat underestimating at the lower end of the scale, and overestimating at the higher end.

The low degree of calibration came as a surprise. Lawyers clearly receive feedback about their predictions, and it is inconceivable that they would not pay attention to it. A possible explanation is that prediction of the outcome of a court trial is too difficult. One indication supporting this hypothesis is the difference between Dutch and U.S. lawyers. In the United States many cases are settled before they are even tried. This custom of plea bargaining is of course mostly used when one of the parties knows that the chances are very poor. In the Netherlands all cases are tried, and lawyers assist defendents even when they are almost certain to lose. Hence in the Netherlands there is a class of predictably lost cases that does not exist in the United States.

The present finding, although tentative, presents an obstacle to the use of expert predictions. In many other fields experts may in the same manner be insensitive to the experience of being wrong. Lichtenstein & Fischhoff (1980) showed that the majority of their subjects could be taught to be well-calibrated. The training was achieved by feeding back information about overall calibration, not by showing the correct answer to each individual problem. We simply do not know whether knowledge about individual true outcomes can have any effect on calibration. The lawyers experiment suggests that it does not.

EXPERIMENT 8: CALIBRATION OF BRIDGE PLAYERS

In this experiment we asked bridge players to assess the probability that a contract would be made. Two groups of subjects were used: 16 highly experienced players from one of the top clubs in the Netherlands, and 28 ordinary players who never participated in tournaments at the national or international level.

The assessments were made immediately after the final contract was reached. The expert players played 28 games each, resulting in 448 predictions. The ordinary players played each 24 games, which resulted in 672 predictions.
Since prediction of the outcome on the basis of incomplete information is the essence of the game of bridge, and since both over- and underestimation are punished by the scoring rule, one may expect that bridge players are well-calibrated.

The results showed that the top players were very well calibrated, considerably better than the ordinary players. One reason for the difference is clear: The less experienced players used the extreme predictions of 100% more frequently than the expert players. The expert players used this categories rarely and, what is more important, correctly. This does not mean that the expert players were just conservative. They doubled 23% of the games, amateurs only 10%. However, the weapon of doubling was used incorrectly only once by the experts, and almost in half of the cases by the amateurs.

The difference between experts and amateurs is not only evident from the way they make predictions, but of course also from the way in which they subsequently play. All the contracts that received a 100% prediction from the amateurs could in fact have been successful, as was shown in a post hoc analysis. Nevertheless 22% of these contracts failed, a thing that never happened to the expert players. This lack of skill in playing was not properly weighted by the amateurs. On the contrary, the amateurs generally thought that they were better players than their opponents. Amateurs produced more high predictions when they bid the final contract, more low predictions concerning the opponents contracts. The experts did not exhibit this optimism about own capacities.

The remaining question is, why were the experts better? The amateurs played not at the top level, but still they could have learned not to overestimate, not to double at the wrong moments, not to underestimate the opponent. Amateurs do not learn this. Top players do, through a process of explicit study and evaluation of their results. Not only the results of each hand or each match, but also the consequences for the bidding system. A bidding system is in fact the formalized model that produces well-calibrated predictions. It is the formal processing of feedback that makes the expert bridge players well-calibrated.

CONCLUSION

What can we conclude from this series of somewhat unrelated experiments? In almost all cases we have seen that confidence ratings and predictions can be considerably miscalibrated. But the underlying mechanisms are quite different. The sources of overconfidence demonstrated in the experiments were the illusory nature of the questions and the spreading of expertise beyond its limits. The sources of overestimation were unavailability of feedback, lack of attention to feedback, and feedback in the wrong format. We conclude from these results that calibration of expert predictions should not be studied by asking subjects to rate levels of confidence.

Experts can be the victims of both types of miscalibration. However, overestimation will occur more frequently, since experts are more involved in prediction tasks and do not usually add a degree of confidence to their own statements. Both types of miscalibration could be reduced by proper training. The problem is that training is not easily accomplished when predictions concern rare events or events that occur in a far future. Meteorologists and lawyers can in principle be well-calibrated because they make short-term predictions and collect many data points. Experts in the field of technology and economics are not always in the position to learn from the outcomes they predicted. Not only because the outcomes are rare or in a distant future, but also because they may attribute errors of prediction to intervening influences that were unknown at the time of prediction. We argue that in reality many expert predictions will be incorrect because proper training was not possible. Since experts must adjust their formal or informal models through which their predictions are produced, training in other areas is of little avail. The many instances in which experts' predictions are suboptimal suggest that it is not always wise to model expert systems after the experts they are supposed to mimic. Whenever these intelligent programs are fed with probabilistic information provided by experts, such as importance weights of factors combined in a linear regression model, one should seriously consider the possibility that experts misjudge these weights. There is little reason to believe that experts possess a mysterious sixth sense, an innate intuition, or an undefinable fingertipfeeling. Experts will be calibrated only after they have learned to adjust the parameters in their prediction model on the basis of a formal analysis. In those cases the same analyses could provide the input to expert systems and other intelligent programs.

The title of this paper was phrased as a question. Does the expert know? The answer is: It depends. Experts are less likely to know when they could not, or did not, compare a large number of predictions and outcomes in a systematic way. The confidence of the expert is not a sign of accuracy, because it is often based on the supposed quality of misapplied inferential reasoning and unjustified spreading of expertise beyond its limits. Predictions by experts can be accepted as expert predictions only when the experts have learned to be well-calibrated. Without that, we cannot be sure that they know.

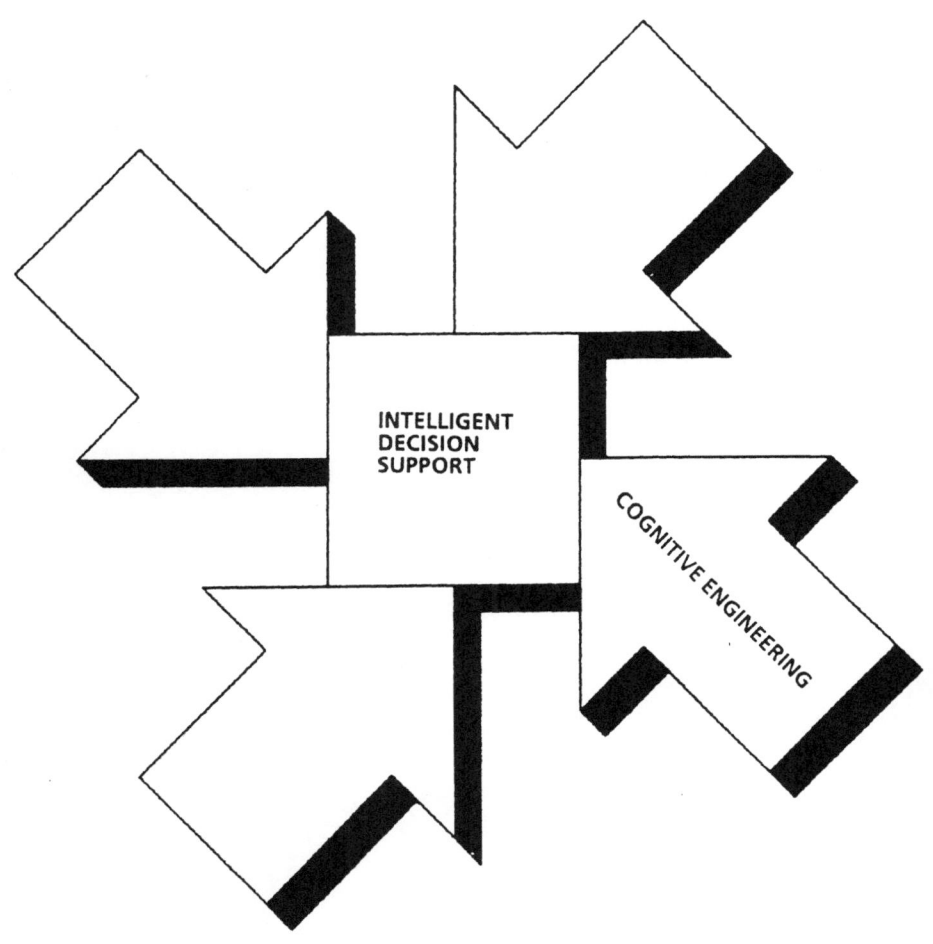

INTELLIGENT
DECISION
SUPPORT

COGNITIVE ENGINEERING

COGNITION AND KNOWLEDGE

THE ELICITATION OF EXPERT KNOWLEDGE

Jacques Leplat
Laboratoire de Psychologie du Travail
de l'Ecole Pratique des Hautes Etudes
Paris, France

The elicitation of knowledge from an expert is practised in several fields and is of interest to various types of specialists: psychologists, pedagogues, cognitivists, engineers. There may be different goals for this elicitation and different uses made of the obtained knowledge. The psychologist, for example, expects to gain a better understanding of cognitive functioning. The elicitation of knowledge may also have practical objectives in terms of cognitive task engineering or intelligent teaching aids, which are not, however, independent of the first goal. The modalities of elicitation may vary according to the goal, but we shall be more concerned here with the general problems involved in this elicitation. We shall often refer to the elaboration of expert systems without, however, neglecting the information provided by pedagogically oriented elicitation. We shall attempt to identify the sources of difficulty in elicitation and then to determine the interest and limits of the methods involved in this elicitation. Rather than developing operational procedures we shall try to specify the difficulties inherent in these methods and examine the most favourable lines of approach. We have voluntarily avoided a direct discussion of the concept of knowledge itself.

I - WHY DO WE NEED TO ELICIT EXPERT KNOWLEDGE ?

One may be surprised that the problem of elicitation has remained so important, pedagogues being still confronted by it. Landa (1984) reminds us of the questions they have to face :

NATO ASI Series. Vol. F21
Intelligent Decision Support in Process Environments
Edited by E. Hollnagel et al.
© Springer-Verlag Berlin Heidelberg 1986

"How should we teach students and non experts to perform at an expert level in any subject area ? In order to be able to teach in such a way, we have to know what is going on in the minds of experts -what mental operations they perform- that enable them to successfully do their tasks and achieve their goals" (p. 237). If these problems persist it is probably because of their difficulty. Lenat (1982) made the following remark which still holds today, that "one current bottleneck in constructing large expert systems is the problem of knowledge acquisition : extracting knowledge from a human expert and representing it for the program" (p. 115). Gallanti and Guida (1985) have confirmed here this point.

A - Expert knowledge : Since we are dealing with elicitation of knowledge from an expert, it is important to specify some general traits of this expert.

- The expert knows how to solve the problems in his field, even those occuring for the first time.

- The expert knows how to state clearly, badly formulated problems and represent the problem in such a way as to allow it to be solved.

- The expert knows how to evaluate the consequences of a deviation from optimal conditions in the solving of a problem.

- The expert builds up his knowledge through practice and may periodically reorganize this knowledge.

- The expert can take decisions without complete information or when information that is usually available is missing (Lenat, 1983, p. 37).

- In certain cases, the expert is also the person who knows how to combine different types of knowledge that are available concerning a particular phenomenon in order to predict its properties or behavior (Hawkins, 1983).

The concept of expertise is very close to those of skill and competence, and similarly corresponds to the act of aiming for a goal. The expert has a particular objective in view and he knows how to find the best path to reach it. The expert is therefore closer to the technician than the scientist in that science aims at developing knowledge whereas technology uses knowledge to achieve fixed goals (Leplat and Pailhous, 1981). This comparison can help us in laying down the problem of expertise. One can also find some useful elements of epistemological reflection in the work of Argyris and Schön (1976) and Schön (1983) concerning the nature of knowledge in practitioners. The questions asked by these authors about practitioners are directly linked to the questions concerning the expert. "We are in need of inquiry into the epistemology of practice. What is the kind of knowing in which competent practitioners engage ? How is professional knowing like and unlike the kinds of knowledge presented in academic text books, scientific papers and learned journals ?" (Schön, 1983, p. VIII).

B - Elicitation and acquisition : The elicitation of expert knowledge may be considered as the inverse process of that which aims at its acquisition during training.

Elicitation aims at externalizing knowledge whereas acquisition aims at internalizing it. If all the knowledge of an expert was the result of controlled acquisition processes, then there would no longer be a problem of elicitation. When the acquisition process is completed the knowledge to be elicited is precisely that knowledge which was intended for acquisition. The only problem here is the control of acquisition through examinations.

Considering the reasons for the lack of symmetry between acquisition and elicitation may help us to improve our understanding of these two processes. The psychology of learning has enumerated many such reasons, we shall examine here only a few of these.

Firstly, one must note that the knowledge acquired at a specific moment for a specific reason is not an isolated, autonomous body, but rather takes position in an already existing system of knowledge, which is different in each subject."... Learning depends on a sub-group of available knowledge and abilities which are effectively mobilised by the subject in a given situation" (George, 1983, p.14) "When an expert learns something new, this something is added to an established and related body of knowledge efficiently organized to receive it" (Hawkins, 1983, p.13). Acquisition is therefore not only local, but is a more general process in which the already existing system of knowledge acquires new properties and may even be reorganized. We believe therefore that knowledge, and in particular the often extremely complexe knowledge of an expert, will always be difficult to specify. It will never be easy to evaluate what is lost when a system of knowledge is isolated from the more general systems in which it is integrated.

A second reason which is fairly close to the one above, concerns the fact that the learner draws from multiple information sources and that these sources can never be totally controlled.

Another reason particularly important in the case of the expert is related to the role of learning through action. The individual is asked to use his knowledge to answer questions, solve problems. He is then given the opportunity to see how well he has performed. This type of learning often plays an important role when the situations are complex and require, for example, the combination of different types of knowledge. A considerable amount of expert knowledge comes from such a source. What is referred to as expert experience come from this origin and therefore, depends partly on the types of situation he had to deal with. As George (1983) has noted, the "informative function of results allows the correction of ineffective acts and therefore the acquisition of new abilities, in other words knowledge about how to act or action procedures" (p. 15).

These comments demonstrate that the acquisition process is a complex one, only partially controlable, and that therefore the knowledge of an expert at a given moment may not be identified with the knowledge that he acquired during his formal training. It follows that techniques of knowledge elicitation for I.D.S.S. must somehow address and retrieve knowledge which will probably not be describable in the same way as the formal knowledge originally acquired by the expert. We can equally understand why the variety of experience before and after the learning period can lead to very large differences between experts.

II - THE OBJECT OF KNOWLEDGE

All work on the elicitation of knowledge necessitates a certain amount of reflection on the object and nature of know-

ledge. A number of the problems which we come across when reflecting on this question are of the same type as those confronted in the sixties by the specialists of programmed teaching.
They are the same as those teaching and it will be useful to
refer to the work done in these fields. This work has shown the
importance of defining the task which the training was designed
for, which leads us to the central question concerning the expert's task. What are the objectives assigned to the expert
which he must respond to and from which the expert system and IDSS will
be conceived ? The answer to this question is not as simple as
it may first appear.

Let us consider, for example, the detection of failures in
a slightly complex system. The general task is to identify the
technical origin of the failure. Although the goal may be clear,
the conditions are not quite so clear, as they may vary enor-
mously. Does one expect the operator to find the origin of all
possible failures using his knowledge of the functioning of the
system, or do we expect him to discover the origin of the most
frequent failures which have been previously identified and for
the detection of which some indications already exist. The operator will function differently in these two situations : in the
first case, he will be more "knowledge-based" and in the second
case more "rule-based". Bisseret (1984) has stressed the particular difficulties encountered in the first case in the elaboration of expert systems. "The unusual unforeseen situations (...)
are of a different nature than familiar situations and that it
is therefore necessary to seek specific assistance solutions
for them" (p. 2).

The definition of the task is therefore a essential step
in the elicitation : it requires a consensus between different
persons involved in the task. It will be particularly important
in the choice of relevant experts and in defining the types of
problems with which they will be confronted in the studies aiming at eliciting knowledge.

III - THE NATURE OF KNOWLEDGE.

Knowledge may be described in various ways and it is often
difficult to dissociate knowledge from the expression of this
knowledge. The operationalization of the concept of knowledge

is by no means an easy task. What do we mean by knowing a machine, knowing how it works ? What do we mean by knowing a
task ? There exists several means of expressing this knowledge,
the relation between these different means not always being simple and their coherence not always being perfect. For example,
differences will be observed if we deduce knowledge of functioning from questions such as "If we manipulate this control what
will be the result ?" or if we deduce it from the execution of
the subject. The operator may, for example, use certain properties that he did not express, or vice versa.

The means of expressing knowledge will be more or less adequate for the processing which they are associated to. We shall
need to distinguish between the expert's expression of his knowledge from the use he makes of this knowledge, and the elabora-
tions of the analyst.

These problems will be approached here via three aspects :
the level of explanation, the acquisition of cognitive skills,
and heuristics.

A - <u>The level of explanation of knowledge</u> : We shall use the
production rules which characterize the formalization of knowledge in expert systems to approach this problem. We shall refer mainly to Clancey's (1983) detailed presentation using some
of his key elements.

Clancey looks first of all at the problem of justifying the
rule. The rule takes the following form "if such a condition,
then such an action" for example in MYCIN "if the patient is
less than 8 years old, don't prescribe tetracycline". Such a
rule does not mention the mechanism by which it may be justified. If we want to show the existence of such a mechanism we
need "to expand the rule" according to Clancey. Taking as an
example the above rule, we could introduce some "intermediate
concepts".
"Tetracycline in youngster
  ➞ Chelation of the drug in growing bones
     ➞ undesirable body change
        ➞ don't administer tetracycline" (p.225).

It is immediately clear in this example that one could go
even further in the clarification process by explaining the

mechanism of chelation. One can talk about depth of understanding in this process : "Conceptually, the support knowledge for a causal rule is a tree of rules, where each mode is a reasoning step that can theoretically be justified in terms of finer-grained steps" (p. 226). As Lenat (1983) noted "the deeper the model, the more costely it is to build" (p. 34).

In the elicitation of knowledge, it is preferable to determine the desired level of explanation beforehand. The operator will spontaneously describe his knowledge at a level that corresponds to his own competence or to the supposed competence of his interlocutor. The level to be obtained will depend on the goals of the elicitation process and the level of knowledge of the future user of the rules. If this person already has a high level of knowledge it will not be necessary to develop the explanation since it will already be known to him. When the user in particular the trainee has seen how the rules are formed, it is necessary to justify them, a process which will help him to memorize them. Providing knowledge of the justification of rules is certainly the factor that follows flexibility in their application, which we have already noted as one of the features of the expert. "Knowing the basis of a rule allows you to know when not to apply it, or how to modify it for special circumstances (...) you can deliberately break the rule because you understand the assumptions underlying it" (Clancey, 1983, p. 241).

In designing IDSS it will be useful to give the operator an opportunity to obtain the justification of the rules if he so wishes. Indeed, it is often observed that when the operator does not know the justification, he tends not to apply the prescribed procedures but to refer to whatever in complete or even inexact knowledge he has about the system.

For example, the operator will have difficulty in applying a rule which entails the momentarily displacement of a variable from its normal position to restore the system functioning. The introduction of such "detours" in a procedure is acceptable by the operator only if he perceives their justification.

B - The acquisition of cognitive skills and elicitation of knowledge : What characterizes the expert is not only an organized system of knowledge but the great availability of this

system, the ease of access to it and the possibility of using
it to complete giving tasks. An expert is characterized by his
skills, and in particular by his cognitive skills. The study
of the acquisition of these skills can help us to understand
better the problems of elicitation Anderson (1982) described
"a framework for skill acquisition that includes two major stages in which facts about the skill domain are interpreted and
a procedural stage in which the domain knowledge is directly
embodied in procedures for performing the skill" (p. 369). The
theory proposed by Anderson deals with "the use and development of knowledge in both the declarative and procedural form
and about the transition between these two forms" (p. 370).

The mechanism involved in the transition from the declarative to the procedural form, known as "knowledge compilation"
allows a better understanding of the nature of the cognitive
skill as well as the difficulties inherent in the elicitation
process. This compilation consists of two basic processes :

- a composition process which reduces two elementary rules
which follow each other in a particular activity into one single rule ;

- a proceduralization process which allows one to leave
aside the declarative information by specifying the practical
application of this information (for example, there exists a
certain number of elementary rules involved in telephoning
abroad : dial the country code, etc... Proceduralizing this
knowledge in the case of a particular person X whom we often
call, transforms the rule into : to call X, dial the following
numbers "1 8 2 4 ..."

This transformation of knowledge with learning has been
described by different psychologists in various languages.
Galperine (1966) following on from Leontiev, described the steps
in the formation of action which are similar to some of the
processes described by Anderson. For example, Anderson's composition process may be compared to Galperine's reduction of
action.

A similar distinction may be found in the work of Landa
(1976) who uses the terms "algorithmic processes" and "algorithmoid processes" ; the first term described a sequence of
instruction which are followed in order, the second described

"processes which occur simultaneously or nearly simultaneously or partially simultaneously" (p. 194). The "algorithmoid processes may be formed in two ways : as a result of the development of an algorithmic process when the consecutive process has been transformed into a simultaneous or nearly simultaneous or instanteneous process" (p. 194) or by being formed fortuitously in this manner from the beginning.

The categories of human behavior proposed by Rasmussen (1983) may also serve to indicate stages in the acquisition of cognitive skills. The behavior is first "knowledge-based" by reference to an internal model with an explicitly formulated goal, then controlled by stored rules ("ruled-based") and finaly "skill-based", in highly integrated patterns, without conscious control.

When we are dealing with an expert, the acquisition process is in its final stage. The rules used by the expert differ from the initial rules, most notably from those of the declarative stage where they were often expressed verbally. They will occasionally have to be redeveloped in order to rediscover the elementary rules. The problem here is to know whether the expert is still capable of expressing his knowledge in a communicable form. What distinguishes the expert from the experienced person may be the greater ease with which he expresses this knowledge.

An important consequence for IDSS of this skill transformations over time, is that adaptative and efficient aids cannot be the same on different occasions. This touches on a general difficulty often met in the use of aids.

- Either these are designed for beginners and are not adapted to experienced operators who consequently make little use of them.

- Or they are designed for experienced people, or by experts who are not familiar with the operator's work, with a content scarcely accessibly to beginners.

The non use of job aids is often attribuable to one or other of these defects. IDSS have to be designed in an adaptative mode, which takes into account the different levels of expertise. They have to present only stricly necessary informations by giving to the user the opportunity to choose easily those which are most useful at the time.

C - <u>Heuristics</u> : Anderson (1982) noted that "all tasks can be characterized as having a search associated with them..." By search he means "that there are alternate paths of steps by which the problem can be tackled and the subject must choose among them" (p. 390). These paths lead to more or less complex and elegant solutions. "With experience the search becomes more selective and more likely to lead to rapid success" (p. 390). Several means of analysing the mechanisms of this search exist. We shall adopt here the perspective offered by one line of research (Polya, 1958 ; Pushkin, 1972) and recently used within the framework of artificial intelligence by Lenat (1982-1983). Lenat has underlined the importance of heuristics which he defines as "pieces of knowledge capable of suggesting plausible actions to follow or implausible ones to avoid" (1982, p. 192), and for which he has provided the operational translations in two programs (Accretion model and Eurisko). He stresses that "the expert must communicate not merely the "facts" of his field, but also the heuristics : the informal judgmental rules which guide him in rapid decision-making. They are rarely thought about concretely by the expert and almost never appear in his field's journal articles, text books or university courses" (Lenat, 1982, p. 195). Every analysis of expert activity aims at demonstrating the existence and use of heuristics. Expertise is acquired only after "a trying period during which they must induce the heuristics of their craft from examples" (p. 195).

Lenat has put forward a certain number of heuristics (for example "analogize : if actor A is appropriate in situation S, then A is appropriate in most situations which are very similar to S" p. 208). These heuristics help to orient oneself in the problem space. The work analyses done on the field often show the existence of such heuristics. Thus a certain number of the " diagnostic rules devised for fault-finding" by Shepherd, Marshall, Turner and Duncan (1977) may be considered as heuristic rules (for example : check all pressure controllers to make sure that they are functioning correctly").

Heuristics formalize the informal judgment "rules of thumb"

which often characterize the activity of the well-experienced expert. They underly what is referred to as the expert's intuition and they are often difficult to specify ("heuristics are not easily elicited from experts", Lenat, 1983, p. 37). The heuristic rules acquire the status of skills when they are completely interiorized and immediately available.

The design of IDSS can profit greatly from an understanding of how these heuristics operate. In particular they can be used to guide the operator's activity. They are specialy useful when the field of expertise is very large and when it is difficult to define an algorithmic procedure.

IV - THE METHODS OF ELICITING KNOWLEDGE

Knowledge is not directly apprehensible and one must find a means of expression which the knowledge will depend on. It will also depend on the situations in which this expression will be recorded. It will therefore always be important to compare the knowledge obtained using different methods. One can only ever hypothesize about the knowledge of one subject and these hypotheses must be verified by various means. In this vast domain we can only describe a few preliminary indications taken from the methodology of work psychology (Leplat and Cuny, 1984).

A - The types of methods : The different types of method used to elicit knowledge may be presented by considering that the expert's activity depends on the characteristics of this expert and the characteristics of the task he has to accomplish. The different methods can be classified in relation to these components.

a - Spontaneous and induced behaviour : Spontaneous behaviour is observed in ordinary conditions of task execution. In particular it participates in the transformation of states who must achieve the goal. Induced behaviour is instigated by the analyst either during or outside of the task (for example, verbalization).

b - Modification in the task : Modifying the task allows the

evaluation of the role played by its different elements. This
is at the basis of simulations, simplified material models of
the task which facilitate the observation of subject activity
and analysis of underlying mechanisms. The modifications may
concern the instructions for the task and the constraints (notably the temporal ones) which control its execution (cf. for example Zakay and Wooler, 1984).

c - Modifications in the subjects : These may concern the same
subject (taken at different levels of learning or competence)
or different subjects with varying characteristics which are
controlled (comparative, longitudinal and cross-sectional methods). Researchers have often successfully compared experts
with beginners. Sweller, Mawer and Ward (1983), with such a
technique, were able to demonstrate the existence of two types
of strategy in the solving of "kinematic problems" and to show
that "the means-ends strategies used by the novices retarded
the acquisition of appropriate schemas" (p. 639).

Modifications in the task and in subjects, which may be
combined, constitute means of revealing the knowledge used by
the expert and of testing the hypothesis concerning this knowledge.

B - The choice of methods : This choice depends on the conditions related to the goals of elicitation and to the characteristics of the situation (defined as the subject task ensemble).

The criterion of choice to be discussed here is based on
the organizational mode of the activity under examination and
in particular on the three types of functioning defined by
Rasmussen (1983).

a - Skill-based activities : The verbalization methods will be
of little relevance with activities of this type as they are no
longer regulated at a representative level and it is difficult
to specify the procedures used.

On the other hand, the methods using the observation of
spontaneous behaviour will be very useful here. The study of visual exploration and the analysis of video recordings will
often provide valuable information on subjects procedures, even

if the inference of cognitive mechanisms and underlying knowledge from observed behaviour is never direct (the same signi-
fier can have several significations, and vice versa, and may
be used in different ways). We will be able to use subject and
task modifications here. An example would be the method of information occultation used by Shepherd et al. (1977).

b - Ruled-based activities : In this case the elicitation of
knowledge may begin by a detailed observation of subject behaviour and also the technique of subject and task modification.
But the privileged methods are the ones that introduce verbalization and which may be associated to the previous ones. They
may be used in many different ways and have been the object of
numerous studies and critical analyses (notably, Ericson and
Simon, 1980 ; Leplat and Hoc, 1981).
    The analysis of the relations between task execution and
its verbalization is one of the classic subjects of psychology
and continues to be a center of interest for research aiming
at specifying the mechanisms that control these relations (Berry
and Broadbent, 1984).
    Here are some of the factors that have a direct influence
on verbalization and the quality of execution :
    - The compatibility between verbalization and the activity
it refers to.
    - The possible interference of verbalization on reference
activity (cf. the concept of reactive measures, Campbell, 1957).
    - The task instructions play an important role in that
they induce the verbalization activity. The instructions can
be used to make the subject verbalize a particular action, procedure or the knowledge underlying the procedure (Leplat and
Hoc, 1981).
    - The verbalization is addressed to the analyst and de-
pends on the representation the expert has of this analyst.

c - Knowledge based activities : In this case, verbalization
plays an even greater role than in the previous one, but the
other methods should not be neglected. Situations may be created here, in which the verbalized expression of knowledge will
be greatly facilitated and much richer. Particularly making
the task concrete in the form of new problems or modifying the
conditions of execution will allow the expression of knowledge
and above all the heuristics used by the subjects.

In conclusion, we wish to underline the fact that the analyst needs to know the different methods of elicitation and
the good and bad points of each one of these.

C - The validation of knowledge.: In conventional examinations
knowledge is validated in relation to the norms of scientific
and technical knowledge. When we are dealing with knowledge elicited from an expert this type of validation is of little use
since it is the knowledge to be validated that is supposed to
be the reference standard for the examination. Validation will
in this situation be determined in relation to the task and
the activities it involves. The fundamental question will be
to verify if the elicited knowledge will allow the expert to successfully complete the tasks in which he is competent, in the domain that interests us. The validation therefore depends partly on the objective of the elicitation. It may only concern
the result, but generally we attempt to validate predictions
concerning the procedure of task processing and we are therefore led to compare intermediate results and to interpret the
deviation between these and the model's predictions.

We shall continually have to ask the question if these divergences are due to the expert only giving part of his knowledge or if that knowledge was incorrectly organized and used
by the analyst. Here are some of the different types of validation that may be considered :

a - Validation by generalization : This consists in applying
the model elaborated from the elicited knowledge to a set of
possible situations belonging to the same class.

b - Validation by contrasting knowledge derived from different
methods : The possible divergences between these different

sets of knowledge are themselves a useful source of information.

c - Interactive validation : The elicited knowledge is used to complete a task. The expert evaluates the results and interprets the divergences. This leads him to complete the knowledge which had been expressed beforehand (a good example is given in Carbonell et al., 1984).

d - Validation by training : We examine if the teaching of elicited knowledge improves the acquisition process and subject competence (Shepherd et al., 1973).

IV - CONCLUSION

The elicitation of knowledge constitutes an important chapter of cognitive systems engineering. Hollnagel and Woods (1983), as well as those involved in the psychological analysis of work, have already underlined this point in stating that "cognitive analysis is needed to understand the cognitive activities required of the MMS..." (p. 592). The elicitation of knowledge is an important part of this analysis. We have presented here some theoretical and methodological questions involved in this elicitation. These questions may act as guidelines for future research in this domain.

The knowledge relevant to the task under consideration has the following important characteristics : it forms part of a more general system of knowledge and it evolves through prolonged activity.

Elicitation is the result of an interaction between the expert and the analyst and depends on the characteristics of both of these. In particular, it depends on the expert's ability to be conscious of its functioning and the instruments available to the analyst. Knowledge may not be considered as raw data, it is reconstructed and organized according to languages and models which are related to the way it is to be used and to the properties of the (cognitive or computer) system that it will be transfered to.

This is why the methodological problems of elicitation are so important. We need to know which tools can help in this elicitation and we need to be able to evaluate the possible influences they may have on the knowledge when being used. Although this point has caused little concern up to now, it would be useful to train experts themselves in the elicitation process and to study the methods best suited to this.

If elicitation of expert knowledge is necessary to design of IDSS, conversely the elaboration of IDSS can give information about the nature of this knowledge, when validation is well done. An IDSS correspond to an hypothesis about the nature of the declarative and procedural knowledge implemented by the expert : to test the efficiency of the IDSS will also be to test this hypothesis, if this hypothesis has been made explicit. So, the design of IDSS and the elicitation of the expert knowledge are too mutually dependent processes which ideally develop in parallel.

NEW VIEWS OF INFORMATION PROCESSING:
IMPLICATIONS FOR INTELLIGENT DECISION SUPPORT SYSTEMS

Donald A. Norman
Department of Psychology and Institute for Cognitive Science
University of California, San Diego
La Jolla, California (USA)

In this chapter I combine several different paths of thought. First, I briefly review developments in the study of human information processing, in particular, those based upon parallel activation structures which operate by a form of generalized pattern matching. Second, I review some issues of human error and demonstrate that the problems are not so much in the making of the error as in the difficulty of discovering the errors, once made. Finally, I conclude with an analysis of the implications for Intelligent Decision Support Systems. The purpose of this chapter is to point out potential issues, problems, and directions of approach.

The PDP Approach to the Study of Cognition

Models of Parallel Distributed Processing (variously called "PDP models," "connectionist models," or "associative models") provide an important new outlook upon processing structures. [1] These models are massively parallel, with thousands or millions of

---

My research is supported by Contract N00014-85-C-0133, NR 667-541 with the Personnel and Training Research Programs of the Office of Naval Research and by a grant from the System Development Foundation. The research on human error was supported by Contract N00014-79-C-0323, NR 667-437 with the Personnel and Training Research Programs of the Office of Naval Research.

1. This section is adapted from my chapter in J. L. McClelland and D. E. Rumelhart (Eds.) (1986). *Parallel Distributed Processing: Explorations in the Microstructure of Cognition. Volume II: Applications.* Cambridge, MA: Bradford Books, MIT Press.

elements interacting primarily through activation and inhibition of one another's activity. Elements are highly interconnected, with perhaps tens of thousands of interconnections. Although the processing speed of each element is slow—measured in milliseconds rather than the picoseconds common in today's computers—the resulting computations are fast, faster than is possible with even the largest and fastest of today's machines. Parallel computation means that a sequence that requires millions of cycles in a conventional, serial computer can be done in a few cycles when the mechanism has hundreds of thousands of highly interconnected processors. These neurologically inspired computational processes place new demands on our understanding of computation, suggest novel theoretical explanations of psychological phenomena, and suggest powerful new architectures for the machines of the future.

Under this new view, processing is done by PDP networks that configure themselves to match the arriving data with minimum conflict or discrepancy. The systems are always tuning themselves (adjusting their weights). Learning is continuous, natural, and fundamental to the operation. New conceptualizations are reflected by qualitatively different configurations. Information is passed among the units, not by messages, but by activation values, by scalars not symbols. The interpretation of the processing is *not* in terms of the messages being sent but rather by what states are active. Thus, it is what units are active that is important, not what messages are sent. In the conventional system, learning takes place through changes in the representational structures, in the information contained in memory. In this new approach, learning takes place by changes in the system itself. Existing connections are modified, new connections are formed, old ones are weakened. In the conventional system, we distinguished between the information being processed and the processing structures. In the PDP system, they are the same: The information is reflected in the very shape, form, and operation of the processing structures.

To ask "why" a particular action was performed or "why" a particular interpretation was placed on an event means to ask "why" a given internal stable state was reached. But there is no simple answer to these questions. To force the questions is to risk confabulation: answers made up after the fact with no necessary connection to reality. In general, *there is no single reason why any given cognitive state occurs.* The system is multiply-connected, multiply-controlled. The states are a result of many different factors all impinging upon one another: The state is simply the best match to all the sources of information.

For me, the work has been important for several reasons. In my personal opinion, many students of cognition have boxed themselves into corners, forcing themselves to develop ever more complex rule structures, ever more complex algorithms for dealing with the behavior of people. Each new experimental finding seems to require a new theory. Clearly, some new approach has been needed. But what?

The PDP approach offers a very distinctive counter-approach. Basically, here we have an adaptive system, continually trying to configure itself so as to match the arriving data. It works automatically, pre-wired, if you will, to adjust its own parameters so as to accommodate the input presented to it. It is a system that is flexible, yet rigid. That is, although it is always trying to mirror the arriving data, it does so by means of existing knowledge, existing configurations. It never expects to make a perfect match, but instead simply tries to get the best match possible at any time: The better the match, the more stable the system. The system works by storing particular events on top of one another: Aspects of different events co-mingle. The result is that the system automatically acts as if it has formed generalizations of particular instances, even though the system only stores individual instances. Although the system develops neither rules of classification nor generalizations, it *acts* as if it had these rules. It is a system that exhibits intelligence and logic, yet that nowhere has explicit rules of intelligence or logic.

The heavy dependence upon particular instances and the lack of explicit generalizations fit a reasonable body of evidence about the importance of specific experiences (e.g., see Kahneman & Miller, 1984; Medin & Schaffer, 1978). These systems try to accommodate themselves to the data by minimizing the overall discrepancy. This leads to categorization and classification of the input signals in a flexible manner—categorization by distance from prototypes, a mode that is consistent with much recent evidence. Here is a system that incorporates learning as a fundamental, essential aspect of its behavior, that makes no attempt to form categories or rules, yet that acts as if it were a prototype-matching, categorization system, that has rules and strategies.

Under this view of things, the common concept of the *schema* must be modified slightly. According to the new view, schemas are flexible configurations, mirroring the regularities of experience, providing automatic completion of missing components, automatically generalizing from the past, but also continually in modification, continually adapting to reflect the current state of affairs. Schemas are not fixed, immutable data structures. Schemas are flexible interpretive states that reflect the mixture of past experience and present circumstances.

Because the system configures itself according to the sum of all the numerous influences upon it, each new invocation of a schema may differ from the previous invocations. Thus, the system behaves as if there were prototypical schemas, but where the prototype is constructed anew for each occasion by combining past experiences with the biases and activation levels resulting from the current experience and the context in which it occurs.

Some years ago, Bobrow and I listed some properties we felt were essential components of the human cognitive system (Bobrow & Norman, 1975; Norman & Bobrow, 1975, 1976, 1979). We constructed the list through observations of human behavior and reflection upon the sort of processing structures that would be required to yield that behavior. We concluded that the system must be robust, relatively insensitive to missing or erroneous data and to damage to its parts. We argued that the system had to work by descriptions rather than precise specifications, by partial information rather than complete information, and by competition among competing interpretations. The properties we listed as essential include: graceful degradation of performance, content addressable storage, continually available output, and an iterative retrieval process that works by description rather than by more traditional search. The PDP mechanisms have exactly these properties.

Problems With Discovery and Correction of Mistakes

Now change the topic quite radically to the analysis of the discovery of error. Later I come back to relate this to the PDP approach. Errors can be classified as *mistakes* or *slips*; I define a mistake as an error in the formation of an intention and a slip as an error in the execution of the intention. The two forms of error are quite different in their causes, in the likelihood that the error will be discovered, and in the steps that one might take to reduce the incidence of the error. This leads to the major points of this section, namely:

- A major class of mistake is that of *misdiagnosis*. There are several consequences that follow from misdiagnosis, perhaps the most obvious and important being that the behavior that follows is then likely to be inappropriate for the situation. There are two issues to consider:

  Why did the misdiagnosis occur?

  What is required to detect the misdiagnosis?

- Detection of misdiagnosis is hampered by a *cognitive hysteresis*, resulting from:

  A bias to search only for confirming evidence.

  The danger of partial explanations.

  The similarity between the actual and the perceived event.

Let me emphasize the need to understand the cause of an error. Suppose that an operator of a Nuclear Power Plant is in the midst of a set of alarms that signify a plant failure. A pressure injection pump turns on automatically, injecting water (under high pressure) into the reactor core. Now suppose that the operator erroneously turns off the pump (as indeed one is supposed to do under certain circumstances). We can ask: What is the likelihood that the operator will detect this error and re-activate the pump? The answer, however, depends upon *why* the pump was turned off. If it was turned off by a "slip" (perhaps because the pump control was either nearby or similar in appearance and operation to a control that the operator intended to work), then the likelihood that the operator will detect the error is reasonably high. Because the operator did not intend to turn off the pump, any evidence that the pump was off — perhaps noting the change in core water level and pressure due to the shutting off of the pump — leads to detection of the error.

The situation is quite different if the pump was turned off by a "mistake"; then, the likelihood that the pump "error" will be detected can be very low. Suppose that the mistake occurs because the operator misdiagnosed the plant state and concluded that the pump should not be on. Because in this situation the operator really did intend to shut off the pump, any evidence that the pump was off would simply be interpreted as confirmation that the intended act had been performed properly.

The difference between these two otherwise identical behavioral sequences is instructive. Although the observed *behavioral error* in both cases is "erroneously turning off the high pressure injection pump," in the second case, the *true causal error* is the plant misdiagnosis. Given the misdiagnosis, the act of turning off the pump is *not* perceived as an error by the operator and so, it cannot be detected by normal monitoring of plant state; it is a natural consequence of the misdiagnosis.

What Is Required to Discover a Misdiagnosis?

It is commonly thought that errors in decision or behavior can be detected by monitoring the situation and noting discrepancies between one's expectations and the actual outcomes. As I have just shown, error correction is not that simple, especially following mistakes in diagnosis. This is where the PDP work comes in. The person is often not even aware of having made any decision. That is, as we engage in our daily, routine behavior, we are continually faced with perturbations of the existing situation to which we must respond. According to the PDP approach, cognitive structures configure themselves automatically, without the need for conscious awareness or thought. It is the rare occasion when we recognize that we are faced with an undiagnosed situation which requires assessment and classification. Most of our activities are done naturally, smoothly, and without conscious effort.

Even when we have deliberately and consciously diagnosed the situation, we do not necessarily evaluate all resulting outcomes to see if they are discrepant from expectations. Many discrepancies are probably not even noted, and of those that are, many are probably not judged to be relevant to the assessment of the situation. To see why the detection of mistakes is not so simple we need to examine two things: first, why the mistake occurred in the first place; second, how a person assesses the situation.

Factors Responsible for a Misdiagnosis

The likelihood that a situation will be misdiagnosed is affected by three different factors, all of which may apply at any one time:

1. Description and capture errors.
2. A priori biases, based on prior experience.
3. Erroneous or incomplete knowledge.

A description error occurs when several situations are similar to one another—their *descriptions* are similar (Norman, 1981)—so that there is a strong possibility that there will be confusion among them and an error of diagnoses may occur. A capture error occurs when another situation is both similar to and more frequent than the current one, and despite recognition of the differences, the stronger event "captures" control, leading to false diagnosis or action (Norman, 1981; Reason & Mycielska, 1982). These errors are subject to the second factor—*a priori* biases based on experience. In general, the more frequently an event or action sequence occurs, the more likely it is to be judged appropriate whenever the current situation looks similar to it. Both errors are likely in PDP systems. In PDP terms, the system settles into a stable state that matches the current experience. Problems arise when this is not the correct interpretation. The PDP mechanism will settle into any state that has a sufficiently stable minimum, even when this is only a *local* minimum, not the *global* minimum that is desired.

There is a bias in favor of events already experienced. This need not be conscious; most people are not aware that they have biases of this sort. Second, from the point of view of Bayesian decision making, this is perfectly sensible behavior. Although a bias in favor of more frequent diagnoses or actions may occasionally lead to error, on the whole it leads to more rapid and accurate assessment of and response to the situation. After all, it is indeed the more frequent event that is more likely. The first two factors indicate that a false diagnosis is apt to occur for situations that appear similar to the actual one. As we see in the next section, this leads to one of the major difficulties in detection of the error.

Factors Responsible for Failure to Detect a Misdiagnosis

Given that misdiagnoses of a situation clearly occur, the interesting and important aspect then becomes why they can sometimes persist for such long durations without being detected. After the event is over, in the brilliance of the hindsight produced by a retrospective analysis, it is difficult to see how the misdiagnosis could have been missed. Well, hindsight always is superior to foresight, and even the most obvious occurrences in hindsight are not so obvious at the actual time of the event (thus, see Fischhoff, 1975, 1977).

In any real situation, there is often a lot happening simultaneously, most of which is irrelevant to the situation of interest. This means that we must learn to ignore the irrelevant events in the environment and attend only to the relevant ones. But how do we determine which is which? In a nuclear power plant, there are numerous indicators in the control room, perhaps several thousand, most of which are not relevant to the major event sequence under analysis. But when a major event occurs it is not obvious how to distinguish the relevant alarms and indicators from the irrelevant. In addition to these problems, there is a major psychological factor that affects how sensitive a person will be to apparent discrepancies: I call this factor *cognitive hysteresis*: the tendency to stick with a decision. There are at least three cognitive factors that affect this tendency:

1. The similarity of the actual event to the perceived event.
2. The danger of partial explanation.
3. A bias to search only for confirming evidence.

1. The similarity of the actual event to the perceived event. As discussed earlier, a major reason for a false diagnosis is the similarity between the actual event and the perceived event; the symptoms overlap. Moreover, once the classification has been made and the appropriate responses to this (false) classification taken, the similarity between the actual and perceived event makes it likely that those initial responses were appropriate for both the false and the correct classification. This similarity impedes the ability to detect the error, for the initial information that arrives after the diagnosis has been made and the initial actions taken are apt to be consistent with the diagnosis, thereby supporting and strengthening it.

2. The danger of partial explanation. In my studies of mistakes and their detection, I have noted one recurring theme: Discrepant information is noted, but discounted because it is "explained away" by finding a reason for the discrepancy. Thus, at Three Mile Island, the operators thought they had closed a Pressure Relief Valve (whereas it was defective and did not close). However, they did notice that the there was a high temperature reading on the output side of the valve, something that could not happen unless liquid were still flowing through it. In this case, the operators remembered that they had previous difficulty with leaks in the valve, which would not produce sufficient flow to be of concern but would give the high temperature reading. As a result of this explanation, they did not check further on the valve (they could have examined the quantity of liquid that was flowing), and this discrepant valve seriously complicated the severity of the incident. Note that the operators' decision was sensible; they had thrown the proper switch and the light indicating a shut valve was on. A secondary indicator— the temperature— indicated some flow of liquid, but there was an explanation for this. The point is that the existence of the explanation stopped the operators from going further with their analysis. It is not clear how much we should expect plant operators to question each observation they make. We could never get very far in our daily activities if each minor variation from expectations had to be explained and analyzed in great detail.

Partial explanations appear to be involved in other situations where discrepant evidence is detected, but "explained away," even though if pursued, the evidence would have pointed to the false hypothesis (see Lewis & Mack, 1982).

3. A bias to search only for confirming evidence. A pervasive bias of humans in decision making or problem solving situations is to search for confirming evidence to their hypotheses at the expensive of disconfirming evidence. This bias holds despite the fact that confirmation is often a weak source of evidence whereas a search for negative evidence would provide quite efficient tests of the hypothesis (see Wason, 1960, 1968). This bias means that the person searches for consistent evidence, often thereby missing disconfirming evidence of the hypothesis.

Cognitive hysteresis: The tendency to stick with a decision. The three factors just described all lead a person to stick with a hypothesis, once it is made. I call this *cognitive hysteresis*: the tendency to maintain a previous state beyond the point where the situation would otherwise warrant it. This situation is well known, both in psychology (where it is called "functional fixation") and in applied settings, where operators are said to get "locked into" their opinions. The situation follows from many of the causes that I have just discussed.

PDP Models and Error Detection.

The PDP approach is consistent with this story of human error. PDP models are characterized by several properties. One is that multiple sources of information combine to yield the final result. The result is a state that best matches the arriving evidence and the prior knowledge of the system. The state can be characterized by its "energy" or discrepancy from a perfect fit. For complex situations, there are many states that give reasonable fits, and the problem of finding the best fit is the traditional hill-descent problem of finding the global minimum in a situation without being trapped in a local minimum. Local minima correspond to capture errors or to misdiagnoses. They are good characterizations of the situation, but they are inappropriate. The problem is that once trapped in a local minimum, it is very difficult to get out. In many PDP theories, the only way to get out is to disrupt the system either by adding noise or by heating up the temperature (the theoretical treatment of PDP systems borrows from statistical thermodynamics, where "noise" and "temperature" are intimately related).

In practical terms, this means that pattern-match systems such as PDP systems can find reasonable, but incorrect matches. When they do, they will suffer from the form of "cognitive hysteresis" I have described. To get out of such a situation requires either a massive amount of new information or significantly different new information. More on this later.

Implications for Intelligent Decision Support Systems

What do we expect of Intelligent Decision Support Systems (IDSS)? The answer depends upon what function is seen to be served by an IDSS. Why do we need IDSS? In part, it must be because both humans and automated systems are flawed. After all, if the human were a perfect decision maker, problem solver, and performer, there would be no need for aids. And if the IDSS were perfectly reliable, fully functional, and technologically practical, then we wouldn't need the person. The fact that we wish for one to support the other indicates that both have problems. This being the case, how do we manage to combine the two—the person and the IDSS—in such a way as to enhance the overall performance? To answer this requires that we know more about both people and systems. In answering the question, we need to be wary about the exact mix of tasks that is devised for the person and IDSS. Just because either of the partners can do one component well does not mean that the division should reflect this. Today, in too many instances, the machine system does whatever it is capable of doing, leaving the rest to the person. This leads to a number of severe problems, some operational, some motivational.

Problems With Intelligent Decision Support Systems

IDSS is another step in the automation of complex control processes, a situations where the person is still in control of the system, but where the person is "assisted" by a support system. Usually, however, this means that the person is really acting as a monitor of the system (although sometimes it is the system that acts as a monitor of the person). The result is to transform the job of the person to that of supervising the system. Many problems with this change are well known. Many are analyzed in the book on *Supervisory Control Systems* edited by Sheridan and Johannsen (1976) and in Wiener (1985) and Wiener and Curry (1980).

One major difficulty arises when IDSS take the person "out of the control loop," resulting in the loss of a feeling of control and general alienation from the task. This results in several things including possible serious error when there is a lack of synchronization between user and person, and a general deterioration of the skills of the person who comes more and more to rely upon the automated system. A second problem comes from the design of these systems. The user needs a clear *system image*, so as to maintain a good mental model of the operation (Norman, 1986).

What Can Intelligent Decision Support Systems Offer?

From the discussion in this chapter—plus pieces hinted at—we can put together some of the properties that an IDSS should offer. From the analyses of the PDP model of cognition and of human error, especially the failure to recognize misdiagnoses, we can draw a number or suggestions for IDSS systems. In brief, the suggestions are these:

- *Aid in correcting misdiagnoses (and the resulting cognitive hysteresis).* The IDSS should help the user continually assess the current state of the system, suggesting alternative explanations and keeping alert for possible critical discrepancies that might otherwise be overlooked or explained away. The problem is to figure out how to do this in a useful way. Nobody likes a backseat driver: The IDSS must earn its trust by the reasonableness of its advice (and perhaps by its accuracy).

  Whenever the user makes a diagnosis, it is important to make sure that the user maintains flexibility as new events occur. The IDSS could take the role of a critic, aiding the user to see alternative possibilities. A major danger is the power of a partial explanation (and the deliberate tunnel vision of hypothesis-guided interpretation). Another way of viewing this is that the IDSS could act as the proper temperature catalyst, always slightly jiggling the system to make sure that the person does not get trapped into the local minimum of a misdiagnoses, but always considers other hypotheses that may lead to better description of the ongoing events.

- *Aid in keeping an accurate mental model of the system.* Although not discussed in this chapter, a critical aspect of human performance is the development and updating of an accurate mental model of the system and its current state (see Norman, 1986; Norman & Draper, 1986). For this purpose, it is essential to present information in such a way that people can readily assimilate it into their own mental models. In general, this will mean graphical presentation in a manner consistent with the thought processes of the person. Problems can arise when the user has an inappropriate User Model, or when the User Model does not reflect the correct status properly.

- *Keeping the person in the control loop.* Studies of overautomation indicate the need to keep the person involved in the operation of the system—in the control loop (Wiener, 1985; Wiener & Curry, 1980). This is especially a problem in cases where the equipment is perfectly capable of taking over large segments of the task. Here, it is important that the person still play an active part in the ongoing activity, not because this presence is required, but because it automatically keeps the person up to date on the current status of the system, the better to respond if an emergency situation occurs. There are several problems here, including the motivational one of keeping the person interested in the task and the informational one of providing the correct support for the task.

- *Real time crisis management.* In major system difficulties, the person can become overloaded, distracted when many events occur in a short time period, either continually changing focus to deal with each new arriving information (data-driven behavior) or focusing too narrowly upon what might turn out to be not critically relevant aspects of the situation (the tunnel-vision behavior caused by excessive conceptually-driven behavior). Here is where an IDSS could play a major role. Again, this is not a topic discussed in this chapter, although it impinges upon many of the issues that have been discussed. Basically, a person's conscious resources are limited and in times of crisis, it is easy to become overloaded. The goal of the

IDSS is to prioritize the information and advice to be given to the person, allowing the major focus to be concentrated upon the major problems, not dissipated among all the minutiae that clamor for attention.

EXPERT KNOWLEDGE, ITS ACQUISITION AND ELICITATION IN DEVELOPING INTELLIGENT TOOLS FOR PROCESS CONTROL

Leena Norros
Technical Research Centre of Finland
Electrical Engineering Laboratory
Espoo, Finland

The essence of expertise in process control

The process operator performs many tasks in his daily duty. When studying process control it is methodologically motivated to try to crystallize the essential demand that characterizes this task. Many of the researchers consider the diagnoses of the process status and the decision on the relevant operative measures the major difficulty in this type of work. This combined activity, judgement, could be defined as the critical activity, the mastery of which essentially determines the quality of the whole work.

However, among researchers there exists some confusion about the nature of this activity. A traditional interpretation of judgement would be, as Buch (1984) points out, that it is a trait that good operators innately posses, or an ability that is acquired as a by-product of working experience. Eventually, good judgement has traditionally developed successfully in that manner.

In the context of enhancing the control of complex systems also the problem of judgement and its traditional interpretation have been reconsidered. The inconsistency of expert judgements in situations were the decision problem is

NATO ASI Series, Vol. F21
Intelligent Decision Support in Process Environments
Edited by E. Hollnagel et al.
© Springer-Verlag Berlin Heidelberg 1986

not represented as a familiar syndrome of correlated attributes would speak of the inadequacy of judgements based on experience (Wickens 1984). Brehmer (1980) formulates the hypothesis that a concept of the object to be learned is necessary for learning from experience. The more complex the object to control the more urgent the need for a concept.

Some researchers take an even stronger position and explicitly differentiate between empirical and theoretical thinking (f.ex. Dawydow 1977). Typically empirical judgement would emphasize accumulation of operative experience. On this basis general features are abstracted and formulated into operative rules and rule hierarchies. The problem that remains is the question whether a particular case would fit a certain rule or any rule, for that matter. The need for this kind of decision manifests the need for a qualitatively different way of thinking. The demand to decide for the first time in a new situation requires an ability to derive the solution from the internal dynamical principles of the object. Thinking, which is oriented towards defining the object as a system and investigates its dynamical internal relations, uses the developmental history of the object as a method to conceive the object. This kind of thinking could be defined as theoretical thinking and would form the basis for theoretical judgement.

The very important distinction between empirical and theoretical thinking can also be made through analysing the role of conceptual tools in both of them. In empirical thinking the tools such as concepts and theories are in a static relation with the object of thinking. In a way they are given ready and steady and can be applied as such. In theoretical thinking there is typically a conscious attempt to form conceptual tools. The object is conceptualized and the concepts serve as further tools. Likewise, the concepts can be transformed objects for the purpose of their development. Thus, instead of a steady relation between the tools and

object, in this case, the two continuously interchange. It is
interesting to note that the basic forms of these two kinds of
thinking can be observed in investigating children's problems
solving (Koslowski & Bruner 1972, Karmiloff-Smith & Inhelder
1975). Furthermore, strange enough, the same difference in the
relation between the tools and object would also characterize
the difference between man and machine (Tichomirov 1975). Thus
man is a theorist, machine an empirist.

Intelligent decision aids as tools for theoretical thinking

Our claim is that the interpretation given for the tools in
the work process is related with two alternative strategies in
developing man-machine -systems. We call these strategies
substitution strategy and developmental strategy; in the
former the methods are adopted as given, in the latter as
changing. We could say that the essential demands of process
controlling activity are very often interpreted and the
operators trained in the empirical spirit. Implicitely or
explicitely the operators are required to find the right
measure or even rule in a particular situation. Also the
diagnostic training tends to orientate the operators towards
the selection of rules not towards understanding of the
phenomena. It would be very natural that also the available
decision aids would be integrated into the whole in the same
manner. This would lead to adoption of the substitution
strategy: the operator is interpreted as doing the same as the
machine can do, thus it is possible to substitute machine for
man.

As an example, it is instructive to look at the implementation
of expert systems into the practice of a physician. The
medical expert systems, such as MYCIN, realize a certain
classification task in the frame of the accepted taxonomy of
diseases. Its validity is completely dependent on the validity
of that taxonomy in the "health" system, which is the object
of the doctor's work. If this work is conceived as making

diagnostical classifications, the MYCIN can be seen as
substituting for him. If this is not the case, it should be
asked how the expert system helps in grasping the health
problems of the people. This problem is usually not raised: in
a study of the use of another medical expert (FIRST-AID) in
the diagnostical consultation, it was investigated whether the
use of the expert would change the normal diagnostic routine
and how. The result was that it did not, which was stated with
content (Brownbridge et al. 1984). No contextual information
of the adequacy of the diagnoses was reported, which indicates
that the problem of expert system use was defined on the
external routine level of the medical activity.

In the last example the machine performs indeed what was
expected of man. The same substitution becomes evident also in
the widespread idea that man and machine are two equivalent
components of a system. Their functions are determined in an
optimal division of tasks. Further it is often asked who does
a particular task better, a computer or a man. Yet, it should
actually be less interesting to know the result of this
comparison directly. Rather it would be reasonable to
investigate how a human expert could do his work better with
computer aid. Thus the human expert or a group of them is
considered as subject mastering the totality of functions.

The model of man that lies behind the idea of substitution
implies that man is provided with a given ahistoric set of
psychological and physiological features. An alternative model
would state that man acquires new functions and abilities
through creating more complicated means of activity.
There is practical evidence for the possibility of an
alternative strategy in developing man-machine systems. For
example, during a validation study of a new alarm system it
was discussed how to interpret the possible result that the
alarm system based on the idea of critical functions were to
be beneficial (Hollnagel et al. 1983, Kautto 1984). The
problem would be: is it the system or the already earlier

developed concept of critical functions that cause the
effect. From the context of our present discussion we can
state that the adoption of the basic concept of critical
functions at least in some intuitive form is a prerequisite
for the successful use of the tool.

In this case the creation of a new tool had already begun in
the praxis as a manifestation of objective needs of
controlling the particular process. To promote such
spontaneous development among the operators and, to strenghten
its power by offering conceptual tools, is the main context of
the alternative developmental strategy. It demands
encouragement of the operators to adopt an experimental
attitude towards the process. This kind of motivational and
cognitive relation means essentially a new type of work and it
is needed for acquiring qualifications of the users of the
complicated technology.

The approach in developing the experts' "intelligent" tools

A computer system which is the essence of any intelligent
tool is a materialized model of the object over which it
should help to have control. It is not independent of but
rather a part of the operator's model structure of the
process. Like all models also the materialized ones should be
continuously developed and enriched. This means that not even
the models are accepted as ready tools but also as objects.
This seems to be an important requirement for a computer
system that should be used interactively and aid in reasoning.
A genuine _interactiveness_ is interaction between the _use_ and
development of the system.

However, it seems that the usual way of creating the model for
a decision aid is not in accordance with the above general
idea. In the words of a specialist, the elicitation of expert
knowledge should be performed through a non-specialist on the

expert knowledge domain. The interviewer should, however, be able to ask clever questions (Brooking 1985). This kind of a superficial approach would encourage the human experts to report their knowledge and procedural skills directly as ready results.

In our opinion, the human experts should be asked to consider their theories, concepts, heuristics, rules, procedures and all possible means as objects of inquiry. The researchers particular role is to formulate <u>special experimental tasks</u> that would open the possibility for the expert to analyse, critisize and develop the whole arsenal of his means. As a result a whole structure of models could be made explicit, one part of it being the potentially materialized system.

The use of particular tasks is methodologically and methodically crucial. They serve as a tool in the process of creating new means. In a sense these tasks are also a didactic method to qualify the operators even further. This is not a minor point when we take into account that often the very expertise we are interested in is in the beginning neither conceptualized nor verbal. The set of tasks serves also as an analytical means for following the development of thinking in the process of acquiring the mastery of the object (Toikka et al. 1985). We could say that the acquisition and elicitation of expert knowledge are in this method integrated, as also training and research. The need for this interaction was also pointed out by Leplat in his article in this volume.

The organizational frames that are required for such developmental activity might differ. Quite evidently old organizational solutions might turn out to be outdated. Essential in the new organizational solutions would be that the two aspects of work, planning and performing are approaching.

In the Electrical Engineering Laboratory of the Technical Research Centre of Finland there are two running studies, one in the power industry and the other in modern metal industry, in which this approach is currently applied and tested. The specific problem of developing expert systems has thus far not been handled but has been planned for the next years. We do not see any principal differences between the process of developing computer tools in general and expert systems in particular.

PROCEDURAL THINKING, PROGRAMMING, AND COMPUTER USE

Steen F. Larsen
Institute of Psychology
University of Aarhus
Risskov, Denmark

Information processing technologies are spreading rapidly
throughout the industrialized world. Can we identify cognitive
prerequisites -- knowledge and skills -- which are particular to
understanding, using, and constructing such devices? If so, we
will be better prepared to deal with present difficulties of
novices and workers, to educate people for future mastering of
the technology, and to foresee psychological and social consequences that the dissemination of these abilities may engender. The assumption behind the talk of "computer literacy" is
that such cognitive prerequisites exist and can be taught,
though nobody seems to agree what they are. I shall begin by
presenting and discussing a proposal about the cognitive contents of computer literacy, put forward at a conference in
Houston a few years ago by B.A. Sheil (1981), a psychologist at
Xerox.

Procedural reasoning

Sheil proposed that "the key idea underlying information technology is procedural reasoning. Procedural reasoning is the process by which one determines the effect of a set of instructions
or, alternatively, the set of instructions that will achieve a
particular effect. ... (It) is the fundamental skill underlying
both programming itself and any appreciation of programmed artifacts" (Sheil, 1981, p. 8; author's emphasis).

NATO ASI Series, Vol. F21
Intelligent Decision Support in Process Environments
Edited by E. Hollnagel et al.
© Springer-Verlag Berlin Heidelberg 1986

The concept of procedural reasoning seems perfectly justified when applied to programming. Since information processing devices are controlled by exhaustively described procedures in their programs, it is evident that programming must involve thinking in terms of procedures. But why should any use of programmed devices require procedural thinking? To make such devices do what one desires, to make them function efficiently, to correct errors, etc., one must have a cognitive model of the way they work, and only a procedural model will be able to make sense of the behavior of the devices. In other words, the user of a technological system must possess a mental model of its functioning which is isomorphic to the model that was used to create the system (though not necessarily equally detailed). Therefore, Sheil (1980) argued that procedural reasoning is becoming an ability of the same general importance as traditional literacy -- a view that he emphasized by coining the term pro-cedural literacy to replace computer literacy.

What are the cognitive characteristics of procedural reasoning? Except for the definition cited above, Sheil is vague and partly self-contradictory. Thus, he makes the strong claim that "Procedural reasoning is a fundamentally new way of thinking" (Sheil, 1981, p.8), in the acquisition of which "users get very little transference from their established skills" (p. 10). However, he also notes that "the use of quasi-procedures such as job instructions, recipes, directions for how to get from one place to another, etc., provides evidence that there is some kind of procedural framework which is widely used to formulate, follow and reason about procedures ... a naive procedural semantics" (p. 24, author's emphases) or pre-existing procedural skills which he urges programming instruction to build on. In other words, procedural reasoning is not a completely new form of thinking, after all. Let us take a closer look at its precursors before trying to determine what is novel and therefore difficult about it.

Implicit and explicit knowledge

I shall argue that the "quasi-procedures" examplified above are only the top of an iceberg of procedural knowledge. The examples all concern procedures formulated in language. But it is clear that to put a job description or a set of directions into words we must know the procedure already. If this knowledge is not in verbal form, we may still be able to describe it by executing the procedure -- by acting or by thinking through it in our head -- and observing ourselves as we do so. Behind explicit descriptions of procedures we may thus assume the existence of implicit, non-verbal procedures that we can only examine consciously when they are actually executed.

That much of cognitive processing is unconscious and non-verbal should not be surprising; it is a basic idea of cognitive psychology (Mandler, 1985). Our ability to retain knowledge which enables procedures to be carried out is generally known as pro-cedural memory, and it is considered to constitute a type of memory different from verbal, episodic or semantic, memory -- perhaps even using different brain mechanisms (see the discussion by numerous researchers in Tulving, 1984). But procedural memory is concerned with knowledge that may only be expressed in actual operations, whether mental or practical; it is implicit or "tacit" (Polanyi, 1966). In contrast, the explicit, verbal knowledge of procedures in Sheil's examples is directly access-ible at will, can be manipulated consciously and communicated to others. Thus, procedural reasoning is thinking in terms of ex-plicit knowledge of procedures.

Note that this distinction between implicit and explicit know-ledge is similar to Winograd's (1975) dichotomy of procedural and declarative representation, and Graf & Schacter (1985) have recently suggested that it may provide an explanation of certain types of amnesia. It is important to realize that it is a dis-tinction of different formats of knowledge whereas the contents -- the procedures that are specified -- may be identical.

Difficulties of explicating procedures

Granted that every human adult may be presumed to possess a vast repertoire of implicit procedures, probably far richer than the limited set of procedures implemented in any programming language, why is so little of it put into words and why is it at all difficult to reason procedurally and to learn to program? Undoubtedly, part of the problem has to do with limitations of cognitive capacity which make it difficult to perform skilled acts and think about them at the same time. But the sheer existence of explicit procedure knowledge shows that this limitation may be overcome.

A more basic problem is concerned with the level of detail of explicit procedure knowledge. Our conscious thinking is usually at the level of actions and the goals they are performed to achieve whereas we do not need to consider the more elementary operations which are employed to reach a goal under the circumstances that happen to prevail; this adapting of our skills (implicit procedure knowledge) to situational parameters is taken care of automatically. (For the distinction between actions and operations, see Leontyev, 1981/1959; it is similar to Marr's, 1982, computational versus algorithmic levels of modelling and to Reiser, Black & Abelson's, 1985, activities versus general actions).

In social situations, moreover, there is usually no need for talking about operations because members of a given culture will generally share knowledge at this level; skills are the common currency of a culture. Therefore, natural language is not developed to specify implicit procedure knowledge in detail and with precision. In general terms, it appears that spoken language is best fitted for describing specific events (as in narratives) and written language for describing general properties and relationships (as in essays; cf. Olson, 1975). As for procedures, a reasonable -- but as yet untested -- hypothesis is that the level of actions is the basic level of description in natural language, similar to the basic level of describing objects

(Rosch, Mervis, Gray, Johnson & Boyes-Brean, 1976). When the
need for a really fine-grained description arises, for instance
when instructing a novice, it is thus more easy and efficient to
demonstrate the procedure than to describe it.

The point of this argument is that explicit description of implicit procedure knowledge is not inherently difficult. The problem is rather that our language is poorly suited for this propose and we therefore also lack appropriate descriptive skills.
However, programming requires a fine-grained, explicit description of the procedures to be carried out by the machine, and
programming languages may therefore offer the descriptive tools
which are needed to explicate procedure knowledge and thus make
it accessible to conscious thinking. Most people who have tried
to program a well-known procedure, for instance the procedure of
an experiment to be run under computer control, will have experienced that the programming made them aware of hitherto un-
noticed imprecisions and alternatives in the process. This "consciousness raising" effect is also one of the scientific advantages of computer simulation of theoretical models.

Implications and limitations of the present argument

If it is true that programming languages enable implicit procedure knowledge to become consciously accessible, then programming skills will by themselves give rise to development of explicit procedural reasoning. Programming does not presuppose
some mysterious procedural reasoning ability; rather, this ability -- that is, procedural literacy -- is fostered by learning
to program.

Let me finally mention one reservation to this hypothesis and
one interesting empirical implication. The reservation is that
existing programming languages may be far from ideal in respect
to knowledge explicating power because of two things: They are
formal and they are machine-dependent. The requirement of formalization is an added difficulty that may very well divert

attention from the task of describing the procedures themselves. As Johnson-Laird (1983) has shown to be the case for logical reasoning, our internal cognitive processing may have a more analogical character. I prefer to talk about <u>procedural thinking</u> rather than reasoning in order to avoid the connotation of formal rule-following that adheres to the latter term. The <u>machine-dependence</u> of programming languages means that arbitrary constraints are placed on which procedures may be expressed straightforwardly and in what way. It would be desirable to have an informal, general-purpose procedural language to use in learning and instruction -- in the direction of the "pseudo-code" recommended in programming textbooks (e.g., Borland International, 1984).

An empirical implication of the ideas presented here is that people with some programming experience should possess a domain-independent ability to think in terms of procedures whereas the procedural knowledge of non-programmers should be domain-dependent. For instance, if a task demands that procedures are detected and remembered, then non-programmers should be able to perform equally well as programmers provided that the domain concerned is well-known. If the domain is unknown to both parties, however, programmers would be expected to perform at almost the same level whereas non-programmers should be clearly inferior. Data from a pilot experiment that I carried out at Emory University, Atlanta, USA, in collaboration with Larry Barsalou suggests that this may actually be the case.

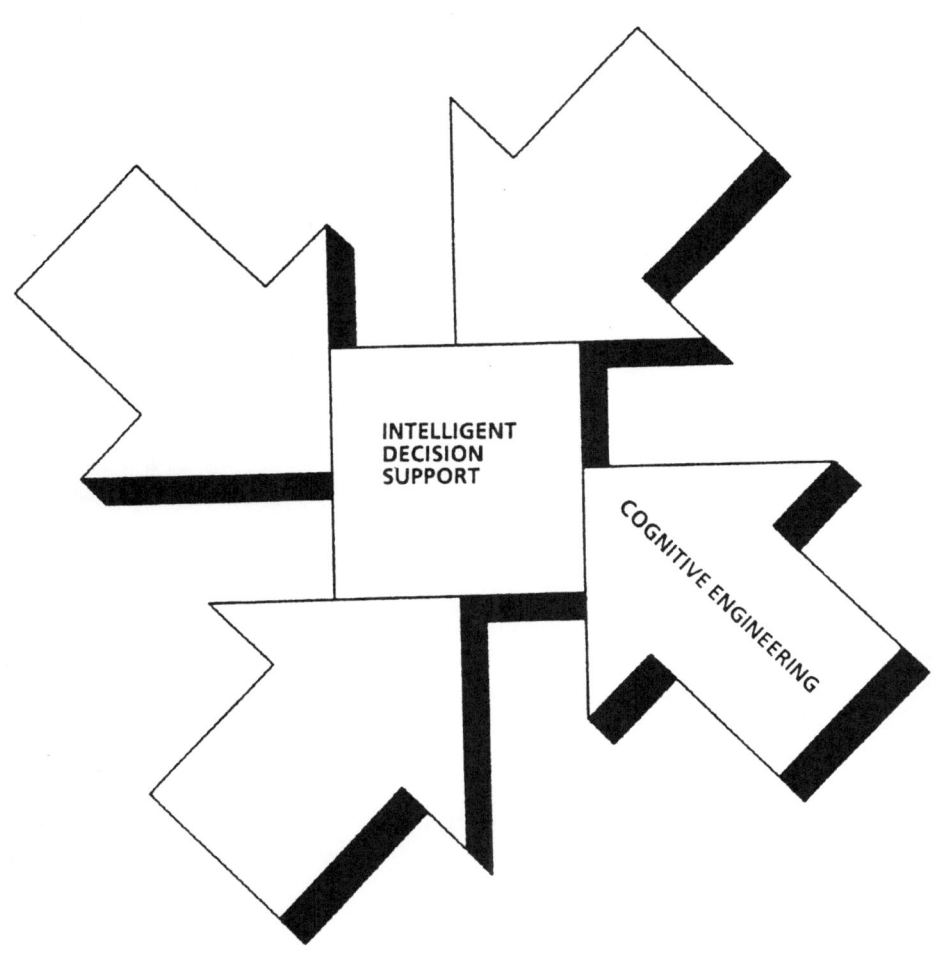

INTELLIGENT DECISION SUPPORT

COGNITIVE ENGINEERING

COGNITIVE SYSTEMS

PARADIGMS FOR INTELLIGENT DECISION SUPPORT

David D. Woods
Westinghouse Research & Development Center
Pittsburgh, Pennsylvania (USA)

Paradigms for Decision Support

Advances in artificial intelligence (AI) are providing powerful new computational tools that greatly expand the potential to support cognitive activities in complex work environments (e.g., monitoring, planning, fault management, problem solving). The application of these tools, however, creates new challenges about how to "couple" human intelligence and machine power in a single integrated system that maximizes joint performance. This paper examines some of the important issues about the use of tools to support cognitive tasks, such as what is useful advice and what is an effective combination of multiple decision makers, that are raised by the capability to produce powerful, intelligent artificial systems.

Tool builders have focused, not improperly, on tool building -- how to build better machine problem solvers, where the implicit model is a human expert solving a problem in isolation. But tool use involves more. Building systems that are "good" problem solvers in isolation does not guarantee high performance in actual work contexts where the performance of the joint person-machine system is the relevant criterion. The key to the effective application of computational technology is to conceive, model, design, and evaluate the joint human-machine cognitive system (Hollnagel & Woods, 1983). Like Gestalt principles in perception, a decision system is not merely the sum of its parts, human and machine. The configuration or organization of the human and machine components is a critical determinant of the performance of the system as a whole (e.g., Sorkin & Woods, 1985). The joint cognitive system paradigm demands a problem-driven, rather than technology-driven, approach where the requirements and bottlenecks in cognitive task performance drive the development of tools to support the human problem solver. As a result, computational technology should be used, not to make or recommend solutions, but to aid the user in the process of reaching a decision. The challenge for psychology is to provide models, data, and techniques to help designers build an effective configuration between the human and machine elements of a joint cognitive system.

NATO ASI Series, Vol. F21
Intelligent Decision Support in Process Environments
Edited by E. Hollnagel et al.
© Springer-Verlag Berlin Heidelberg 1986

A Hypothetical Computer Consultant

In order to understand the joint cognitive system paradigm for decision support, let us examine an alternative way to provide decision tools that is exemplified in many current expert systems. In this paradigm, the primary design focus is to apply computational technology to develop a stand-alone machine expert that offers some form of problem solution. The *technical* performance of this system is judged on the question: are the solutions offered *usually* correct (e.g., Yu et al., 1979). This paradigm emphasizes tool building over tool use and questions about how to interface human to machine are secondary to the main design task of a machine that usually produces correct decisions.

A typical encounter with a hypothetical *intelligent* computer consultant developed in this fashion consists of: the user initiates a session; the machine controls data gathering; the machine offers a solution; the user may ask for *explanation* if some capability exists; the user accepts (acts on) or overrides the machine's solution.

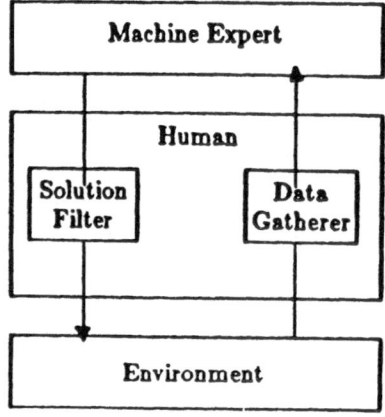

Figure 1. The joint cognitive system implicit in a hypothetical computer consultant where design focuses primarily on building a machine expert that outputs usually correct solutions; the human's role is to gather data and to filter any poor machine solutions.

What is the joint cognitive system architecture implicit in this hypothetical machine expert? The primary focus is to apply computational technology to develop the machine expert. In practice, putting the machine to work requires communication with the environment -- data must be gathered and decisions implemented. Rather than automate these activities they are typically left for the human (Figure 1). Thus, interface design for the hypothetical consultant is not so much how to interface the machine to the user, but rather, how to use the human as an interface between the machine and its environment. This results in a user interface design process which focuses on features to aid the user's role as data gatherer (e.g., Mulsant et al., 1983) and features to help the user accept the machine's solution. One consequence of this emphasis is that control of the interaction resides with the machine. However, human factors work in person-machine systems has established that a machine locus of control can have strong negative effects on user and total system performance (e.g., M. Smith et al., 1981; Hoogovens Report, 1976).

A related characteristic of this paradigm is that user acceptance and the machine expert's *technical* performance (again, in the sense of offering problem solutions that are usually correct) are seen as independent issues (e.g., Shortliffe, 1982). *Excellent decision making performance does not guarantee user acceptance* (Langlotz & Shortliffe, 1983, p. 479). Thus, lack of user acceptance is a problem in the user which must be treated by measures outside of the essential characteristics of the machine expert. One proposed technique is to embed other, useful support capabilities in the same computer system that implements the machine expert, e.g., data management functions such as computerized data entry forms or standard report generation (Langlotz & Shortliffe, 1983). Some designers of machine experts go so far as to suggest that systems *provide the physician [i.e., user] with the ability to report the facts he considers important (even if they are not used internally) [i.e., by the machine expert]* (Mulsant & Servan-Schreiber, 1983, p. 11-12; italics added). The cognitive system viewpoint suggests, on the other hand, that problems with user acceptance are very often symptoms of an underlying deficiency in the *cognitive coupling* (Fitter & Sime, 1980) between the human and machine subsystems (e.g., machine control).

This paradigm emphasizes user acceptance of the machine's solution. However, since these systems are imperfect, output typically consists of some form of confidence or likelihood estimate over a set of possible diagnoses. The user is expected to act on the machine's solution, but what is the machine's solution: the highest likelihood category? likelihood weighted by consequences? by some form of expectation? for dynamic domains, are temporal fluctuations important? what if there are several high likelihood options or no high likelihood options? Choosing a solution to act on is further complicated because of the non-standard procedures that are typically used to compute likelihood estimates. Due to the method used to generate the confidence values, the likelihood data usually rests on an ordinal measurement scale. However, they are often represented as interval scales to the user which carries a high potential for misleading the human decision maker and complicating his task. These examples point out that computing likelihood over limited categories underspecifies the cognitive activities underlying diagnosis; likelihood is only one element of decision making under uncertainty and risk. Failure to recognize this can lead to error prone links in the joint cognitive system (but see Schum, 1980 and Einhorn & Hogarth, 1985 for treatments of how evidence supplied by one system should be used by a subsequent decision maker).

The hypothetical computer consultant outputs some form of problem solution (recommended solution; solution categories ranked by some form of likelihood). However, it is the human user who has responsibility for the outcome. Of course, the user has the authority to override the machine, in principle, i.e., to filter the expert machine's output. This form of "cognitive coupling" (Fitter & Sime, 1980) between cognitive systems has several strong implications. First, does the user really have the authority to override machine output in practice. Since the user's only options are to accept or reject system output, there is great danger of a responsibility/authority double bind where he either always rejects machine output (perhaps by finding or creating grounds for machine unreliability) or he abrogates his decision responsibility (the user may not override the computer, regardless of circumstances, if the cost of an error in overriding is too high). When people refer a problem to a human specialist, they generally pass on both authority and responsibility together (e.g., Miller, 1983); thus, a specialist who is called in on a case typically acts as the primary problem solver and not as a consultant to another problem solver. The responsibility/authority double bind has been observed with non-AI decision aids that output solutions (e.g., Fitter & Sime, 1980) and with increases in control automation that fail to address the operator's new role as supervisor of automated resources (e.g., Hoogovens Report, 1976).

Second, how good are people at discriminating correct from incorrect machine solutions, and how does discrimination performance vary with user expertise and over different types and depths of explanation? Very little is known about what factors affect human performance at filtering another decision maker's solutions (I am addressing here the person's ability to filter machine output <u>after</u> the machine expert has been developed and deployed; current, typical methods for constructing a machine expert are extremely dependent on human experts/knowledge engineers' ability to detect and correct erroneous machine output). What level of expertise is needed to recognize erroneous machine output or when a situation is beyond the capabilities of the machine (machine experts are at best only usually correct)? Can people use syntactic cues (this output looks funny for this type of problem) or experience-driven associations (in this situation, the machine often screws up) to filter erroneous system output?

A related issue is the question of loss of skill. Some degree of expertise would seem to be required to filter machine output; what factors determine if the user of a machine expert can develop or maintain that expertise. Work in process control emphasizes that knowledge only becomes operationally effective if it has occasion to be used (De Keyser, this volume). Learning by doing applies to cognitive as well as to perceptual-motor skills:

> It has to be recognized that in giving up the interplay between knowledge and its regular practical exercise, we are departing from the only conditions we know for the successful development of art and science (Council for Science and Society, 1981, p.68).

Issues on the loss of cognitive skill are closely related to the loss of skill questions that arise in control automation: how will the human acquire or maintain the manual (or cognitive) skill to take over or adjust control when automation breaks down or when disturbances occur that are beyond the capability of the automation (e.g., Hoogovens Report, 1976; De Keyser, this volume)? does a man-in-the-control-loop (or decision loop) architecture improve human fault management performance in highly automated environments (e.g., Ephrath & Young, 1981; Wiener, 1985)?

When the machine expert offers some form of problem solution, the user could perform the filtering task by computing his own problem solution and then comparing it with the machines's output. This strategy results in a redundant, as opposed to diverse, joint human-machine cognitive system architecture, and, analogous to equipment reliability, the design issue is whether a redundant or some diverse architecture results in more reliable overall performance.

The critical question is what is the criterion for judging an effective system. In the paradigm represented by the hypothetical consultant, the "system" is defined as the machine expert and "effective" means usually correct machine solutions. However, an alternative approach is to define the system as the combination of human and machine (the human-machine cognitive system) and for effective to mean maximizing joint performance, i.e., performance of the whole should be greater than the performance possible by either element alone (see Hollnagel, this volume, for a treatment of the implications of this for the evaluation of intelligent systems).

What kind of decision tool is the hypothetical consultant?

> one of the big problems is the tendency for the machine to dominate the human ... consequently an experienced integrated circuit designer is forced to make an unfortunate choice: let the machine do all the work or do all the work himself. If he lets the machine do it, the machine will tell him to keep out of things, that it is doing the whole job. But when the machine ends up with five wires undone, the engineer is supposed to fix it. He does not know why the program placed what it did or why the remainder could not be handled. He must rethink the entire problem from the beginning. (Finegold, 1984, p.115)

Given limited user participation in the problem solving process, the danger of a responsibilty-authority double bind with support systems that offer solutions rather th'an informative counsel, the potential loss of cognitive skill, and the potentially difficult and unsupported task of filtering poor machine solutions, the impoverished joint cognitive system implicit in the hypothetical computer consultant does not represent an effective model for the use of decision tools, i.e., for decision aiding.

Joint Cognitive Systems

Effective cognitive system design requires, first, a problem-driven, rather than technology-driven, approach. In a problem-driven approach, one tries to learn what makes for competence and/or incompetence in a domain (i.e., cognitive task analysis; Woods & Hollnagel, 1986; Rasmussen, this volume), and then to use this knowledge to provide tools which help people function more expertly. If the problem-to-be-solved is dangerous environments, then an automated decision system is a viable solution category. If the problem is human inconsistency or memory lapses, then a redundant cognitive system architecture may be one appropriate path. It is insufficient to say human diagnostic performance is not as good as I would like (even by experts), therefore I will build a machine for diagnosis. One must ask what aspect of the diagnostic performance of the current person-machine system is the bottleneck. Studies of cognitive performance in work environments have shown person-machine performance problems such as:

- fixation or perseveration effects in applied problem solving (Woods, 1984; Montmollin & De Keyser, 1985; Norman, this volume);

- weaknesses in counterfactual reasoning: would y have occurred if x had not?

- data sampling/information acquisition problems: can the user find and integrate the "right" data for the current context and task (e.g., Woods, 1986).

Can machine experts that offer problem solutions counteract any of these problems? For example, one characteristic of fixation effects is early termination of the evaluation of alternative hypotheses; therefore, a good joint cognitive system should support a broader exploration of solution possibilities. Would a machine expert offering its own solution broaden the evaluation of alternatives or narrow evaluation and exacerbate fixation problems? If failures of attention are the underlying problem, then a decision aid that helps the human problem solver focus in on the relevant data set for the current context is the goal; computational technology supplies the means to build real systems that embody techniques to achieve this goal. When tools dominate, rather than constrain, the design of a decision support system, the designer runs a strong risk of the error of the third kind (Mitroff, 1974): solving the wrong problem.

Second, if joint cognitive system design is to be effective, there is a need for models and data that describe the critical factors for overall system performance. Sorkin and Woods (1985) contains of an analysis of joint cognitive systems modelled as two decision-makers-in-series (cf. also, Schum, 1980; Einhorn & Hogarth, 1985). The first stage consists of automated subsystems that make decisions about the state of the underlying process. When alerted by the first stage system, the subsequent human decision maker uses evidence provided by the automated subsystems and his own analysis of input data to confirm or disconfirm the decision made by the automated monitors and to decide on further action. These analyses show, first, that the performance of the joint system can be significantly enhanced or degraded relative to the performance of the machine element alone depending on interactions between the characteristics of the subsystems (primarily the response criterion of the automated subsystem and the user's workload and monitoring strategy). Second, the value of the output of the first stage is better thought of as information, in the sense of evidence or testimony, to be used by the person to reach his decision, rather than as a proffered solution to be accepted or rejected. Third, the inferential value of the information provided to the human decision maker is highly sensitive to the characteristics of the joint system; for example, the value of the evidence provided by the automated subsystem degrades rapidly if it exhibits a bias for or against possible events, even if it is a sensitive detector alone.

Empirical studies of human-human advisory interactions can also help to determine what is good advice. Alty & Coombs (1980) and Coombs & Alty (1980) found that unsatisfactory human-human advisory encounters were strongly controlled by the advisor. The advisor asked the user to supply some specific information, mulled over the situation, and offered a solution with little feedback about how the problem was solved. While a problem was usually solved, it was often some proximal form of the user's real problem (i.e., the advisor was guilty of a form of solving the wrong problem: solving a proximal case of the user's fundamental or distal problem). The advisor provided little help in problem definition. There is a striking parallel between these characteristics of unsatisfactory human-human advisory encounters and the characteristics of the joint cognitive system implicit in the hypothetical computer consultant analyzed earlier.

On the other hand, Alty & Coombs found that in more successful advisory encounters a partial expert (experiencd computer user with a domain task to be accomplished) consulted a specialist (local computer system expert). Control of the interaction was shared in the process of identifying the important facts and using these to better define the problem. In this process each participant stepped outside of his own domain to help develop a better understanding of the problem and, as a consequence, appropriate solution methods.

These studies (cf. also, Pollack et al., 1982) reveal that good advice is more than recommending a solution; it helps the user develop or debug a plan of action to achieve his goals (Jackson & Lefrere, 1984, p. 63). Good advisory interactions aid problem formulation, plan generation (especially with regard to obstacles, side effects, interactions and tradeoffs), help determine the right questions to ask, and how to look for or evaluate possible answers. This means good advice must be more than a solution plus a solution justification; it must be structured around the problem solving process to help the user answer questions like: what would happen if x, are there side effects to x, how do x and y interact, what produces x, how to prevent x, what are the pre-conditions (requirements) and post-conditions for x (given x, what consequences must be handled).

Studies of advisory interactions reveal another important characteristic of joint cognitive systems: the relationship between the kinds of skills and knowledge represented in the human and in the machine (as opposed to relative skill levels). The human user of decision tools is rarely incompetent in the problem domain. Instead, the human and machine elements contain partial and overlapping expertise that, if integrated, can result in better joint system performance than is possible by either element alone. Today, no one expert in any field can keep up with the amount or rate of change of information. The result is the generalist-specialist problem. Most real world problems require the integration of different specialists each of which contributes a unique point of view (e.g., Hawkins, 1983; Coombs & Alty, 1984). One kind of expertise is the ability to evaluate and integrate specialist knowledge in some real problem context. As a result, the designer of a joint cognitive system must address the question of what is or what should be the relationship between machine expertise and human expertise in a given domain: is the person a generalist who manages and integrates input from various machine implementations of specialist knowledge?; is the human one specialist interacting with the knowledge of other specialists to deal with a problem at the junction of the two fields? Effective interactions between these kinds of partial, overlapping experts (which I suspect will be a dominant type of joint cognitive system) requires that knowledge from different viewpoints is integrated during the decision process including problem formulation and plan evaluation.

Fundamentally, the difference between the hypothetical consultant analyzed earlier and the joint cognitive system approach to decision support is a difference in the answer to the question "what is a consultant." One operational definition of a consultant (operational in the sense that systems purported to be consultants are built in this fashion) is some one (thing) called in to solve a problem for another, on the asssumption that the problem was beyond the skill of the original person. Given this definition, the important issues for building decision aids is to build better automated problem solvers and to get people to call on these automated problem solvers (the acceptance problem). The joint cognitive system perspective, on the other hand, defines a consultant as a reference or source of information for the problem solver. The problem solver is in charge; the consultant functions more as a staff member. As a result, the joint cognitive system viewpoint stresses the need to use computational technology to aid the user in the process of solving his problem. The human's role is primarily to achieve total system performance objectives as a manager of knowledge resources that can vary in "intelligence" or power (e.g., Sheridan & Hennessy, 1984). To build effective human-machine cognitive systems, there is a need for techniques and concepts to determine what knowledge resources are needed by a domain problem solver and to integrate and communicate these results as appropriate during the problem solving encounter.

Systems are beginning to be developed which embody these characteristics. One path is to build machine advisors that critique the human problem solvers plan (Coombs & Alty, 1984; Langlotz & Shortliffe, 1983; Miller, 1983). These systems consist of a core automated problem solver plus other modules that use knowledge about the automated problem solver's solution and solution path, knowledge of the state of the user-computer interaction, and knowledge of the user's plans/goals to warn the user, to remind him of potentially relevant data, and to suggest alternatives. I will not explore this path for building joint cognitive systems further, other than to comment that it imposes strong requirements on the power of the tools needed to produce advisors of this type. Instead, I will briefly sketch a second path towards joint cognitive systems based on what has been called direct manipulation (Hutchins, Hollan & Norman, 1985) or graphic knowledge systems.

Towards Joint Cognitive Systems: Graphic Knowledge Systems

The Significance of Data

Studies of human performance in complex domains reveal that performance failures can often be linked to problems in information handling -- data overload, getting lost, keyhole effects, tunnel vision to name but a few. All of these information handling problems represent manifestations of an inability to find, integrate or interpret the "right" data at the "right" time, or the significance of data problem (cf., Woods, 1985a). The significance of data problem occurs whenever large amounts of potentially relevant data must be sifted in order to find the significant subset for the current problem context. Failures in this cognitive task are seen when critical information is not detected among the ambient data load, when critical information is not assembled from data distributed over time or over space; and when critical information is not looked for because of misunderstandings or erroneous assumptions. This conceptualization points out that information handling problems are rarely caused by a lack of a data but rather by its overabundance (i.e., most human information handling tasks are, in the language of Communication Theory, information compression tasks) and that attempted fixes which focus only on increasing the available data will exacerbate the problem.

The significance of data problem arises because a particular datum gains significance or meaning only from its relationship to the context in which it occurs or could occur. In other words, there are no facts of fixed significance. For example, you hear on the radio that today's pollen count is 23. This number has no meaning, or information value in itself. This piece of data can explain why you are or are not sneezing often today, or help you evaluate whether the tradeoff between symptom relief from an anti-histamine and the side effect of drowsiness is or is not justified, or help you to decide if you should attend the picnic planned for that afternoon, only if you also have data about the pollen measuring scale (e.g., linear or logarithmic), limits, normal or expected values, recent history, seasonal referents, and how your hay fever symptoms vary with pollen levels.

The important point for the development of effective decision support systems is the critical distinction between the available data and the meaning or information that a person extracts from that data (e.g., S. Smith, 1963). The available data are raw materials or evidence that the observer uses and evaluates to answer questions (questions that can be vague or well-formed, general or specific). The degree to which the data help answer those questions determines the informativeness or inferential value of the data. Thus, the meaning associated with a given datum depends on its relationship to the context or field of surrounding data including its relationship to the objects or units of description of the domain (what object and state of the object is referred to), to the set of possible actions, and to perceived task goals (after Gibson, 1979, what that object/state affords the observer). The process is analogous to figure-ground relations in perception and shows that information is not a thing-in-itself but is rather a relation between the data, the world the data refers to, and the observer's expectations, intentions, and interests. As a result, informativeness is not a property of the data field alone, but is a relation between the observer and the data field.

Take the case of a message about a thermodynamic system which states that valve x is closed. Most simply, the message signals a component status. If the operator knows (or the message also states) that valve x should be opened in the current mode of operation, the message signals a misaligned component. Or the message could signify that with valve x closed, the capability to supply material to reservoir H via path A is compromised. Or given still additional knowledge (or data search), it could signify that, with valve x closed, the process that is currently active to supply material to reservoir H is disturbed (e.g., data such as actual flow less than target flow, or no flow, or resevoir H inventory low). Furthermore, the significance of the unavailability or the disturbance in the material flow process depends on the state of other processes (such as, is an alternative flow process available or is reservoir H inventory important in the current operating context). Each interpretation is built around what an object affords the operator or supervisor of the thermodynamic system, including an implicit response: correctly align component, ensure capability to supply material (or take into account the consequences of the inability to do so), repair the disturbance in the material flow process (or take into account the consequences of the disturbance), or discount these messages based on current objectives.

The significance of data problem refers to breakdowns in the cognitive task of filtering/segregating relevant from irrelevant data or failures of attention. This cognitive task is critical in most naturalistic problem solving situations (e.g., Dorner, 1983; Selfridge et al., 1984; Montmollin & De Keyser, 1985). Woods (1984b) contains results on operator performance in nuclear power plant emergencies which reveals the major role played by attention failures when operational problems occur. For example, the lessons learned task force review of the Three Mile Island accident found that "it seemed that although the necessary information was, in general, physically available, it was not operationally effective. No one could assemble the separate bits of information to make the correct deductions." (Joyce & Lapinsky, 1983). Lees (1983) and Woods (1986) found the significance of data problem at the heart of a variety of deficiencies in the identification and response to abnormal conditions with conventional alarm systems in process control domains. Similarly, military history reveals that intelligence failures such as Pearl Harbor (Wohlstetter, 1962) and the Yom Kippur war (Shlaim, 1976) are not due to a lack of data, but rather to an inability to filter and integrate from disparate sources the relevant subset of data in noisy environments.

It would seem that a solution to the significance of data problem is straightforward: analyze the problem domain to determine the set of relevant data and then collect and present only that data. However, the critical constraint on performing the relevant/irrelevant discrimination is that what is relevant depends on the context and changes over time in dynamic decision making situations. As a result, the set of potentially relevant data for all possible cases becomes very large relative to the set of data that is actually relevant at any point during the problem solving process. The task, then, is how to restrict the possibly relevant data set while avoiding errors of overfocusing (tunnel vision), failures to track shifts in relevance (fixation or perseveration), and errors of too little focus (inability to differentiate the relative importance of data) and while preserving flexibility, range and generality of response to potentially changing task demands and circumstances.

Person-machine system performance problems are often associated with attention failures, but in many situations people exhibit a remarkable ability to focus in on relevant data. Examination of the psychological mechanisms that operate in these situations can reveal ways to enhance, support and perhaps extend human abilities. First, in natural environments, people generally perceive or pay attention to the meaning or significance of signals (which is a relationship between observer, signal and observed world) and only secondarily to the signals themselves (e.g., Yates,

1985). When observers scan a visual scene or picture they tend to look at "informative" areas, i.e., the ability to recognize data of high potential relevance as context varies (cf., Woods, 1984a). "Visual search is an active interrogation of the visual world during which people systematically detect and use meaningful patterns of relationships to decide where to look first and in what sequence to seek further information (Rabbitt, 1984, p. 273).

Second, one characteristic of human expertise in a particular domain is the ability to quickly recognize what is potentially important or promising and to progressively focus in onto the critical data for the current context (e.g., DeGroot, 1965). This ability is related to the fact that human experts are able to see problems in terms of meaningful high level structures rather than individual data elements. For example, the chess master sees a web of threats, attacks and vulnerable points (e.g., unbalanced pawn structure) where the non-expert's unit of analysis is individual pieces (e.g., DeGroot, 1965; Chase & Simon, 1973), or the experienced pilot perceives his position in terms of a landing envelope, stall conditions, and the possibility of a go-around. K. Duncan (1981) even uses the relevance of the data requested as problem solving unfolds in time to the ultimate problem solution, as a measure of fault diagnosis performance.

It is important to keep in mind that what data are relevant depend on where one is in the problem solving process and what has and could happen. Given hindsight or the position of an omniscient observer, one can specify what data are needed for the ultimate solution, but this misses the cognitive task of focusing in on that relevant subset that is critical from the point of view of the person in the problem solving situation. It is attention failures or breakdowns in this "focusing in" process that produces performance problems in tasks whose solution seems obvious to hindsight or to outside observers.

The mechanisms of perception and attention which underlie these abilities to focus in on relevant data provide the basis for effective decision support through enhancing the communication of the significance of data (see e.g., Kolers & Smythe, 1979; Norman & Bobrow, 1979; Johnson-Laird, 1983; J. Duncan 1984; Kahneman & Treisman, 1984; Rabbitt, 1984 and Yates, 1985 for descriptions of some of the relevant psychology). The goal is to assist the human's decision process by aiding his ability to focus his attention on relevant problem aspects and to switch problem views, in other words, to help the human find the relevant data set. The central obstacle to accomplish this is that a particular datum gains significance in relation to its context, hence, the need for techniques to display data in context.

The capacity to <u>see</u> domain situations as a web of objects, relations and purposes, rather than as collections of data elements per se, is fundamental to the ability to progressively focus in on relevant data. People use knowledge of meaningful objects and events, knowledge of the relationships between objects and possible actions, and knowledge of relationships between perceived goals and objects to intelligently direct search behavior and action selection. As a result for person-machine environments, improved joint system performance depends on a shift in emphasis away from display of the data as signals (albeit with proper design for data accessibility) towards a focus on communicating what is signified by the data. If available data are organized and displayed so that the user can directly see the state of task-meaningful objects, then natural mechanisms for focusing in on the relevant data for the current context will be more effective. Consider again the thermodynamic system example. In this case valve position is one element of multiple data sets that communicate about the availability and performance of a particular material flow process. The human problem solver should then be able to refer to that material transport process as an object with attributes, e.g., inactive-available, inactive-unavailable, active-should be active, active-performing correctly, etc., and to see the state of that process in the context of related goals and processes such as the state of goals that are or can be affected by the operation of this process.

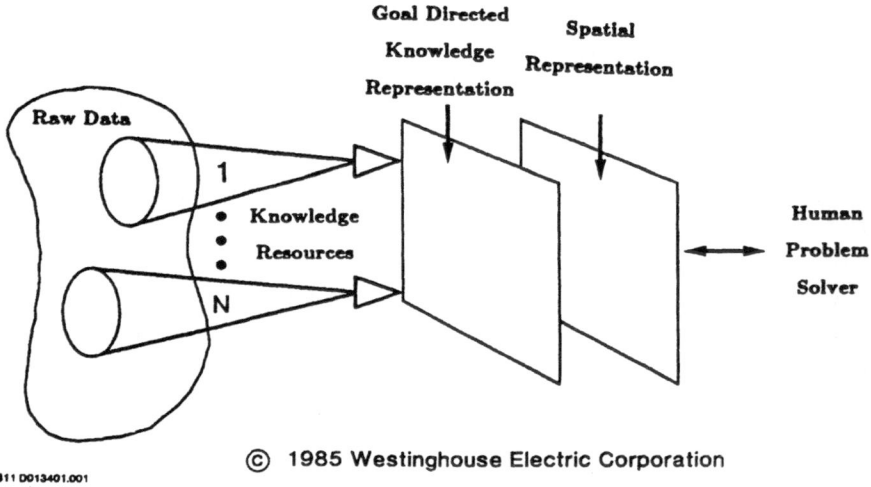

811 D013401.001

Figure 2. Graphic Knowledge Systems.

Enhancing the perceptability of the significance of data as a mode of decision support, then, depends on

1. identification of the task-meaningful units or objects and relations, or a knowledge representation for the domain;

2. a map of the relationship between available data and the units/relations in the knowledge representation, or the domain data-information correspondences;

3. a spatial representation or visualization of the knowledge representation and data-information correspondences;

or, what might be called, building a graphic knowledge system (Figure 2).

Goal-Directed Knowledge Representation

The units of description and the web of relationships between those units that are appropriate for domain tasks can be constructed by structuring the data field (the field of available data) in terms of the goals to be accomplished, the relationships between goals (e.g., causal relations, constraints) and the means to achieve goals (how a process functions, resources, options, pre-conditions, post-conditions). The methodology for producing this description of the cognitive activities and the requirements for cognitive activities in a domain is described in Woods and Hollnagel (1986). In this approach to building knowledge representations, an analysis of the joint human-machine cognitive system is partially driven by the cognitive demand characteristics of the domain (derived from analysis of what the designed system accomplishes and how it accomplishes those objectives, cf., Warfield, 1973; Lind, 1981; Pearl et al., 1982) and partially by the psychology of the human problem solver as relevant to the demands imposed by the environment. This analysis generates the questions that may need to be answered in some particular situation: what is the current goal-of-interest? is that goal satisfied or violated? is the process that achieves that goal working correctly or is it disturbed? is an alternative process available if needed or is it unable to perform? what constraints imposed by related goals govern the operation of a given process to achieve the goal-of-interest? what are the physical systems of components and their interconnections that implement the relevant processes? It is the answers to this bounded but generic set of questions about inter-goal/constraints, processes-goal relationships, inter-process relationships, and about how a process functions that organizes available data and provides informative counsel to the domain problem solver.

The resulting knowledge representation is object (not data) based; therefore the same data point may be relevant to the characterization of many objects. Objects are organized around goals-to-be achieved; this means the mapping is function or issue based as opposed to event or scenario based (e.g., a failure tree). The goal organization captures data on the relationships among goals and on how a goal is achieved (Lind, 1981). Thus, possible actions are organized around the goals that they can affect, and data about pre-conditions, post-conditions (effects), constraints (side effects or inter-goal couplings) and alternative means are captured. There are multiple levels or granularities to the object topology (Rasmussen, 1983; J. Duncan, 1984) in the sense that there is no one, fixed unit of description. Rather, there are related levels of representation with a unit of analysis appropriate to that level of the domain of interest. The result is a nested series of contexts (e.g.. the example of a valve closing in a thermodynamic system discussed earlier). Because of extensive cross connections between goals in actual work environments (at least, for the domains investigated to date with this technique), the topology is not strictly hierarchical and cannot be seen in a single, birds-eye view.

Data-Information Correspondences and Knowledge Resources

The knowledge representation specifies how the problem solver should see the domain. Data are the raw material or evidence to answer questions about the status of the units and relations in the knowledge representation, i.e., the data-information correspondences. For example, what parameters, targets, and limits indicate that a goal is satisfied or what data (e.g., tank levels, valve positions, pump status, flow rates) indicate that the functions performed by a process are working correctly (e.g., a material flow process consisting of sources, transport mechanisms or delivery points). In effect, a datum generates multiple messages or interpretations. At a high level, it signifies or activates a unit in the knowledge representation, e.g., $x > y$ signifies a violation of goal A; at another level, the datum signifies that, of the set of things that represent violations of goal A, the particular signal is $x > y$.

The ability to map from the field of available data to the state of task meaningful objects and relations defines one kind of informative advice. Using machines, or what I will call knowledge resources, to help the human problem solver move between raw data and the state of domain objects/relations constitutes one form of decision aiding (cf., Hutchins, Hollan & Norman, 1985). These knowledge resources can vary from weak to powerful (or, if you prefer, vary in *intelligence*). At a minimum, there is a need to collect the sets of data that testify to the state of an information category in the knowledge representation; the person provides the integrated assessment. With more machine power, knowledge resources can perform some of the data integration tasks; for example in thermodynamic systems, sets of rules are needed to evaluate derived parameters (is a goal satisfied?), active/inactive status of processes, available/unavailable status of processes, and when an alternative process should or should not be active. Finally, knowledge resources can function more intelligently to compute the status of high level objects and relations (e.g., warnings of prerequisite violations, reminders about post-conditions, reports of potential side effects or consequences) and report the result to the human problem solver.

This model of decision support explicitly recognizes that the knowledge resources needed to support problem solving in complex, as opposed to toy worlds, are heterogeneous, diverse in type and level of power, and incremental in development and validation. The knowledge representation (and therefore the human's cognitive activities in the domain) provides the motivation (in a systems design approach) for what knowledge resources are needed, and it provides the framework to meaningfully integrate the results provided by heterogeneous and evolving knowledge resources.

Spatial Representation

Visualization of a problem aids comprehension and conceptualization. The priority of space as an organizing principle is so compelling that non-spatial data are often given a spatial representation to improve comprehension (e.g. Haber, 1983). For example, people use Venn diagram representations to assist syllogistic reasoning (Johnson-Laird, 1983), and the implications of a set of data are found more easily in a cartesian plot or other graphic form than in a data table (Chambers et al., 1983). Similarly, showing how some device or system works (not how it looks) can aid a person to control or troubleshoot that device (e.g., Lieberman, 1985). For example, Brooke & Duncan (1981) found, in studies of fault-finding performance, that display format affects "the ability of the diagnostician to perceive what is relevant and what is not" (p. 188). and Payne et al. (1984) found that perceptual complexity affects problem solving performance in a variety of tasks.

The knowledge representation annotated with data sets specifies how the expert problem solver should see the domain; to effectively communicate this information to the user, the display system should reflect or should be an analog of the knowledge representation (cf. also, Norman & Draper, 1985). In an analogical representation, the structure and dynamics of the representation contains information about the structure and dynamics of what is represented. The correspondence between symbol and referent is not arbitrary, is not one of surface or pictoral similarity (i.e., icon or pictograph), and is not completely isomorphic. The topology of domain objects and relations specifies what about the domain should be reflected in the representation presented to the problem solver (see Sloman, 1978 and Kolers & Smythe, 1979 for further discussion of the defining characteristics of analogical representations).

Analogical representation produces economies of processing in problem solving by reducing memory, referencing and computational demands (this is not to argue that the problem solving power of analogical representations or depiction is always higher than descriptive representations; the choice of representation depends on the domain of problems to be addressed and on the goals of the representation builder). Lenat & Brown (1984) is a fascinating example, in machine problem solving, of the power of appropriate analogical representations.

- If A is an analogy of S, then relationships within S are represented in A without being explicitly named in A. Thus, modifying a diagram also and automatically changes the relationship between the modified object and all other objects in the picture. The effects or consequences of the change do not have to be calculated; they can be directly observed.

- Using a map one can get to all of the relationships involving a certain place through a single access point; by contrast, each part of the region would have to be referred to many times, in a large number of statements, if the same data set was expressed as a description (where the structure of the representation is independent of the structure of what is represented). As a result, it is easier with a map to access any part of a mutually relevant data set given any one element of the set.

- In good analogical representations, non-useful possibilities are inherently difficult to follow and useful possibilities are easy to follow (for example, it is harder to invent a drawing of an impossible object than it is to describe an impossible object) because constraints in the semantics of the domain (including causal and consequential relations) are expressed as constraints in the syntax and allowable transformations of the representation (cf. also, Hutchins, Hollan & Norman, 1985). As a result, for example, messages and alarms about how to manipulate the interface per se disappear and are transformed into messages and alarms about domain issues.

- All of the above help reduce the dependence of performance on user working memory capacity because the analogical representation acts as an external memory or provides contextual retreival cues (e.g., Norman & Bobrow, 1979; Williams, 1984). In this view, memory limitations are symptomatic of, rather than causally related to, deficiencies in interface and decision support facilities.

The concept for decision support is to capitalize on the power of analogical representation in human attention and problem solving by building spatial representations of data worlds that are not inherently spatial. One technique to support the user's view of relevant subsets of data is multiple perspectives (Williams, 1984). The significance of a given data point varies with the observer's point of view, and a perspective is a description of an event or device from a particular viewpoint. The concept of multiple perspectives is based on a metaphor that examining a data base is like a person standing at some point in a topology. The portion of the data topology that is viewable from a user specified vantage point corresponds to the data of high potential relevance given the user's interest (as expressed by the requested vantage point). This field of attention is constructed and manipulated through horizon limits, landmarks, height of vantage point (level of summation), granularity of the view. The use of multiple perspectives depends on the analysis of domain objects and relations in order to define what perspectives are meaningful and what portion of the topology is of high potential relevance in that context; e.g., from the perspective of some goal of interest, data about goal satisfaction, inter-goal constraints, process performance, and process availability form data subsets of high potential relevance. The user can then recognize interesting or important conditions within this data field to focus on or pursue. The display of the potentially relevant data set helps the user progressively refine and focus in on the important data for the current context.

Display selection now becomes a process of selecting an object and selecting a perspective or context. Conceiving of data selection in terms of objects and perspectives helps build interface transparency because the user's data queries are formulated in terms that are meaningful relative to his/her domain tasks rather than in terms unique to the interface itself. Because perspectives act as filters about what is relevant or irrelevant given a vantage point, they provide implicit control over the portion of the database available to the user at any time (from the data base point of view, the problem of how much and what kind of data to present to the user of large, heterogeneous data bases or, from a psychological viewpoint, the problem of specifying the user's normative field of attention in varying contexts, Moray, 1981).

A second graphic technique to visually represent portions of the knowledge structures is object or integral displays, displays that communicate the state of a high level unit in a single perceptual object. There has been a large amount of work to develop various object display formats and to research their benefits. Wickens (1984) contains an overview of the psychological basis for integral display formats, and Goldsmith &

Schvaneveldt (1984), Wickens et al., (1984), and Carswell & Wickens (1984) contain research results that empirically confirm the theoretical justification. For example, in a fault detection task on a simple dynamic system (where available data was related through the dynamics of the system and needed to be integrated to determine if the system was working normally) Wickens et al. (1984) found that there was a significant performance advantage with an object display (geometric pattern) over seperate display of the data (individual bars). Finally, practical experience with integral displays in several applications has confirmed the benefits, for example, contact-analog displays in flightdeck applications (Roscoe & Eisele, 1976), in statistics to aid in the interpretation of the results of multivariate analyses (Chambers et al., 1983), and displays that help one see how a system, program, or device functions (e.g., Lieberman, 1984). Building a successful object display for a particular application requires that the designer knows the relevant portion of the knowledge representation and associated data-information correspondences and uses this to select, scale, and assign data to the graphic form in a way that enhances the perceptability of the information to-be-communicated (cf., Woods, 1985).

The graphic knowledge system concept for decision support is based most simply on the empirical finding that visualization enhances human problem solving. Hence to support visual thinking, there is a need (a) to map the meaningful units of knowledge and their relationships and (b) to use this knowledge representation to guide what information (and therefore what knowledge resources or aids) the human problem solver needs, and (c) to build an analogical representation of the important domain units/relations (or to make the knowledge structures visible).

Conclusion

The issues explored in this paper show that questions of tool use cannot be treated as a secondary design problem, to be handled as only an interface issue that is relevant late in the development of an automated problem solver. Rather the characteristics of the joint person-machine cognitive system (implicit or explicit in the design) have a fundamental impact on the ultimate effectiveness of the new system in the actual work context and on the definition and architecture of the tools themselves.

Criteria for effective decision support are best met by using AI tools to support the user in the process of solving his own problem by providing informative counsel such as warning the user of potentially dangerous conditions, reminding him of potentially relevant data, suggesting alternative response strategies, performing overhead information processing tasks (massage the data into a form to more directly answer the questions the human problem solver must face), and by using direct manipulation, graphic techniques to enhance the user's information extraction, problem structuring and thinking (visualization aids comprehension). From the joint cognitive system perspective, the intelligence in decision support lies not in the machine itself (the tool), but in the tool builder and in the tool user.

Acknowledgements

Portions of the section on paradigms for decision support are adapted from my article "Cognitive Technology: The Design of Joint Human-Machine Cognitive Systems," AI Magazine, 6, in press.

Thanks to all the members of Westinghouse's Man-Machine Functional Design group for their confidence to try this approach in decision support applications, the valuable feedback they provided to refine the concepts, and their skill in wielding these conceptual tools to develop actual intelligent decision support systems.

Special thanks are also due Erik Hollnagel, Jens Rasmussen, Morten Lind and, in particular, Emilie Roth for the many valuable discussions on decision support and helpful comments on earlier drafts that aided in the distillation of these ideas -- they provided a model of informative advice.

A FRAMEWORK FOR COGNITIVE TASK ANALYSIS IN SYSTEMS DESIGN

Jens Rasmussen
RISO National Laboratory
Roskilde, Denmark

ABSTRACT

The present rapid development of advanced information technology
and its use for support of operators of complex technical systems are changing the content of task analysis towards the
analysis of mental activities in decision making. Automation
removes the humans from routine tasks, and operators are left
with disturbance control and critical diagnostic tasks, for
which computers are suitable for support, if it is possible to
match the computer strategies and interface formats dynamically
to the requirements of the current task by means of an analysis
of the cognitive task.

Such a cognitive task analysis will not aim at a description of
the information processes suited for particular control situations. It will rather aim at an analysis in order to identify
the requirements to be considered along various dimensions of
the decision tasks, in order to give the user - i.e. a decision
maker - the freedom to adapt his performance to system requirements in a way which matches his process resources and subjective preferences. To serve this purpose, a number of analyses at
various levels are needed to relate the control requirements of
the system to the information processes required and to the
processing resources offered by computers and humans. The paper
discusses the cognitive task analysis in terms of the following
domains: **The problem domain**, which is a representation of the
functional properties of the system giving a consistent framework for identification of the control requirements of the
system; **the decision sequences** required for typical situations;
the mental strategies and **heuristics** which are effective and
acceptable for the different decision functions; and the cognitive control mechanisms used, depending upon the level of skill
which can/will be applied. Finally, the end-users' criteria for
choice of mental stategies in the actual situation are considered, and the need for development of criteria for judging
the ultimate user acceptance of computer support is discussed.

NATO ASI Series. Vol. F21
Intelligent Decision Support in Process Environments
Edited by E. Hollnagel et al.
© Springer-Verlag Berlin Heidelberg 1986

INTRODUCTION

Since some system designers may be fed up with discussions about task classification and the tedious requirements of the "time-line studies" of traditional task analysis, the present discussion will be opened by a short review of the development of task analysis as depending upon the technology applied for the human - work interface. Hopefully, it will be clear that not only has the content of task analysis changed, but also that the dependence of the success of a design upon the quality of a task analysis will no longer be a question of degree - which some designers might find could be reached by proper training anyhow - but that lack of task analysis for a design based on advanced information technology may very well lead to fatal failure.

The need for cognitive task analysis is rapidly growing as new information technology is introduced in the interface between humans and their work substance. Task analysis has been an important element in the Human Factors of systems design during several decades. The nature of task analysis has, however, changed significantly through this period. Several phases in this need can be identified, depending on the technology applied in the task interface.

During the early phases of the industrial development, operators were interacting directly and physically with the machinery. Instruments were sparsely used, and task analysis was not considered; performance depended on the general technical master-apprentice training, and work conditions could be arranged empirically by behaviouristic time and motion studies.

As the size and complexity of industrial installations grew and electrical instrumentation technology was developed during the thirties and forties, centralised control rooms appeared. Large numbers of indicators and alarm annunciators together with operating controls were necessary, and this situation led to human factors analysis of the interface design. Still, however, there was no great need for task analysis. All primary sensor-based data were available at all times, and only the physical appearance and arrangement of interface components had to be chosen in relation to features such as the frequency of tasks, and the relevance of the various indicators and control keys for individual tasks. Operators were still trained as apprentices on older plants. Therefore, system designers only had to consider human factors as a final polish.

This situation changed with the advent of electronic instrumentation which made high levels of automation and very sophisticated equipment technically feasible. In particular for military equipment, reliability requirements during missions pointed at the need for analysis of operator tasks and work load. Especially when process computers were applied in industrial control systems for data presentation, the need for task analysis became obvious. Data then had to be selected and grouped for presentation by means of visual display formats which were designed .to match the different operating conditions and operator tasks. However, what frequently happened was an introduction of new instrumentation technology without major changes in the basic operating philosophy. Thus, the primary sensor data were still presented, but the requirements of the individual tasks and situations had to be considered on the basis of a task analysis, which could also be used for task allocation, i.e. the distribution of tasks between operators and automatic equipment. However, this task analysis could in general be done in terms of requirements of the technical system for control operations, because it primarily serves to select data to be included in the individual display formats. Typically, very little attention was paid to the question about what the task implied in terms of human characteristics, except with respect to rather general categories of operations which were related to training and selection criteria. This kind of task analysis was described systematically by Miller (1963).

The newest stage is the use of modern information technology for preprocessing of the measured data in order to integrate basic data into messages which match an operator's needs when involved in critical decision making during disturbed system operation. The automatic equipment takes care of normal operation and all planned sequence operations on the system. The major function of the operating staff, and, therefore, the task shaping the design, is the task as supervisory controllers during emergencies. This preprocessing of information cannot be planned without an intimate knowledge of the decision strategies applied by the operators, of the subjective goal adopted, and of the mental representation of plant properties used for the task. This means that the development of a framework for cognitive task analysis is necessary, as also discussed by Hollnagel and Woods (1983) in their arguments for "cognitive systems engineering". This kind of cognitive task analysis will have to be focused upon functional data processing aspects of the cognitive activity.

A further development in the application of advanced information technology in human task interfaces will be the introduction of means for interactive decision making which, in the present context, involves a highly integrated cooperation between the system designer and the user with a computer as the means of communication. To the more functional information processing aspects of cognitive task analysis are then, as a result, added questions of affects and attitudes such as criteria for accepting advice, for trusting messages, for understanding a decision partner's inference. The question of task allocation changes towards allocating responsibility and authority. Decision task elements are not "allocated"; rather the role as either active decision maker or concurrent monitor is dynamically allocated to the agent with the proper resources; both partners will probably, at a more or less intuitive or rational level, have to keep track of the proceeding task simultaneously, in order to cooperate, which means to be able to take over authority and mutually to trust each other.

In the following sections, a framework for cognitive task analysis from the functional, information processing point of view is discussed.

THE CONTEXT OF A COGNITIVE TASK ANALYSIS

In the typical case, when designing human-computer systems the problem is the need to design systems for support of decision making which is also effective during situations which have not been foreseen during design, and which are not familiar to the user even if the system is operated by a group of well trained, professional users. In other cases, the problem is to design information systems which are to be operated by a wide variety of users whose background and formulation of needs are poorly known. In such cases, the design cannot be based on a detailed quantitative model of the actual information process; instead a model or conceptual framework must be used which describes the interaction in terms of related categories defining the boundaries of a design envelope within which users can generate effective ad hoc tactics suiting their subjective preferences.

A widespread academic conception of the design process is a more or less orderly progress from a statement of the goals through several levels of functional formulations until, finally, the material implementation is settled. In general, iterations between the phases are included, but by and large the top-down design is accepted as the formal model.

This model does not correspond to realities, in particular not during periods with rather stable technology. Design is then largely a "horisontal" process. The previous system design is updated to be adapted to slightly changed needs or means of implementation. During such periods designers and users have a common conception of merits and difficulties by the present design and the modifications to introduce.

This approach will not be effective during periods of rapid technological changes. Designers and users will not have the same intuition, if any, and design should therefore be a more orderly top-down process. However, the benefits to be gained from new technology can only to a limited extent be formulated as design goals by future potential users. In addition, systematic top-down system synthesis is not part of many engineering curriculae, which are mostly considering analysis of systems which are supposed to be the result of the "art of engineering". The odds are that the coming successful designs will be based on "bright ideas", "inventions", etc. The evolution of new industrial practice will typically be based on mutual industrial competition and survival of the "fittest design". But even if the design itself is not based on an explicit conceptual framework in terms of a cognitive task analysis, such a framework will be needed for proper evaluation and documentation of the design. This evaluation and documentation will have to consider two aspects.

The first is the internal functional consistency in the design concept, i.e. whether the functional characteristics and limitations of the various subsystems and pieces of equipment are both necessary and sufficient for the overall functional requirements. For the normal technical functions this is no major problem since it can be done empirically by prototype experiments or simulation facilities. At present, however, such empirical evaluation is not feasible for an assessment of the correspondence between the potential risk of a large-scale industrial installation and the safety targets established for risky, but rare, events. In addition, empirical evaluation of the quality of the human-machine interface in support of operators' decision making during such events will not be feasible. For both of these features, an analytical assessment of the internal consistency of the design is necessary. This implies a check of the design concept against a conceptual model of the function of the total system which is based on a decomposition of system function into elements at such a level that either empirical data can be collected, or the judgement of subject matter experts can be used.

As an example of the need to rationalise the consistency of industrial practice by explicitation of the design concept in a period of technological change, consider the typical alarm system of a large process plant. In a plant designed by bringing together standard equipment and subsystems, the alarm system is typically the conglomeration of the warning signals which the individual suppliers consider important for the monitoring of the state of their piece (as the basis for performance guaranties). Hundreds of alarm windows presenting the state of the individual physical parameters may overload the operators' information capacity during disturbances. When the information capacity of process computers became available to control system designers, the use of alarm analysis and reduction was typically preferred to a reconsideration of the alarming philosophy and a design of an integrated control concept.

The other is the evaluation of the correspondence between the design concept and the conditions of the ultimate operation. This match has two sides. The designer can seek to predict the conditions of the final operation in his design, or the users can try to match the conditions to the design basis. In both cases, an explicitly stated model of the cognitive task underlying the conceptual design will be necessary. This cognitive task analysis and, in particular, its preconditions in fact specify the operating conditions to be accepted by the user, and will be the only rational reference for judgement of the causes of eventual malfunctioning. Unfortunately, the conceptual design basis is frequently very implicitly communicated. This leads to a high degree of ambiguity in the evaluation of the match between a design concept and real life conditions, in particular for the human-system interaction. The following discussion of a frame-work for cognitive task analysis attempts to formulate a model which may serve to bridge the gap between engineers' analyses of control requirements and psychologists' analyses of human capabilities and preferences.

THE DOMAINS OF A COGNITIVE TASK ANALYSIS

For design it is necessary to structure the great variety of real life work conditions into domains which correspond to design decisions. By the use of a multi-facetted description system it is possible to represent a great variety of conditions by a rather low number of categories in each domain, related to general features. The distinctions drawn between different domains of analysis resemble the distinctions of Leplat (1981) between analysis in terms of tasks, activities, and processes. In the present discussion, however, the denotation of task will be goal-directed activities of work in general, irrespective of the level of description. Leplat also stresses the need in system design to realise the difference between the prescribed task and the actual activity of operators. This is implicitly done in the present approach, since the aim is design of systems in which the operators can adopt activities and mental processes according to their individual preferences while performing the tasks required by the system.

From this point of view, the following dimensions of a conceptual framework for description of a cognitive task appear to be relevant for system design:

The problem domain. The first domain of an analysis which will serve to bridge the gap between the purely technical analysis of a system and the psychological analysis of user resources should represent the functional properties of the system in a way which makes it possible to identify the control requirements of the system underlying the supervisory task. This is an engineering analysis in technical systems terms and will result in a representation of the problem space. An appropriate representation should reflect the varying span of attention in the part/whole dimension, and the varying level of abstraction in the means/end dimension. Change in representation along both dimensions is by designers and operators in order to cope with the complexity of a technical system, depending upon the situation and the phase of a decision task (Rasmussen, 1984).

The means-end dimension is illustrated by the abstraction hierarchy of Figure 1, in which representations at low levels of abstraction are related to the available set of physical equipment which can be used to serve several different purposes. Representations at higher levels of abstraction are closely related to specific purposes, each of which can be served by different physical arrangements. This hierarchy is therefore

LEVELS OF ABSTRACTION

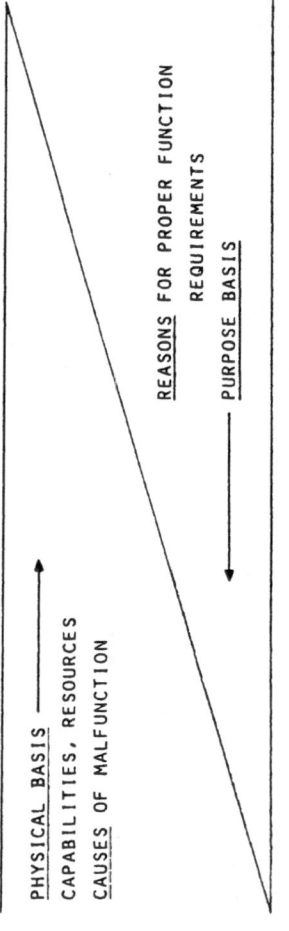

FUNCTIONAL PURPOSE

    PRODUCTION FLOW MODELS,
    CONTROL SYSTEM OBJECTIVES ETC.

ABSTRACT FUNCTION

    CAUSAL STRUCTURE, MASS, ENERGY &
    INFORMATION FLOW TOPOLOGY, ETC.

GENERALISED FUNCTIONS

    "STANDARD" FUNCTIONS & PROCESSES,
    CONTROL LOOPS, HEAT-TRANSFER, ETC.

PHYSICAL FUNCTIONS

    ELECTRICAL, MECHANICAL, CHEMICAL
    PROCESSES OF COMPONENTS AND
    EQUIPMENT

PHYSICAL FORM

    PHYSICAL APPEARANCE AND ANATOMY,
    MATERIAL & FORM, LOCATIONS, ETC.

Figure 1. The functional properties of a physical process system can be described at several levels of abstraction, representing the functional purpose and the physical implementation in different degrees in a means-end hierarchy. (Reproduced from Rasmussen (1983) with permission from IEEE).

useful for a systematic representation of the mappings in the purpose/function/equipment relationship which is the context of supervisory decision making. The relationship is a complex many-to-many mapping without which there would be no room for choice or supervisory decision making. According to Ashby's (1960) law of requisite variety, this mapping will have to be increasingly complex as the variety of abnormal situations which must be controlled successfully is increasing as it is the case in modern large-scale industrial operations. The increasing complexity makes it more and more likely that an operator may test an erroneous but quite reasonable hypothesis by actions upon the system (Rasmussen, 1984) and thereby change a minor incident into an accident. Similar lines of reasoning have led Perrow (1984) to introduce the notion of the "natural accident".

Considering a control task at any level of the hierarchy, information about the proper function, target states, and answers to the question "why" is obtained from the level above, while information about present limitations and available resources, i.e. answers on the question "how", can be obtained from the level below. In the context of supervisory control, the focus is upon those functions in human-machine systems which are related to correction of the effects of faults and other disturbances. Operational states can only be defined as disturbed or faulty with reference to the planned or intended functions and purposes. Causes of improper system functioning depend on changes in the physical world. Thus they are propagating - and explained - bottom-up in the hierarchy. In contrast, reasons for proper system functions are derived top-down in the hierarchy from the functional purposes considered during design.

During plant operation, the task of the supervisory controller - human and/or computer - will be to ensure that proper actions are taken to match the actual state of the system with the target state specified from the intended mode of operation. This task can be formulated at any level in the hierarchy. During plant start-up, for instance, the task moves bottom-up through the hierarchy. In order to have an orderly synthesis of the overall plant function during start-up, it is necessary to establish a number of autonomous functional units at one level before they can be connected to one function at the next higher level. This definition of functional units at several levels is likewise important for establishing orderly separation of functional units for shut-down and for reconfiguration during periods of malfunction.

During emergencies and major disturbances, an important supervisory control decision is the selection of the level of abstraction at which to consider the control task. In general, the highest priority will be related to the highest level of abstraction: First consider overall consequences for plant production and safety in order to judge whether the plant mode of operation should be switched to a safer state (e.g. standby or emergency shut-down). Next, consider whether the situation can be counteracted by reconfiguration of functions and physical resources. This is a judgement at a lower level representing functions and equipment. Finally, the root causes of the disturbance are sought to determine how it should be corrected. This involves the level of physical functioning of parts and components. Generally, this search for the physical disturbance is of the lowest priority, aside from the role which knowledge about the physical cause may have for understanding the situation.

The actual content of the information about the design basis which is necessary to enable operators to make the appropriate decisions should thus be identified from the abstraction hierarchy:

For prediction of responses of the system to control inputs in supervisory control decisions, knowledge about functional relations at each of the levels in the hierarchy is necessary. This includes knowledge of plant anatomy and spatial arrangements at the lowest level of physical form. At the level of physical function, important information is the description of the functioning of equipment, for instance, in the form of pressure-flow-rpm charts for pumps, reactivity-power equations for nuclear reactor cores, etc. Possibilities at the level of more generic functions are phase plots for water-steam systems ("steam tables"), heat transfer characteristics of cooling circuits, and control strategies for automatic controllers. More general characteristics in terms of power and inventory balances will be typical of more abstract functional requirements. Finally, at the level of functional purpose, the production requirements and the specifications of risk targets and limits for dangerous releases are given.

This kind of information, describing relationships within each level of the hierarchy, will in general be immediately accessible in engineering manuals and system descriptions. Such information as well as descriptions of the functional mapping upwards in the hierarchy are typically related to established and well documented methods for engineering analysis. This is not the case for information describing the downward mapping which represents the design decisions; i.e. the reasons behind the chosen implementations. This information is typically implicitly found in company or engineering practice or is based on the designer's personal preferences and seldom finds its way to the operators. This may be crucial for control decisions when over-ruling of a design requirement, e.g. an interlock protection, has to be considered during critical situations. In this way, reasons for fatal operator decision errors may be propagating downward in the hierarchy. Traditionally, much effort is spent on presenting operators with analytical, bottom-up information about the system. Only little attention has been paid to the need for top-down, intentional information on reasons for the implementations chosen during the design. To give access to such information, ad hoc advice facilities are typically established in the form of technical supervisors on call and - in the nuclear industry - "resident technical advisors" and "technical support centers". This kind of information should be directly available to the operating staff, probably in a kind of "expert system" computer-based tutoring system.

The lack of information on reasons will probably not be a prob-lem in systems of moderate size and risk levels, for which only the rather frequent operational states have to be considered since the reasons for these will be immediately and empirically known to the operating staff, because their effects are fre-quently met. This is not the case for large systems where safety specifications also have to consider rare events. In such sys-tems, reasons for infrequent yet important functions may be much more obscure to operators and special means may be required to make them understood. The information can be difficult to collect, once the design has been completed. It is a frequent experience for operating organisations that questions to system suppliers concerning their design bases are hard to have answered; typically, minutes from project meetings have to be retrieved since the man having the knowledge has moved to another position. Information representing reasons for design choices, for production and safety policies in a company will have the character of heuristic rules which are verbally stated, and an information base in the form of an "expert system" and an "expert knowledge acquisition" program to collect such infor-mation may be a useful tool for alleviating these difficulties.

The decision sequence. The next domain of analysis to consider is related to the decision process which has to be applied for operation upon the problem space. It is generally accepted that the decision process can be structured into a fairly small number of typical decision processes representing the various phases of problem analysis and diagnosis, evaluation and choice

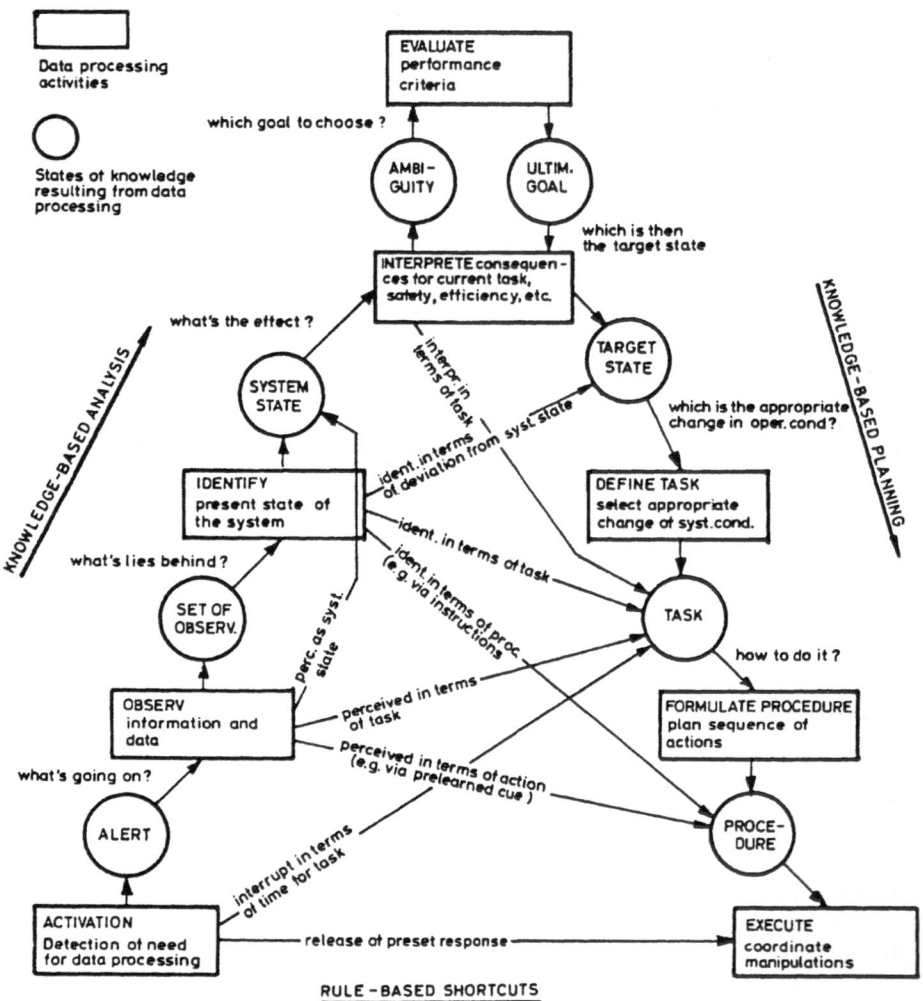

Figure 2. Schematic map of the information processes involved in a control decision. Rational, functional reasoning connects the "states of knowledge" in the basic sequence. Stereotyped processes and heuristics can by-pass intermediate stages. (Adopted from Rasmussen (1976) with permission from Plenum Press).

of goal priority, planning of resources, and, finally, execution and monitoring, see Figure 2.

Human mental economy can be achieved by the partitioning of a complex process in subroutines connected by more or less standardised key nodes representing states of knowledge useful for linking different processes, for bringing previously successful subroutines into use in new situations, and for communicating with cooperators in decision making. This is important since the task will be shared by the operator, the computer, and the systems designer. The task analysis will have to consider whether the decision is to be based on the designer's analysis a priori, and stored in the system by means of operational instructions and/or computer programs in order to ensure proper treatment of, for instance, rare event risk considerations, or whether the decision must be left for an on-line evaluation by operators and/or computers, see the example in Figure 3.

At this level the task analysis will be in implementation-independent terms, and will include an identification of the cognitive control mechanism required; i.e. which of the subroutines that can be preplanned routines called from memory, and which must be organised ad hoc. The decision ladder model of Figure 2 can be used as a scratch pad to represent the allocation of the different phases to the designer, the computer, and the operating staff.

Mental strategies and heuristics. An analysis in this problem domain can serve to identify those information processing strategies which are effective for the different phases of the decision sequence in order to identify the required data, control structures, and processing capacities. It is generally found that a given cognitive task can be solved by several different strategies varying widely in their requirements as to the kind of mental model and the type or amount of observations required (see for instance for concept formation: Bruner et al., 1956; for trouble shooting: Rasmussen et al., 1974; and for bibliographic search: Pejtersen, 1979). An analysis of the task is therefore important in order to identify the different strategies which may be used to serve the different phases of the decision sequence, and to select those which are considered effective and reliable.

This analysis, which is related to operations research rather than psychology, identifies the information processes required in implementation-independent terms as a basis for a subsequent human-computer task allocation based on demand/resource matching. It may be difficult to identify the useful, possible

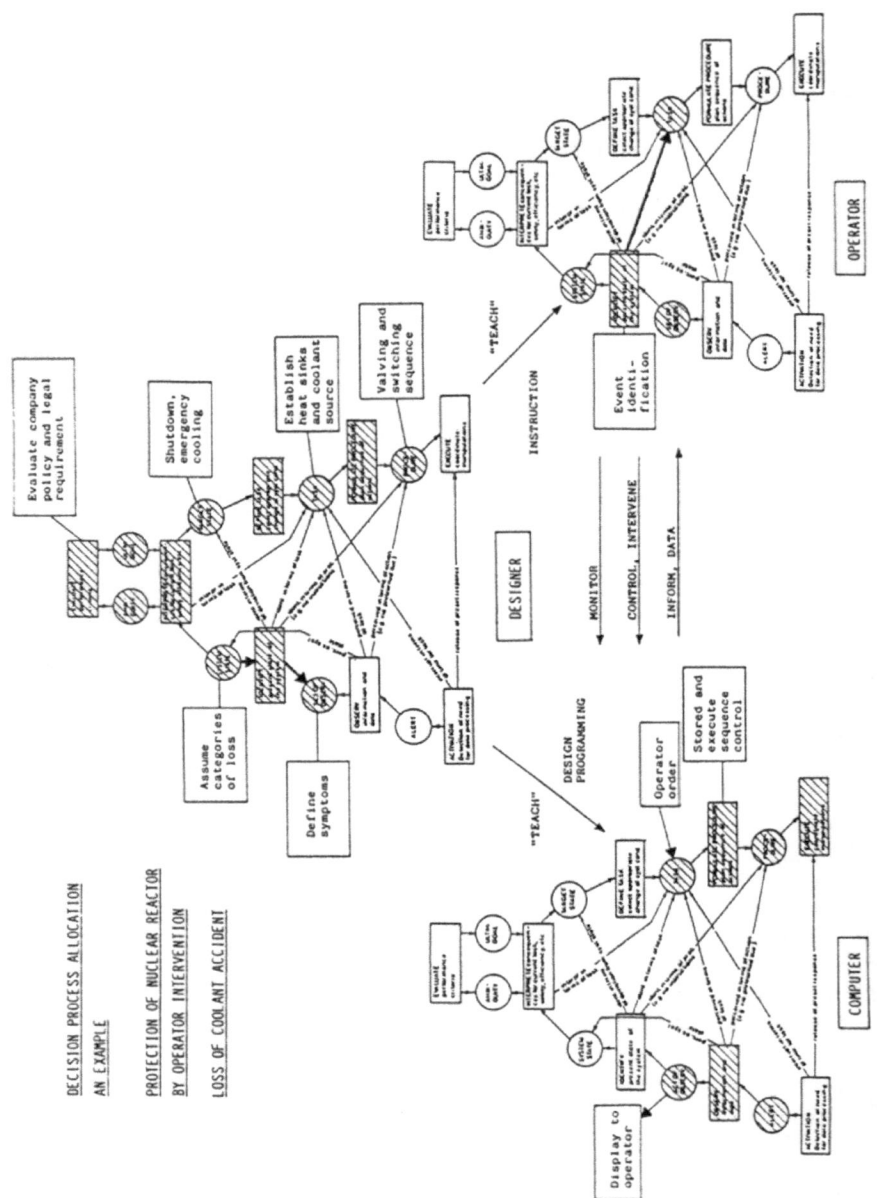

Figure 3. The figure illustrates the role of the designer, the operator, and a computer in a control decision which has only been partly automated. The designer has planned a repertoire of protective seqences which are stored in a computer but left the diagnosis to an operator. In addition to the decision functions, the designer, operator, and computer can support each other in different inform/advice/teach/learn functions.

strategies by rational analysis, but since users are very inventive regarding identifying clever tricks, the strategies may be identified by empirical studies of user behaviour. This, of course, requires psychological as well as domain expertise.

An important part of the analysis is an identification of the general resource requirements of the strategies in terms of data about the actual system state, on the basic functional relations, the processing capacity needed, etc., and of the consequences of errors. The results will be the basis for matching these requirements with the resource characteristics of the designer, the computer, and the operating staff for planning of their roles in the interactive decision task.

In addition, guides can be obtained for selection of suitable support of the mental models required in terms of information content of suitable display formats and the data integration required. It is the design objective to match the displays to the mental model which will be effective for the task and to choose interface design, operator selection and total job content in a way which will guide operators' subjective preferences in that direction. It should be noted that different strategies will have very specific needs with respect to type of useful data, to support of the mental model etc., and that in "intelligent" decision support systems it should be possible to let a computer analyse the user queries in order to identify the user's strategy and then to supply the required support in displays and messages of the proper form. The system should develop a model of the user (Hollnagel and Woods, 1983).

Cognitive control domain. While the information **content** which should be included in the display formats to support the decisions of a supervisory controller is identified from a task analysis, the **form** of the displays should be selected from consideration of human cognitive control mechanisms. This is necessary in order to have an interface design which will lead the future users to develop mental models and heuristic rules which will be effective for the information processing strategies chosen as the design basis. The model of human information processing used should be able to represent the different cognitive control mechanisms which are active at different levels of training, i. e. during routine as well as unfamiliar situations. The level of training will determine the domain of cognitive control and, thereby, the nature of the information used to control the activity and the interpretation of observed information. Control may depend on a repertoire of automated behavioural patterns, a set of state-action production rules, or problem solving operations in a symbolic representation. And, consequently, the same display configuration may be interpreted as a pattern of **signals** for control of direct sensori-motor manipulation, as **signs** which may act as cues for release of heuristics, or as **symbols** for use in functional inference. If the same display configuration is to support the skilled expert's heuristics as well as the analysis of the novice, careful analysis is necessary of the potential for interpretation of the displayed information as signal, signs, or symbols related to the task (Rasmussen, 1983).

Another important aspect of the model of cognitive control structures will be its use to identify mechanisms leading to errors in human performance. It seems to be possible to account for a majority of the frequent slips and mistakes by a limited number of cognitive mechanisms closely related to learning and adaptation (Reason, 1982, 1985; Rasmussen, 1984). In short, such mechanisms are related to interference due to similarities between cue patterns releasing stereotyped actions, between automated movement patterns, etc. An important part of a cognitive task analysis can be to screen the cognitive activities related to crucial decisions in order to identify sources of interference from "similar", typically more frequent task sequences. This points to the fact that cognitive task analysis may not only have to consider selected critical tasks; i.e. the general background of "trivial" activities is important. Such tasks are the source of the general level of training, of the large repertoire of automated routines which may be the raw material for composing special procedures, and which is with certainty the source of interference with less familiar activities. A model which seems to be promising for this kind of analysis is presently being developed by Reason (1985).

In a consideration of the cognitive control structures, it will be important to treat facilities for error recovery in a cognitive task analysis. The control domain applied for control of a current activity may not be the one needed for detection and recovery from own errors, and this will have ramifications for interface design which must both support the control task and enhance error detection and recovery. A special analysis should be applied for evaluation of error recovery features.

This kind of analysis depends on a model of the cognitive behaviour of a human in a complex interaction with the environment which depends upon experience across time and tasks. Since such a model should be able to predict the relationships among elements of behaviour which are typically studied by separate branches of psychology, it has to be a model at a higher level than the usual psychological models, as Hollnagel and Woods (1983) argue, "the whole of a man-machine system is more than the sum of its parts." There are, of course, limitations in the present analytical models of human behaviour as it will also be demonstrated in the discussion in the sections below, but this should not prevent the use of the existing models to guide designers in the direction of more user-friendly and error-tolerant systems, by identification of features defining boundaries of the domains within which it is possible to generate acceptable performance.

The analyses discussed so far have been considering the elements of the stepwise transformations from the functional properties of the physical system to the operator's interpretation of the messages presented by the decision support system, which are illustrated by the left column of Figure 4. In addition to the effects of this path of interaction, an operator's responses will be depending upon his mental models and strategies achieved from his professional training and general technical background. To ensure compatibility between the mental models and strategies adopted for interface design and those developed during training, a cognitive task analysis should also be the basis for planning of training programs, whether it be class room or simulator training. Development of schemes for training diagnostic skills within petrochemical industries, based on analysis of expert troubleshooters' strategies, seems to indicate a potential for great improvements compared with the present tradition (Shephard et al., 1977).

Finally, stop rules for the level of detail to include in a cognitive task analysis may be derived from consideration of the cognitive control domain necessary. Model-based analysis will be necessary only to the level of automated subroutines, which may be much more reliably planned from empirical studies and human factors handbooks. The stop rule in task analysis for planning of training programs proposed by Annett et al. (1977) seems to be useful also in the present context. According to their stop rule, task analysis will be terminated when the product of the frequency of a subtask and the cost of error in its execution are below a certain value, since analysis can then be replaced by observation, or benefit from improvements will be marginal.

Criteria for choice of mental stategies. The requirement that the interface design should lead operators to form effective mental models and adopt proper strategies presupposes that it is possible to characterise the different strategies with respect

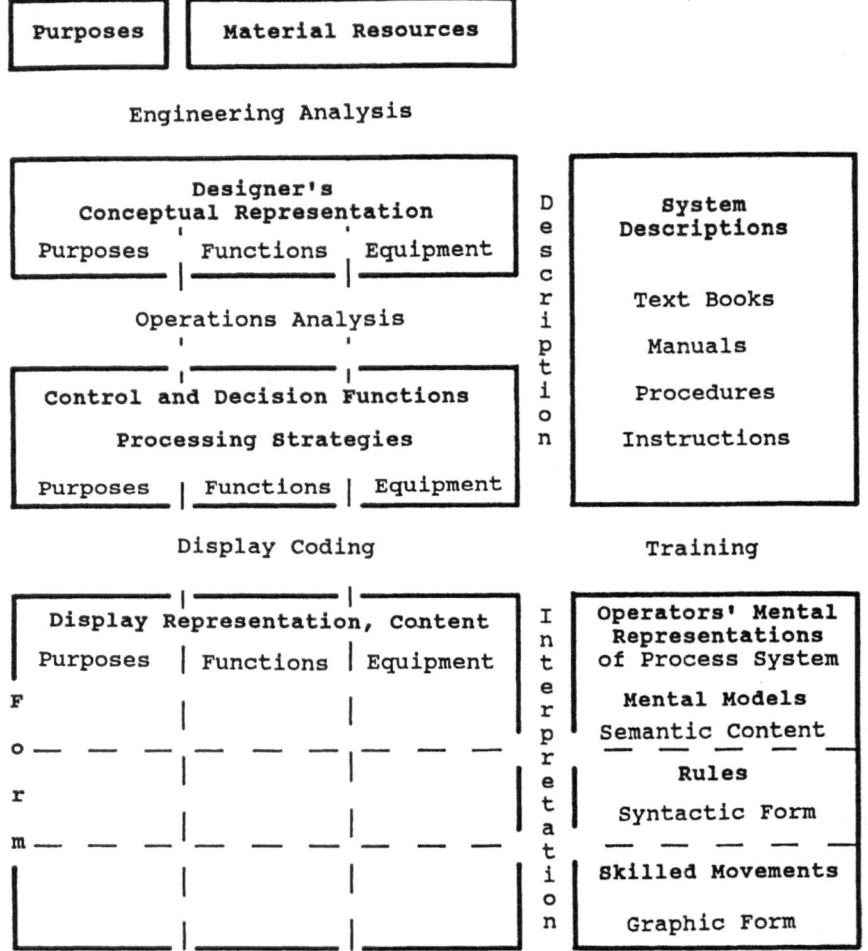

Figure 4. A schematic map of the different representations of a process system which are relevant for a cognitive task analysis. The aim of such an analysis is an integrated consideration of these representations and their mutual compatibility. During design, analysis along the path represented by the left hand column should ensure proper content of the displayed information. The form will depend on the level of cognitive control aimed at. This will also depend on the communication through the right hand training path.

to features that are related to the users' subjective criteria for choosing a given strategy as the basis for their performance. The criteria for this choice will frequently depend on properties of the information process itself, rather than its result. This requires an analysis in the borderline between operations research and psychology, and represents an area where laboratory experiments and generalisations from various real life analysis can be very fruitful. Some results are available, indicating important process criteria, such as cognitive strain, load on short-term memory, cost of observation, time available, data and resource limitations in general, cost of mistakes, etc. (Bruner et al., 1956; Rasmussen et al., 1974). More research in this area is needed. Interface design and computer support of decision making will only be successful if based on a proper knowledge of the performance criteria and subjective preferences the user will apply for the choice of strategy in the actual situation.

Role allocation. In addition to the support given in the problem solving process proper, there will be a need for exchange of background information. The designer, computer or the end-user may possess knowledge or data useful in the other partners' running task because of better memory characteristics, other information sources, etc. What does the designer/computer/user know which the partner needs?

Interactive decision making is sometimes considered to be based on a - static or dynamic - task allocation, where the cooperators are allocated functions and exchange results at suitable "states of knowledge". This is a realistic picture of the communication during rather familiar situations. However, when unusual situations occur, it will be necessary to have a much more intimate cooperation. During such situations human cooperators typically engage in a discussion to elaborate on preconditions of interpretation of observations, on the status and origin of their mental model, relevance of previous experience, etc., a conversation which seems to be a prerequisite for understanding messages and accepting advice. Such interaction takes place whenever the course of events or the partners' responses do not conform with intuitive expectations.

This means, in fact, that for systems intended for interactive decision making or "intelligent" support, it is not a question of allocating information processing functions, but of allocating authority and responsibility - to designer or user. The decision partners on site, the operator and the computer, who are the representatives of the designer, and who have to complete the design of control strategy for the particular unforeseen event, in reality have to operate in parallel during the process in order to be able to communicate meaningfully. The question only is who is responsible and has to be consistent and complete, and who is allowed to be satisfied with the intuitive hunches and approximate estimates sufficient to understand and trust the partners' messages. The problem for cognitive task analysis is to determine what kind of information should be communicated, and which "states of knowledge" in the information process of the computer are suited for communication to match the processing and expectations of the user and prepare him to take over. Typically, the information processes suited for the computer and for the human decision maker will be very different and, therefore, the necessary key nodes for exchange of information should be very carefully chosen. This communication will include messages in the whole range from neutral messages concerning background information, over advice or recommendations on useful approaches, to instructions and strict orders. Which are the criteria for the users' acceptance of the messages in the sense they are meant? What constitutes the difference in the conditions leading to cooperation, compared with those leading to rejection of support and to a competitive interaction?

Very little research is available in this area. The discussions in existing general textbooks focus, however, on aspects of preconditions for advice acceptance which are relevant for design of decision support systems. Nowell-Smith (1954) discusses different categories of communication which are relevant in this respect, such as: Learning, instruction, advice, exhortation, and command. He also mentions a number of ways in which a piece of advice can go wrong: The adviser deliberately deceives, he mistakes facts or conditions, he is inappropriately assuming that the advisee has the same perception of the problem as he himself has, the advice is excellent but unforeseen conditions emerge and, finally, he did not know what the problem is. Except for the first, these are all conditions which may serve to decrease an operator's confidence in a designer's attempts to help. In his discussion of possible criticism of advice, Gauthier (1963) notes: "The misuse of advice is subject to the law of diminishing returns; extend it too far, and no one will seek or attend to advice". This statement could have been addressed to some of the alarm analysis concepts mentioned earlier. A designer's pre-analysis of alarm patterns will probably be trivial to operators in frequent situations, and wrong in the more complex ones.

Howland et al. (1976) analyse trustworthiness of communication and find that "an individual's tendency to accept a conclusion advocated by a given communicator will depend in part on how well informed and intelligent he believes the communicator to be. However, a recipient may believe that a communicator is capable of communicating valid statements, but still be inclined to reject the communication if he suspects the communicator is motivated to make non valid assertions. It seems necessary, therefore, to make a distinction between 1) the extent to which a communicator is perceived to be a source of valid assertions (his "expertness") and 2) the degree of confidence in the communicator's intent to communicate the assertions he considers most valid (his "trustworthiness)". These appear to be key questions when a designer by means of his analysis of hypothetical scenarios attempts to assist an operator unknown to him during a complex situation in a distant future.

In general, research in this area seems to be related to activities such as social counselling, educational guidance, advertising and political propaganda, which means that research related to computer-based decision support systems should be considered a high priority research area.

CONCLUSION

In this framework for cognitive task analysis, the important feature is not the detailed form of the models underlying the analysis at the various domains. The most important aspect is the attempt to identify a set of dimensions in a multi-facetted description of the intelligent task a human and a computer are supposed to perform in cooperation. The concepts used in the dimensions should be able to bridge the gap between an engineering analysis of a technical system and a psychological description of human abilities and preferences in a consistent way, useful for systematic design. At the same time, the concepts should be compatible with those used for analysis of control and computer systems.

The present discussion of cognitive task analysis is only considering the concepts related to the **content** of the communication to be considered in the design of intelligent decision support systems, i.e. the representations of the problem domain, the control decisions to be made, and the decision strategies which will be applicable. For a successful design of the **form** of displays and other means of communication, including tools for effective information retrieval, models of human preferences and criteria for choice of the ultimate approach by the users are necessary (Woods, 1984). Development of such models requires analysis of human performance in real decision situations, either from actual accident scenarios (Pew et al., 1981; Reason, 1982; Woods, 1982) or during complex simulated scenarios (Woods et al., 1981).

DECISION MODELS AND THE DESIGN OF KNOWLEDGE BASED SYSTEMS

Morten Lind
University of Aalborg
Aalborg, Denmark

ABSTRACT

When developing knowledge based systems for realistic
domains the designer is faced with a complex task of planning
the control structure of the system. At present most systems
are developed bottom-up leading to programs which are difficult
to understand and maintain. It would be desirable to be able
to develop KBS's top-down from specifications of systems
functions before considering the implementation in software.

In the present paper we will consider the use of a model
of human decision making developed by J. Rasmussen at Riso
National Labs as a tool for specification of the functions of
KBS's and their organization in several levels. We will apply
the decision model in two ways.   The first deal with the
decisions to be performed by the KBS in terms of the domain
requirements. The other application of the decision model
describe the decisions to be made by the KBS in terms of
manipulation of the knowledge in the KBS knowledgebase.   The
relevance of this application appears from the observation that
the basic control cycle of the interpreter in a production
system has some striking similarities with the decision model.

BACKGROUND

The work reported in the present paper is part of an
effort to apply systematic approaches to the design of a KBS
for realistic i.e usually complex domains. In most cases KBS's
are developed with a strong orientation towards analysing the
domain in terms of the mechanisms supported by existing
programming paradigms.   The dangers of this tool oriented
approach is that the system designer is not encouraged to
analyse and specify the task to be solved in a way independent
of the actual implementation.   As a result the process of

NATO ASI Series, Vol. F21
Intelligent Decision Support in Process Environments
Edited by E. Hollnagel et al.
© Springer-Verlag Berlin Heidelberg 1986

acquiring knowledge from the expert becomes very difficult and may even be inadequate. Another result may be lack of transparency of the final KBS architecture.

Top down methods has gained a certain popularity for the development of traditional software and the question is whether a similar approach is possible for KBS design. By adopting a top down approach to the KBS design problem we need concepts which are suitable for the specification of KBS functions. We have considered the use of decision or task models of the type described by Rasmussen (1985), Breuker and Wielinga (1985) and by Kepner and Tregoe (1960) as possible candidates for guiding the development of KBS's. We will select the decision model developed by Rasmussen as the basis for our discussion. The model is shown in fig. 1 and it is assumed that the reader is acquainted with the model.

Decision models are interesting because they are very general, suggestive and apply on many task levels. Generality is a desirable feature but create problems when the models are applied in KBS design because the task context should be precisely defined. Most people would intuitively understand the meaning of a decision model like the one presented in fig. 1. because it has so many possible interpretations. However, this make it difficult to distinguish what is actually represented in the model and what is inferred or implied by a person looking at the model.

We have attempted to use Rasmussens decision model as a design tool in the development of a KBS for diagnosis of process plant. In this study we considered the decision model as representing control knowledge which could be used by the KBS if properly identified and formalized. The diagnosis system is based on the use of Multilevel Flow Models of the process plant to be diagnosed (Lind, 1982) and require a fairly elaborate control structure to guide the problem solving processes. Concepts which could aid in the specification of control structures would be very usefull for this design task.

As part of this work we have identified some inherent ambiguities in the decision model shown in fig. 1. An analysis of the problems encountered has led to the identification of some fundamental problems with the decision model. These problems should be solved if the model should be used as a formal tool for specification of control structure and KBS functions. It is on of the aims of the present paper to describe these problems for later clarification.

199

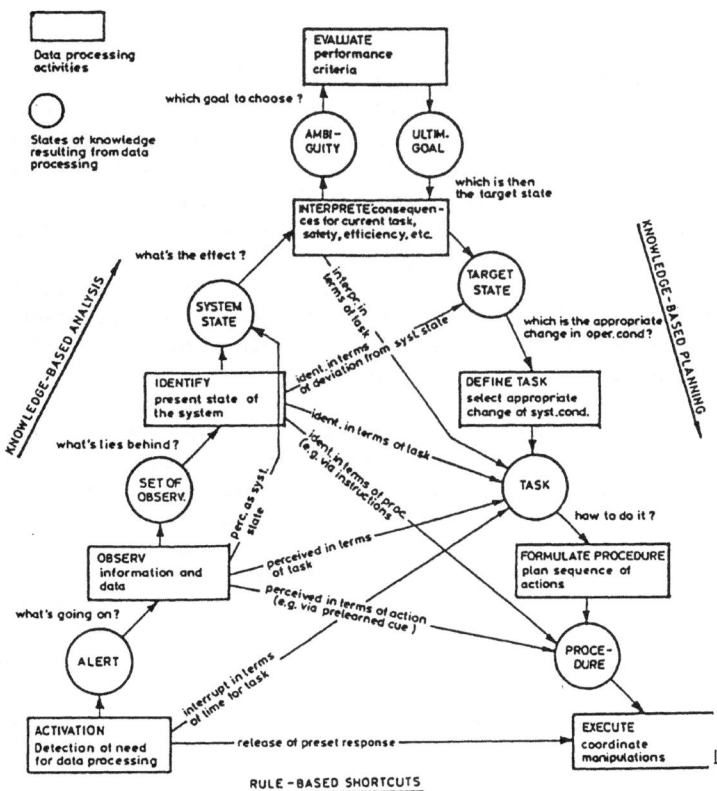

Fig. 1. The decision model (Rasmussen, 1985).

THE DECISION MODEL

The decision model define a framework for the planning of
different problem solving activities in a KBS for dealing with
systems diagnosis and control. Initially the model was
developed for describing problem solving of operators in
process plant control rooms. But as shown in the present
paper, it can also be used as a usefull conceptual tool in the
design of advanced KBS architectures. In such an application
it is used as a normative model i.e for prescribing the
functional structure of the system. The decision tasks
included in the model are included in the taxonomy of expert
tasks presented by Stefik(1982). The model in fig.1 show how
these tasks can be considered as subtasks in systems control

and how they relate to each other.

A unique feature of Rasmussens model is the distinction between socalled knowledge based problem solving corresponding to the use of deep knowledge of the system in diagnosis or planning and the use of rulebased problem solving based on shallow knowledge. In this way it make a distinction between the way a novice would solve a problem (by reasoning through a rational sequence using deep knowledge) and the way an expert would shortcut the rational sequence. By integrating these different types of problem solving into one framework the decision model may be used for planning the use of different types of knowledge in problem solving.

We will use the model mainly for discussing decision making in diagnosis and control of a system. The decision tasks relate accordingly to the environment of the decision agent (human operator or KBS). However, another interpretation of the decision model is possible which relate to the description of the decision processes going on internally in the computer i.e. describe how the KBS inference machine manipulate symbol structures representing domain knowledge in the knowledgebase. The two interpretations are related by the way the domain knowledge is represented in terms of knowledge representation formalisms. These relations will not be considered here.

HOW CAN THE DECISION MODEL BE USED IN KBS DESIGN?

The decision model can be used in two ways for the design of KBS's. It can be used normatively as a conceptual framework for specifying the tasks to be done by the KBS or it can be used descriptively for the interpretation of human problem solving performance.

When using the decision model as a normative model it is used to specify the tasks involved and relations between tasks via states of knowledge. This can be considered as a high level informal specification which can guide the further development of the system. However, if the model is formalized (which is the core topic of this paper) it can be used as knowledge represented explicitly in the KBS and used for controlling the problem solving.

The use of decision models or similar models representing tasks networks for interpretation of human problem solving performance has been described by e.g. Breuker and Wielinga (1985). They consider the use of socalled interpretation models similar to the decision model for the analysis of verbal data as part of the process of knowledge acquisition.

DECISION TASKS AND INFORMATION PROCESSES

In the following we will mainly deal with the decision tasks on an abstract level as we will consider the decision model as describing a network of tasks related by knowledge states. But although we consider each task as an abstract entity we will mention briefly how a task relate to the actual information processes which are required to solve the task. The basis for solving a task is knowledge about the system controlled (be it a physical system such as a process plant or a symbol manipulation system). Each decision task can in general be performed in using many different categories of knowledge and they may be implemented as KBS's. A KBS for solving a specific task in the model can accordingly be considered as a resource which is available for achieving the task in question and the choice between information processes as a resource allocation problem. Rasmussen (1985) discuss different strategies for diagnosing system faults based on different types of knowledge of the system diagnosed. These strategies define different information processes with different requirement to information about the plant and processor resources (e.g. memory size and complexity of inferences).

WHAT DOES THE DECISION MODEL REPRESENT?

Although the decision model is very suggestive and would provide a usefull tool for the design of knowledge based systems it is not directly suitable for this purpose in its present form. The problem is that many important aspects are left unspecified or implicit. In the following we will analyse the decision model with the aim of uncovering this implicit information. The analysis is a step towards the formalisation of the model. Only when the model is formalised it can serve as a tool for KBS design.

There are two main problematic aspects of the decision model which we will discuss. First, it should be noted that the model does not describe fully the information flow in decision making but only describe the flow of control from one task to the other. As an example we could mention that the "DEFINE TASK" subtask require both the system state and the target state as input information (see fig. 2). This fact does not appear explicitly from the diagram although it is implicitly assumed. It is not made explicit that each decision task in the general case can be based on all plant data both measured or computed values of plant parameters available or models of plant structure and function. This fact do not appear from fig. 1 and the knowledge states in the model corresponds therefore to different stages of elaboration of the plant information. Each knowledge state represent the new information which is created by performing the previously decision task. Another example illustrating the same problem occurs because of the inherent circularity of the diagnostic process (Lind, 1984). The circularity occurs because it is necessary to know the goal (protect system, protect operation or locate fault) before a proper level of state identification can be performed. But on the other hand the goal can only be chosen when the system state is known. This means that the goal (which is one of the knowledge states on the top of the decision model) may provide information to the observation or identification tasks (see fig. 3). These dependencies are not made explicit in the present form of the model.

Secondly, it should be noted that the knowledge which provide the basis for the decisions made in a given task change with time. This is obviously the case when the state of the plant controlled change. However, it is more important to realize that the result of a decision task may also depend on the state of the decision making processes themselves. Thus, if the same type of decision task is performed several times within a longer task sequence (e.g. in a diagnosis - planning - execution - diagnosis cycle) the information processes involved may be different at the different steps. Another way to state this fact is that a decision maker may chose the information processing strategies on the basis of the current knowledge acquired. A way to cope with this aspect in a formalized way is to consider each decision task in the model as a task category or class which have different instantiations (corresponding to different information processes) which applies for different kinds of knowledge available.

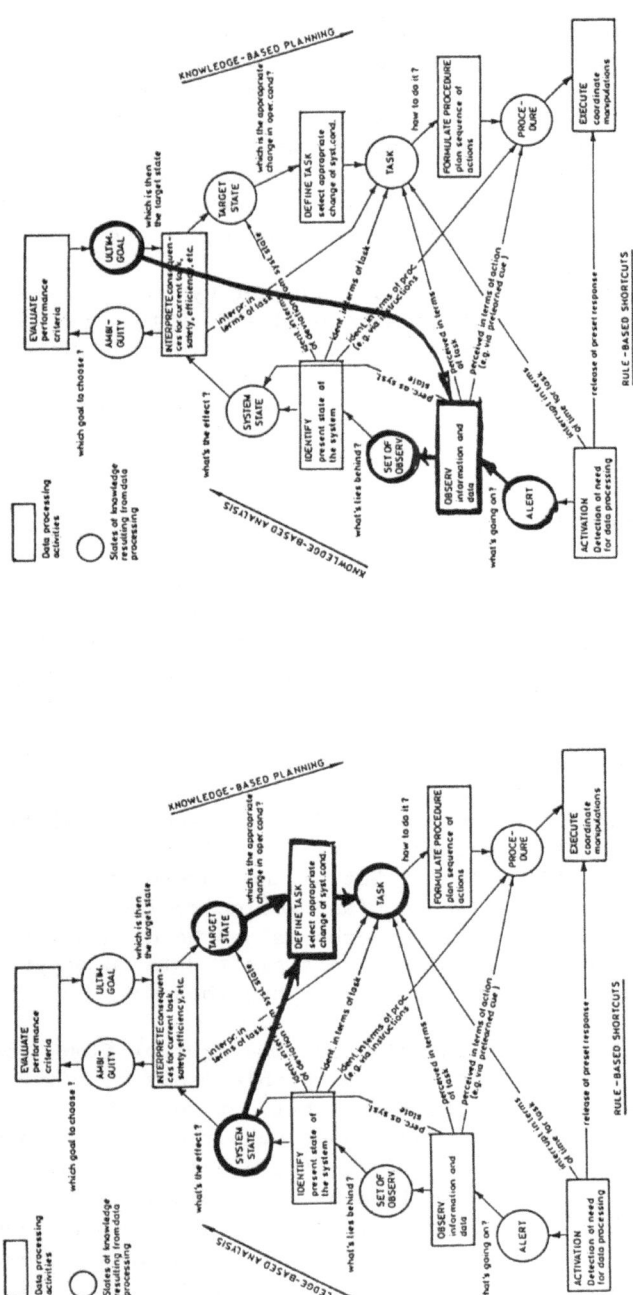

Fig. 2 The definition of task is dependent on system state. The dependence is shown as an additional link.

Fig. 3 The observation task depend on the goal chosen. The dependence is shown as an additional link.

The distinction between rule based and knowledge based decision making deal with a part of the problem mentioned above but it does not take into account that there may be different "shallow and deep" decision processes which can be applied under different conditions of knowledge available. The problem can be described in general terms as a lack of means to represent the flow of data or information during problem solving. This deficiency make it impossible to describe the dynamic aspects of decision making described above.

However, we need to consider another dynamic aspect of the decision processes. This aspect deals with the problems occurring when a decision task cannot be completed due to lack of sufficient information or decision criteria. This is illustrated below by a small scenario describing a simple diagnostic process.

Example scenario:

We assume that the initial situation in the diagnosis is that the decision maker is waiting to be alerted by plant signals indicating that the plant is disturbed and a diagnosis of plant state is required. When alerted the decision maker begin to observe plant parameters. After the collection of a set of observations has been completed plant data are interpreted in order to identify the system state. Up to this stage the sequence of tasks performed fits nicely into the sequence prescribed by the decision model on the left branch. But as the interpretation of plant state is abiguous (as would be the case when multiple competing fault hypotheses has been generated) the decision maker collect more plant data in order to be able to discard one or several hypotheses.

It is realized that in this case the control flow reverses and it is determined by the actual problem solving state inside the interpretation box. The fact that it was impossible to distinguish between two hypotheses make the decision maker to collect more observations i.e to initiate a task which is not in the direct sequence as described by the decision model. We can therefore conclude from this example that a problem occurring inside one of the boxes will invalidate the sequence prescribed by the decision model in fig. 1. The sequence described by the decision model presume the that tasks are solved so that there is no need to change the control flow.

THE DECISION MODEL REPRESENT A PROBLEM SPACE FOR DECISION PLANNING

The first step in solving the problems with control flow in the decision model is to declare that the model does not prescribe a unique sequence of tasks but it represents the problem space for planning of task sequences. This means that the model does not imply in itself any specific ordering in time but descibe only the logical dependencies of the individual tasks involved. However, removing the sequence information does not solve our problem directly it only makes it possible to reformulate the task sequencing problem as a planning problem. Before we examine this decision planning problem we will explore the consequences of considering the model as a problem space for decision planning. We will consider the left branch of the decision model in fig. 1 i.e the tasks involved in diagnosis.

There may be several alternative ways of doing state identification using e.g deep knowledge of the system structure and function or shallow knowledge such as fault symptom-cause pairs. This means that we can expand the decision model (the part we consider for the present purpose) by decomposing the individual tasks as shown below. The tasks may be distributed among separate problem solvers (KBS's). In fig. 4 we have decomposed the identification task into three subtasks. Each subtask produce one or several state hypothesis and the SELECT task deduce the system state from these hypotheses using knowledge about the relations between the IDENTIFY subtasks. The observation task may also be decomposed into several tasks if the individual identification tasks are based on different sets of observations. We get in this case the structure depicted in fig. 5.

DECISION PLANNING

The task network can be used to plan the execution of the individual tasks as it describe how the different states of knowledge are connected via decision tasks. Assume for example that observed (measured) plant data are given then we can plan a sequence of tasks which may lead to a goal state where the state of the system diagnosed is known (the top node in the graphs shown above). The plan (sequence of tasks) is generated by finding the path(s) which lead from the initial node to the goal node (forwards reasoning in the task network). However, more elaborate structures may be used which apply a combination of

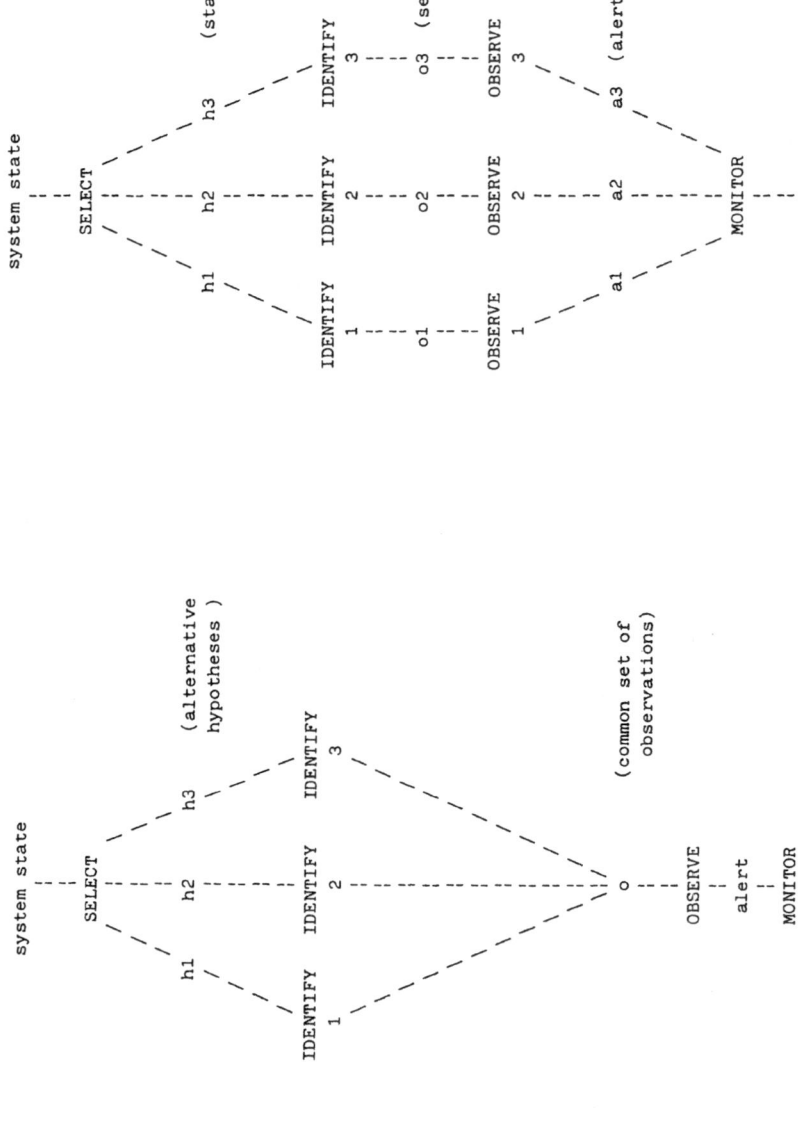

Fig. 4. A decision network with three identification tasks.

Fig 5. A task network with three identification and observation tasks.

forwards driven and bacwards driven reasoning in the task network. The top node may be taken as the goal node and the possible plans may be generated by backwards reasoning. The two cases correspond to two different ways of controlling the decision processes. In the first case it is data driven and in the other case it is goal driven leading to two possible architectures for implementation of the expert system.

There may be tasks which relate states with observations such as e.g. a verification task which from a hyphotesis about system state can derive the observations necessary for hypothesising the given system state.

We can consider the planning of the execution of decision tasks as a metalevel decision problem. This means that we can apply the decision model again but we need to define a new context in which the decision planning can be formulated. This context is defined by the decision model considered as a task network. In this context "system state" will represent the states of knowledge reached on the base level (i.e. how far the base level decisions has got) and plans and actions will relate to base level decision task sequences which has been selected by the (meta level) decision maker in order to change the state of the base level descision process. Accordingly, the metalevel decision processes may include both diagnostics and planning functions. In every situation a problem occurs in one of the information processes on the base level a shift may be made to the planning (meta) level to evaluate the possible actions to be taken.

We can summarize the description of the metalevel decision making in terms of three questions which can be made about the problem solving on this level.

The state of problem solving on the meta level:

Where am I? (what is known at the base level)

What are the goals? (what is the current knowledge
state to be achieved on base level)

What are the options? (what base level information
processes are available)

THE DECISION MODEL AND THE CONTROL CYCLE OF AI PROGRAMS.

In the analysis of the decision model we have up to now been concentrating on decision planning as seen from the point of wiev of the external task. We have indirectly assumed that the task was related to the diagnosis and control of a system in the environment of the KBS or the decision maker. But we can also apply the decision model for describing the internal operations of the computer. Adopting this wiev, we consider the decision tasks as related to the manipulation of the symbol structures in the computers memory.

This interpretation of the decision model is interesting because the control cycle of production systems (and basicly of all AI programs according to Nilsson (1979)), can be mapped directly into the decision model as a specific choice of sequence of tasks and intermediate knowledge states (see fig. 6.). This is presumably not a mere coincidence and raise some interesting questions related to the conceptual basis of the production system cycle or to the origin of the decision model.

CONCLUSIONS

In this paper we have analysed a specific decision model from the point of view of KBS design. We have identified some problems of what the decision model actually represent as the model only implicitly deal with control and information flow.

Two major results have been obtained. The decision processes have been separated into three distinct categories or contexts, one (base level) related to decisions in terms of the external tasks, a meta level decision process related to planning the sequence of decision tasks on the base level. The third context is an interpretation of the decision model in terms of the control cycle of a production system. As this production system may implement one of the decision tasks we have demonstrated that the decision model applies on at least three functional levels of a KBS.

Furthermore, the identification of the close similarity of the decision model and the basic control cycle of AI programs raise some basic epistemological issues in systems theory because there is a one to one mapping between production systems and turing machines. It indicate a possibility to use the decision model to extend the control cycle of AI programs to

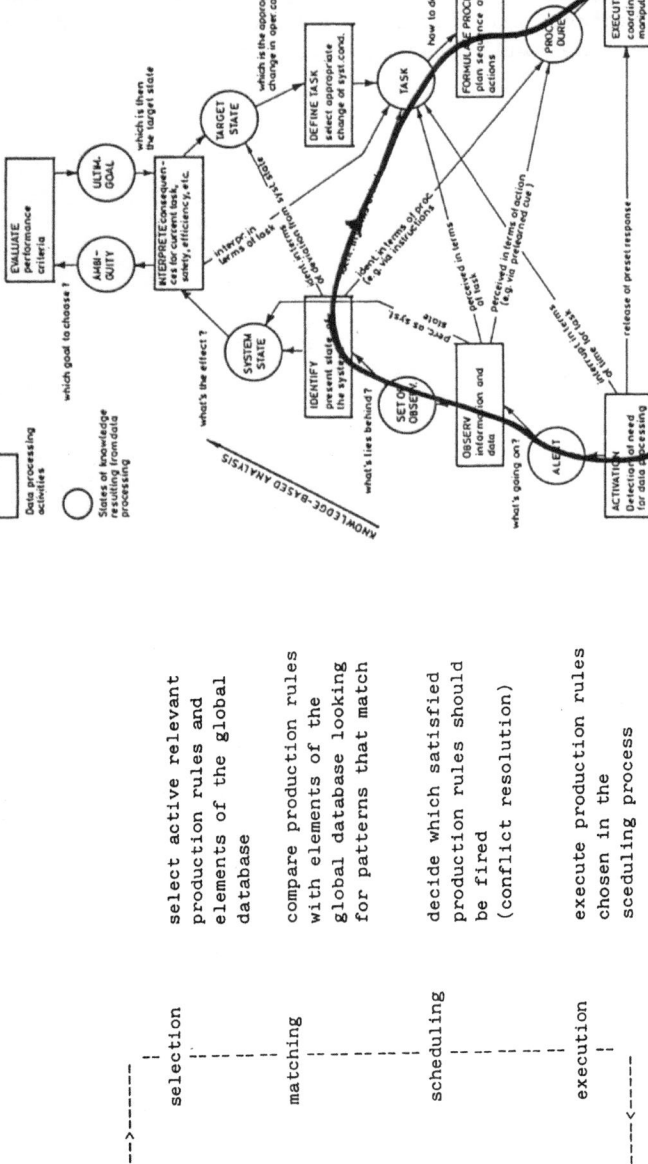

Fig 7. Control cycle of production system mapped into the decision model.

selection — select active relevant production rules and elements of the global database

matching — compare production rules with elements of the global database looking for patterns that match

scheduling — decide which satisfied production rules should be fired (conflict resolution)

execution — execute production rules chosen in the sceduling process

Fig 6. Control cycle of production system

include the top levels in the decision model. The significance
of this extension is a topic for further research. AI
architectures (e.g. Blackboard based systems) which can support
the dynamic control and information flow structures identified
in the decision model will also be investigated.

COGNITIVE SYSTEM PERFORMANCE ANALYSIS

Erik Hollnagel
Computer Resources International A/S
Copenhagen, Denmark

Introduction

For the purpose of this paper an Intelligent Decision Support System (IDSS) is defined as a computer based system designed to collect, organise, process and present information to support decision making in dynamic process environments. A general system evaluation should include: (1) the quality of the system's decisions and advice, (2) the correctness of the reasoning techniques, (3) the quality of the human-computer interaction, (4) the system's efficiency, and (5) its cost-effectivness. I will consider the situation where there is a need to analyse the performance of an IDSS to determine the quality of its functioning, corresponding to (1) and (2) above. To do so will require a description not only of how the IDSS functions but also of how humans make decisions, since it is the functioning of man and machine together that is in focus. It also requires a characterisation of the critical aspects of IDSSs - particularly of what sets them apart from conventional (i.e. non-intelligent) systems.

IDSSs as Technological Systems

Man-Machine Systems (MMS) are generally considered as non-intelligent technological systems, hence described in essentially mechanistic terms. Performance analysis therefore normally refers to the quality of the system product. There are, however, several reasons why this approach is inadequate.

Time Dependency. First of all it may even in principle be impossible to evaluate the quality of a decision at the time it is made, because the verdict depends on the actual consequences of the decision relative to the expected or predicted consequences. This fact is recognised in many dynamic decision models by closing the loop with a feedback from the implementation of the decision to the decision maker. But it basically means that the quality of the decision, in terms of the product, only can be assessed post hoc - and often in a manner that is far removed from the ideal of unbiased scientific methods (cf. Fischhoff, 1975).

Defining a Metric. It may be difficult to find an appropriate metric for the decision. It is clear that conventional single-attribute characterisations of decision outcomes, e.g. in monetary value, will be insufficient but there are no obvious alternatives. The problems are magnified for decision making in process environments, because real life conditions always require multidimensional descriptions.

Shortcomings of Formal Operator Models. Apart from the fundamental difficulties of modelling human action and cognition (which, however, has only discouraged few) practical experiences have not been too encouraging. As we shall see later there is a growing realisation that formal models may have to be relaxed to include more 'soft', psychological concepts. On the whole it can be argued that the mechanistic nature of technological descriptions make them unsuitable for characterising human performance (Hollnagel, 1983). The reason for this is the inability of mechanistic explanations to account adequately for human psychological functions, cognitive as well as affective. Even if we only consider cognitition, the basic characteristics of intelligence cannot be captured in a mechanistic explanation, despite the galant attempts made by e.g. information processing psychology.

IDSSs As Cognitive Systems

Together these reasons mean that it is practically inappropriate (and fundamentally incorrect) to describe IDSSs as technological systems, because the focus is on the product (decision outcome) rather than the process (decision making). If the attempt nevertheless is made, one consequence will be the inability to make adequate performance analyses. An alternative system description must therefore be found. The logical approach is Cognitive Systems Engineering which describes the IDSS as a cognitive rather than a technological system. The fundamental difference is the recognition of the intelligence inherent in an IDSS, which provides a more adequate background for characterising it in an appropriate manner. A short definition of a cognitive system is the following:

> A cognitive system produces "intelligent action", that is, its behavior is goal oriented, based on symbol manipulation and uses knowledge of the world (heuristic knowledge) for guidance. Furthermore, a cognitive system is adaptive and able to view a problem in more than one way. A cognitive system operates using knowledge about itself and the environment, in the sense that it is able to plan and modify its actions on the basis of that knowledge. It is thus not only data driven, but also concept driven. (Hollnagel & Woods, 1983, p. 589)

From this definition a number of other less obvious characteristics of cognitive systems may be derived. For instance, capacity for self-reference (Hollnagel & Lind, 1983), ability to model the environment and other cognitive systems, context dependent knowledge interpretation, ability to use incomplete knowledge or adapt different points of view, etc. These are, however, not strictly necessary for the analysis of cognitive system performance. We need to identify the necessary and sufficient set of criteria for an intelligent interaction between two cognitive systems, i.e. the minimum defining characteristics that make a machine an artificial cognitive system. That may very well also be the upper limit for what computers can achieve in the way of intelligence, at least at present. On the other hand, humans, being natural cognitive systems, certainly bring a host of additional characteristics to the task. But it would be a mistake to apply the full richness of natural communication to an MMS limited by the capacity of an artificial cognitive system.

Natural and Artificial Cognitive Systems

Man is obviously a natural cognitive system. A machine (computer) may be an artificial cognitive system, depending on what features it has. If we consider IDSSs in particular, the following observations may be made:

Goal Oriented Functioning. Any decision making system must by definition be goal directed, i.e. concept as well as data driven. Even though an IDSS may not actually make decisions, it must be capable of providing decision support (advice) that is goal directed, hence goes beyond simple data processing. (Logically it will, of course, have to make hypothetical decisions to give adequate advice.)

Heuristic Knowledge. Since an IDSS functions in a dynamic environment, it must be able to use a constantly changing knowledge base. Part of that may be the result of inferences made by the IDSS itself, which would qualify as truly heuristic knowledge.

Adaptivity. This is a basic characteristic of cybernetic control systems, and both analogue and digital systems may easily be endowed with adaptive mechanisms. A special consequence of adaptivity is multiviability. It means literally that the system can reach its goal in more ways than one, and it implies that the choice of a particular way is based on knowledge of the characteristics and requirements of the current situation, rather than being random. The ability to choose an appropriate way to the goal means that the system is intelligent (in any reasonable sense of the word), and multiviability is a fundamental characteristic of cognitive systems in general. Accordingly, how a decision is made becomes just as important as what decision is made.

Multiviability is a further reason why it is insufficient to consider only the product of the IDSS in performance analysis, as technological descriptions do. It is, of course, essential that the product (the decision) is correct. But since there always are several ways in which this can happen, the process is just as crucial as the product. This also agrees with the general notion that intelligence characterises the means rather than the end.

Ability to Plan. This capacity is required of an IDSS, because it has to function in a dynamic environment. The amount of data that must be processed and the (relatively) slow computing speed makes a straightforward algorithmic approach infeasible (Garbolino, 1985). Instead, processing has to be planned and modified according to heuristic computing strategies.

Together this means that an IDSS matches the basic features of an artificial cognitive system. Conversely, a description that does not capture these features will be inadequate for both system description and performance analysis. It therefore makes sense to assume that an IDSS is an artificial cognitive system. That assumption has some consequences for how the functional structure of an IDSS is modelled and how performance analyses are made and furthermore enables us to describe also the joint MMS, i.e. the operator together with the IDSS, as a cognitive system.

Reasons for Performance Analysis

Any performance analysis must begin by considering the nature and purpose of the investigation in relation to the nature and purpose of the IDSS. A performance analysis may be needed for a number of different reasons. The most important of these are: (1) as an integral part of system design, (2) as quality assurance to ensure proper system performance, (3) to improve and extend system performance, (4) to troubleshoot faulty systems, and (5) for purely scientific reasons to increase our knowledge and mastery of the artifacts we create.

Performance Analysis As Troubleshooting

Preferably, all systems should be analysed and debugged during design and development and delivered with a reasonable certainty that they function as intended. This, however, presupposes an ideal situation that may only rarely occur. Instead, once an IDSS has been implemented a performance analysis is generally only made as troubleshooting when it fails, i.e. when the decisions or advices turn out to be incorrect. One perfectly good reason is that a latent malfunction is detected only when something goes wrong. Logically a correct decision (product) does not imply a correct decision process. But it is pragmatically defensible to analyse only cases where a malfunction has been detected, since that at least indicates that something has gone awry, even though this approach will only capture a subset of all possible faults.

Another prominent reason for performance analysis is design verification. This may follow the accepted principles for system verification and validation, particularly with regard to the software modules. Design verification is important to ensure the proper functioning of the parts (and to a certain extent also the whole) of the IDSS, but there is a fundamental limit to this approach. It tries to confirm that the system complies with design specifications based on tests which, however, must be artificially well-defined and which often are produced specifically for the verification. Beyond that it is generally not possible from the design specifications to predict how the system will perform in a realistic situation, and accordingly there is no independent basis to compare observed performance with. On the contrary, the consequences are often predicted from a simulation of the design specifications (rapid prototyping or explorative programming).

In contrast, when performance analysis is part of troubleshooting, one must in addition to design verification also provide an explanation for an observed undesirable condition. This explanation must be consistent with the model for cognitive systems function and decision making that is inherent in the IDSS, i.e. the identified cause must make sense vis-a-vis the model. Accordingly, this provides a much better check of the power of the model and of the design

specifications (cf. also Gaschnig et al., 1984). I shall
consequently only consider cases where an incorrect decision
has been detected. This provides a natural limit on the scope
of the analysis, as well as some requirements to both system
and process description.

Why Decision May Be Incorrect

Troubleshooting requires a tentative classification of the
possible causes why a decision may be incorrect (and
conversely also for why it may be correct). These may belong to
either of two groups. First, the decision criteria may have
been inappropriate or the set of alternatives may have been
incomplete. Secondly, the decision process may have gone wrong.
In the first case the incorrect decision is attributed to
inappropriate knowledge rather than inappropriate decision
rules. In the second case the cause is found in the rules
rather than in the knowledge (though both, of course, may be
the case.) We may sometimes be able to determine directly that
knowledge was incorrect, for instance if the IDSS missed some
essential facts (whether or not we know how this happened) or
if it used incorrect knowledge (false propositions). It is,
however, generally impossible a priori to assume one or the
other, and the question therefore becomes how one can determine
which is the case.

Since an IDSS has been explicitly designed, we know how it
is supposed to work. But it may in practice be difficult to
find out if the design specifications were followed exactly.
One reason is that the IDSS is so complex that a detailed
step-by-step performance trace would be incomprehensible. We
must instead rely on a reconstructed trace based on a set of
discrete observations/recordings. Although this method has
proved to be useful in cognitive simulation (Newell & Simon,
1972), it contains a considerable element of interpretation.

One possible solution is to equip the IDSS with a
mechanism that automatically provides a trace of what knowledge
it used and what rules it followed. This kind of 'explanation
facility' is a common feature of many expert systems, but it is
not without problems primarily because the explanation facility
is an added feature, hence only reflects what the designer has
anticipated. Another problem is that it is not enough to know
what the IDSS did and that it performed according to the design
specifications. We must also know that they were correct. The
latter is, however, far from trivial because it requires an
independent point of reference. Since I have assumed that an
IDSS can be described as an artificial cognitive system, the
basis for comparison must be a model description of decision
making in natural cognitive systems.

The Prerequisites of Performance Analysis

It is a truism that in order to understand a phenomenon
one must have some assumptions about it. This is so whether we
are concerned with the humanistic or the technological domain
(Hollnagel, 1978). In systems theory we often refer to what we
want to understand as the object system, while the description

is referred to as the model (or model system). The prerequisite then is to have a reasonable initial approximation of the model system.

Psychological analyses can start from our privileged knowledge of how we think, remember, solve problems, make decisions, etc. (Morick, 1971), which provides us with an intuitively correct understanding of how natural cognitive systems function. Technological analyses can refer to reference descriptions of the system as found, for instance, in the design specifications. But analyses of artificial cognitive systems can use neither approach directly. Instead, the obvious prerequisites are a clear description of decision making imbedded in a description of action and control, and a description of how cognitive systems function. In other words, the design specifications must refer to a model of decision making in natural cognitive systems. We must specifically be able to account for how knowledge of the situation (context and agents) is used to select and implement appropriate decision strategies.

Understanding How A System Functions

In psychological terms the problem is how we can understand the functioning of an intelligent system. As far as understanding natural cognitive systems goes we may, in addition to introspection, draw upon the repertoire of psychological methods, e.g. for knowledge elicitation (Leplat, 1985). This can, however, not be done for cognitive systems in general because we cannot make the same a priori assumptions for artificial systems as we do for humans. In the technological sphere the task is rather similar to the debugging of symbolic executions (Coombs & Stell, 1984). In this case one has a user who provides a set of specifications to an interpreter, e.g. a PROLOG machine. If the outcome does not match the user's expectations he must debug the symbolic interpreter, i.e. understand how its process differs from his own. It is usually assumed that it actually is the user who is in the wrong, and that the symbolic interpreter provides the correct execution of the specifications. (The goal is accordingly to teach the user where he made a mistake.) But there is one crucial difference between symbolic debugging and IDSS performance analysis. In the former case it is assumed that we have a correct understanding of the execution, being the machine's version of it. In the case of an IDSS, we have no such a priori correct interpretation of our specifications on which to base the comparisons.

Performance Analysis As Backtracking

Performance analysis is basically an exercise in backtracking from existing data to probable causes by means of assumed principles for system functioning (system mechanisms, decision rules). This is particularly clear in the case of human errors, which start from an observed system malfunction and backtracks through the preceding events. In these cases the search generally stops when the first acceptable candidate for a cause has been found. The main reason is that our models of natural cognitive systems and MMSs are inadequate to explain human errors (partly because errors not are intentional). There is therefore no basis for determining whether the identified cause is indeed the correct one. If we consider incorrect decisions in the same way as human errors, we run a similar risk of oversimplification in decision performance analysis. This emphasises the importance of having an adequate model to support backtracking and analysis. The requirements to this model can be defined by outlining two basically different approaches to performance analysis, Black Box Analysis and Model Based Analysis.

Black Box Analysis

In black box analysis no assumptions are made about the functional structure of the system, i.e. what possible mechanisms it may contain. The performance analysis is simply concerned with the identifiable patterns in system input-output, possibly compared to some reference system (which is assumed to perform correctly). No efforts are made to describe the functioning of the system which, to all intents and purposes, is regarded as a black box.

A good example of black box analysis is Turing's test (Turing, 1950). Although this was originally proposed for a specific purpose (to dissolve the philosophical problem of whether machines could be said to think), it has often been used as the basis for assessing the performance of AI systems (Alty & Coombs, 1984). There have also been more formalised efforts to qualify it as a method of performance measurement (Colby & Hilf, 1974). It is nevertheless fundamentally inadequate as an approach to performance analysis, because it is concerned only with the gross surface characteristics of performance.

It is interesting to note that several systems have been build to pass Turing's test, notably ELIZA (Weizenbaum, 1966). In these cases the goal was to produce system performance what would 'fool' observers, i.e. that would appear intelligent (or whatever quality was in focus). But there was deliberately no attempt to make the mechanisms of the artifact resemble those of the object system. The mechanisms were rather chosen to provide the most impressive performance under given circumstances. For such systems black box analysis is, of course, adequate - in fact, to do anything else would be a mistake.

Black box analysis thus does not try to go beneath the surface manifestations of the system and is therefore not suitable for backtracking. The determination of what the relevant criteria are is based on an intuitive (expert) understanding of the system's functioning, rather than a detailed functional analysis. It is thus concerned more with appearance than function, hence with face validity rather than content validity. Face validity is necessary for a good IDSS, but it is not sufficient. A method that mainly emphasises face validity is inappropriate for performance analysis because an IDSS is designed to <u>function</u> as an artificial cognitive system and not just to <u>look</u> like one.

Model Based Analysis

The opposite of black box analysis is obviously an analysis based on a model of the object system. Here we may make a distinction according to the type of model used, whether it is a formal, normative model, or a more intuitive, descriptive model.

Analysis Based On Normative Models. A normative model makes some specific assumptions about the internal functioning of the system. It is both important that the product resembles that of a natural cognitive system, and that the process used to achieve that is appropriate for the domain. The mechanism chosen is, however, determined by available techniques more than knowledge about how natural cognitive systems function in the same task. Thus appropriate mathematical / computational models are used for the system's internal workings even though this may not match our privileged knowledge.

Examples of such systems are easy to find in AI, e.g. MYCIN and PROSPECTOR. Examples from a different domain are finite-state machines such as a Bayesian Decision Machine (cf. Garbolino, 1985). Even though this model in principle is based purely on mathematics, it is suggested to provide it with an 'information processor' to overcome some of the problems with the more formal approach, in this case the combinatorial explosion. Along the same lines is a proposal by Dubois & Prade (1985) to relax the axioms of qualitative probability by introducing concepts from fuzzy set theory and possibility measures. Similarly, Stassen (1985) makes the cogent point that formal models can best be established for situations that are so well-described that they are unsuitable for the study of imprecisely defined phenomena such as decision making. In other words, the situations are chosen to fit the model rather than the other way around.

Normative models are inadequate as a basis for IDSSs analysis because their goal is to obtain the best possible fit between observed performance and model behavior, without considering whether the model makes sense for the system it describes. Using this approach is tantamount to considering the IDSS as a technological system and therefore does not address the central problems. To perform intelligently an IDSS must exhibit the same characteristics that cognitive systems do, i.e. it must be an artificial cognitive system. Performance analysis can therefore not be separated from knowing how cognitive systems function.

Analysis Based On Descriptive Models. This leaves us with analyses based on descriptive models. Here the emphasis is on having models that are appropriate for the domain and which comply with our knowledge of how natural cognitive systems function, i.e. of how decisions are made. Descriptive models do not provide an absolute basis for evaluating performance products in the way normative models do but rather present a frame of reference for analysing the underlying processes, particularly in trobleshooting. By incorporating knowledge of decision making, the model may indicate whether the explanations we find are the proper ones, in the sense that they adequately explain the incorrect decisions. We thus do not object ourselves to the weakness found in human error analysis, i.e of not knowing whether or not a found cause was appropriate.

To support us in a backtracking type of performance analysis, we must assume (1) that the IDSS is designed according to the model, and (2) that the model is adequate for human decision making, i.e. that it is a functional model for natural cognitive systems. This means that IDSSs must be designed as artificial cognitive systems modelled on natural cognitive systems (because we have no better definition of what intelligence is). We must therefore begin by understanding how natural cognitive systems function, i.e. how they make decisions and behave intelligently. The model will be descriptive in the sense that it is based on what we know about natural cognitive systems. This knowledge is, of course, expressed by a formalism which presently is taken from computer science and systems analysis. That is nevertheless relatively unimportant.

A formalised model does, however, not have to be structured in, for instance, the flow diagrams that are normally used in graphical representations. This technique may certainly aid an understanding of the model, but it may also be grossly misleading (cf. Herriot, 1974). It is far more important that the model adequately represents the <u>functional structure</u> of the IDSS, i.e. defines the functions and the network of control that must be considered in a performance analysis. If these aspects can be defined, we can also derive the measurements we need. In this respect it is essential to realise the difference between a functional and a structural description. Even for common physical processes, such as blood circulation or central heating, functional and structural descriptions may differ widely (Lind, 1985). In particular, there may not be a one-to-one mapping from one description to the other. Similarly, a functional description of an IDSS or expert system need not map one-to-one to the physical description of its implementation in hardware (inference engine, knowledge base, etc.).

A Model of an Intelligent Decision Support System

If we consider expert systems as an example of IDSS, the normal modelling approach is to describe the basic functions in terms that refer to implementation details. Most expert system will thus need a data acquisition part, an inference engine, a knowledge base, a communication module, etc. There may be further details added for the various parts, depending on the application. However, the structure of these descriptions clearly show that their origin is in computer science and AI rather than in systems analysis. This particular way of describing the functional structure of the system is therefore inadequate for understanding its function as an artificial cognitive system.

The Cognitive System

An alternative system description, based on Cognitive Systems Engineering, would emphasise the functional elements required for intelligent action, and might lead to the following list:

Goals. The IDSS must have some way of representing the decision goals. Since decision making normally is recursive the system will have multiple goals. It is essential to know the relative importance of these goals, for instance in relation to changes in the environment (external pacing).

Knowledge. The knowledge of the IDSS may be divided into several categories, e.g. knowledge of the process, knowledge of the user (interacting cognitive systems), knowledge of itself, etc. Much of this knowledge will be in the form of models which must be continuously revised to reflect the current state of the environment.

Plans. Depending on the goal and knowledge of the current conditions the IDSS must select a plan that can bring it closer to the goal. This involves a comparison of the different possible plans, and a continuous monitoring and revision on the strategical, tactical and operational levels.

Status. The current status is an extension of the system's model of itself. Knowledge of the current status must include some sort of time horizon so strategies can be revised in relation to probable future developments. To achieve this requires at least a primitive form of self-reference, hence contains the dangers of infinite recursion.

This rough view of the functional structure of an IDSS does not refer to the same elements as expert systems normally do, e.g. inference engines, data acquisition modules, etc. These parts must, of course, be available but they refer to the implementation of the IDSS rather than its functional structure. The model is consistent with our (formalised) knowledge of how natural cognitive systems function. What is presently missing from it is a more explicit description of the control network, i.e. how the overall decision making is controlled and how it is related to the independent events in the environment.

The Decision Process

The second essential part of the IDSS model is the representation of the decision process. This is normally done by describing decision making as a number of stages that are logically necessary, without assuming that they must be performed in a fixed sequence (cf. Fischhoff, 1985; Hollnagel, 1984; Rasmussen, 1984; Rouse & Rouse, 1983). A generic set is the following:

Option Generation: Identifying possible courses of action (and inaction).

Value Assessment: Evaluating attractiveness or aversiveness of the consequences of each action.

<u>Uncertainty Assessment</u>: Assessing the likelihood of each consequence actually happening.

<u>Option Choice</u>: Integrating these considerations using a defensible (rational) decision rule to select the best (optimal) action.

For the purpose of IDSS performance analysis it is of minor importance how many steps the model contains and what they are called. The essential relation is between the individual steps in the decision making and the control network of the cognitive system.

For the purpose of a performance analysis we must be able to specify this relation in two ways. First of all one should arrive at an overall description of how the resources of the cognitive system relate to the steps in decision making. For instance, changes in the environment may induce changes in decision goals, and this will have consequences for how the decision process is carried out. Similarly, carrying out a step of the decision will have consequences for the status of the cognitive control network (changed goals, knowledge, plans, etc). A particularly important function is that of 'inheritance', i.e. how a change in one part of the cognitive system would affect other parts (e.g. a newly established subgoal may inherit information from preceding goals and decisions).

A proper representation of the decision process must thus not only describe the steps in decision making, but also the metadecision making, i.e. the monitoring and control of decision making. Two of the most important metaprocesses are the following (cf. Reason, 1985):

(1) To decide if the situation is relevant for the ready application of existing rules.

(2) To direct limited cognitive resources (attention & memory) toward critical aspects of the problem space.

This corresponds well with the functional description of a cognitive system given above. The control of when and how existing rules should be applied (entry-exit conditions) corresponds to the selection of plans based on goals and knowledge. And the monitoring and allocation of cognitive resources corresponds to the status monitoring of a cognitive system.

The metaprocesses are just as essential as the rest of decision making. In relation to IDSSs it may be argued that the metaprocesses are the more important, and that the intelligent support should focus on these rather than decision making as such. This would certainly be consistent with the viewpoint of CSE. And if an IDSS shall more than just replace steps in the decision process (which might not be very intelligent) it must entail some form of monitoring or modelling of the decision maker and his current status. The model of the IDSS must be explicit about the functional structure of the artificial cognitive system, the logical sequences in the decision process, and in particular the interaction between the two. Such a description would be independent of the implementation and serve to supply us with the frame of reference for a performance analysis. In order to analyse a specific IDSS it may also be necessary to describe the mapping between the functional structure and the implementation.

Method Overview

When it comes to available methods for evaluating artificially intelligent systems there is little in the way of a standard repertoire. The traditional approach is based on modifications of Turing's test. The alternative is to apply the more rigorous methodology from the field of MMSs.

Turing's Test

The basis for Turing's test is the indistinguishability between the performances of a human and a machine (or between natural and artificial cognitive systems). This was originally proposed as a way of deciding whether or not machines could be said to think (Turing, 1950), but has gradually become the accepted way of evaluating the performance of artificially intelligent systems. It may be disguised and elaborated in various ways (Gaschnig et al., 1984), but the principle remains the same. Performance data are collected from the system in question as well as from a suitable reference system. The data are transformed to a common format (often a quantitative measure) to eliminate obvious cues; in case of qualitative evaluations the protocols from the reference (natural) system are often reduced to match those from the tested (artificial) system). The performances are evaluated 'blindly' by a group of experts, using suitable reliable methods, and the results are used to determine whether the tested system meets the specified criteria.

The Turing Fallacy. Since Turing's test compares the performance of two systems it is concerned with products rather than processes. This is so whether one compares the final product or the intermediate steps. Turing like tests are therefore unsuitable for cognitive system performance analyses. The basic fallacy is that the test should be done 'blind', i.e. without knowing which system is which. In the original version the quality of thinking (in the evaluator) is used to determine the quality of thinking in the system under examination, and this obviously does reveal very much about what thinking is. (Turing may actually have proposed this test tongue-in-cheek.) Applied on AI systems it relies on human intelligence to determine the presence of artificial intelligence, but does not improve our understanding of intelligence as such. Performance evaluation is no substitute for a proper performance analysis, particularly not in cases of troubleshooting.

Man-Machine System Analysis

The number of methods that can be used to evaluate the

performance of MMSs is quite large and methods vary in scope and complexity from simple laboratory experiments to full-scale simulations (Hollnagel, 1985). Several of them can be applied to cognitive systems provided a meaningful relation can be established between performance measurements and system function characteristics. This obviously depends on a clear description of the characteristic functions of the system in terms of cognitive (intelligent) functions.

One useful way of classifying the methods is with regard to how the system is represented in the evaluation. The basic distinctions are here: (1) a conceptual system, which exists as a set of descriptions and specifications only, (2) a static system, where fundamental aspects of the system are tested from a static representation, and (3) a dynamic system, which generally refers to a running implementation of the system in part or whole, the extreme being the actual finished product.

Since I have argued that performance analysis is most efficiently done as troubleshooting, we need only consider the methods where the system is represented dynamically. The other cases are certainly not without interest but are better suited for earlier stages in system design and development. The dynamic simulations should be used to exercise the system to detect (or provoke) faults. Depending on the application and the stage of system development, a particular method may be chosen to collect the appropriate data for performance analysis, for instance from the set described in Hollnagel (1985).

The specific method will determine the richness and complexity of performance data. The basic procedure will be to identify the task segment where something went wrong. The next step is to determine whether the design specifications were followed - assuming that they were correct. This should include consideration of both the direct decision rules and the metacognitive processes. If nothing is found wrong with the decision rules, it is reasonable to assume that the knowledge was incorrect, and to investigate that in more detail.

The reason for proceeding in this way is the following. If the knowledge used in decision making was incorrect or incomplete, it may or may not lead to an incorrect decision depending on how crucial the missing knowledge was. But if the decision rules were incorrectly executed, the decision could only have been correct by chance, i.e. it then really does not matter whether the knowledge was correct or not. It is therefore more important to consider the decision rule, independent of whether the analysis proceeds in the forwards or backwards direction.

A Systematic Approach to Performance Analysis

Performance analysis is often carried out as a bottom-up or data driven activity, in the sense that one works from a set of performance recordings, and tries to draw some conclusions about the system vis-a-vis the purpose. As I have argued here, this must necessarily be based on an adequate model of the phenomenon under investigation, i.e. a model of decision making in artificial cognitive systems. In other words, the bottom-up approach must be supplemented with a top-down approach.

The essence of the bottom-up approach has been captured in the description of the general principles of human performance analysis (Hollnagel et al., 1981). These may, of course, be extended to cover performance analysis for cognitive systems in general, including IDSSs. The formalised description covers the steps from the raw data to the highly elaborated competence theories. The description assumes a simultaneous top-down approach, but does not describe this explicitly. The two aspects of performance analysis are shown in Figure 1 below. The following is a description of the two aspects.

From Data to Competence Descriptions

The bottom-up approach begins with the raw data (performance trace elements). These are the categories of observation that have been defined prior to the analysis. Or, if the analysis was not planned in advance (as might well be the case for trubleshooting), they are the categories of data that are naturally recorded by the system or that one can get access to post hoc (cf. Pew et al., 1981).

The next step is the aggregation of the raw data into a detailed description of the actual performance. The importance is here on establishing a common frame of reference (e.g. time) and to translate the different data types to a common format. Following that the performance description is 'translated' from the technical terms of the observation to a generalised trace in the formal syntax of the analyst. With the current frame of reference that means a translation in terms of cognitive systems engineering and the steps in decision making.

This is followed by the more elaborate analysis which considers all the generalised traces from the same event and tries to combine them into a prototypical description. It is thus essentially a performance model for a specific type of event, hence the level where the results from the performance analysis may be most useful for troubleshooting and fault identification.

The final step is going from the combined, prototypical description to the level of competence theories. This leads to highly elaborate descriptions that may be of little relevance for a particular system, but which contains the generalised kowledge necessary for overall system design and evaluation.

From Competence Descriptions to Data

The top-down approach naturally starts on the level of theories, in this case Cognitive Systems Engineering, theories of action and control, etc. These provide the necessary background for working in the domain and provide the general directions for both system design and analysis. They thus lead directly to the following level of conceptual descriptions. In the case of IDSSs this contains the detailed reference descriptions of decision making in natural cognitive systems, i.e. the model for the design of the articifial cognitive system.

From that, and the design specifications, the critical performance aspects can be identified. These are necessary as a background for analysing the generalised performance trace, i.e. for deriving the prototypical performance. The critical performance aspects, often expressed in terms of model derived variables, are so to speak the filters that identify the salient aspects of system performance. Critical performance aspects may further be refined into essential task segments which are the parts of the combined performance trace that are subjected to further analysis. Both the essential task segments and the critical performance aspects serve to effectively reduce the amount of data that has to be analysed. But it is, of course, crucial that this reduction does not lose anything essential. That is why it must build on a reference model for the IDSS.

The final step in the top-down analysis is the specification of relevant independent and dependent variables. This can obviously not be done without an explicit model of the IDSS and detailed knowledge of the situation. The final measurement method must be based on these definitions and avoid the danger of reversing the relationship. In the case of troubleshooting one does not always reach this stage, because the situation is not that well controlled. This makes it all the more important that the preceding steps are elaborated as far as possible, so there is a firm background for the performance analysis

Summary

The line of reasoning given here can be summarised as follows. (1) IDSSs cannot be adequately described in mechanistic terms but must be considered as artificial cognitive systems. In particular, the minimum conditions necessary for artificial cognitive activity can be specified to be use of heuristic knowledge, adaptivity, and the ability to plan. (2) Performance analysis is best done as troubleshooting, i.e. to explain an observed malfunction. Other types of performance analysis requires a well-defined reference which may be hard to justify. Performance analysis in troubleshooting, however, requires an adequate model of the process in question. Since we are concerned with decision making in artificial cognitive systems, we must have a model for decision making in natural cognitive systems (which may be different from a theory of human decision making). (3) Performance analysis is basically backtracking from an observed phenomenon (fault) to an explanation. The quality of the backtracking and the explanation depends on the model used. I argued that the model should be a descriptive model for cognitive systems, rather than a black box model. The details of such a model were proposed, with special emphasis on the metacognitive processes. (4) The accepted performance evaluation of AI systems is based on Turing's test. This is, however, basically a performance evaluation rather than a performance analysis, hence inadequate for our purpose. Instead we must consider the methods from the field of MMSs, particularly those dealing with dynamic processes. It was argued that one should first consider whether the decision rules were followed, and then whether the knowledge used was correct. (5) Finally the dual nature (simultaneous top-down and bottom-up) of performance analysis was discussed as a prerequisite for specifying performance data and analysis criteria.

The fundamental rules of cognitive system performance analysis are: (1) develop a detailed model of the phenomenon in terms of decision making in natural cognitive systems, and (2) concentrate on performance analysis rather than performance evaluation. In addition to this one should, of course, also observe the more general principles for design analysis and evaluation.

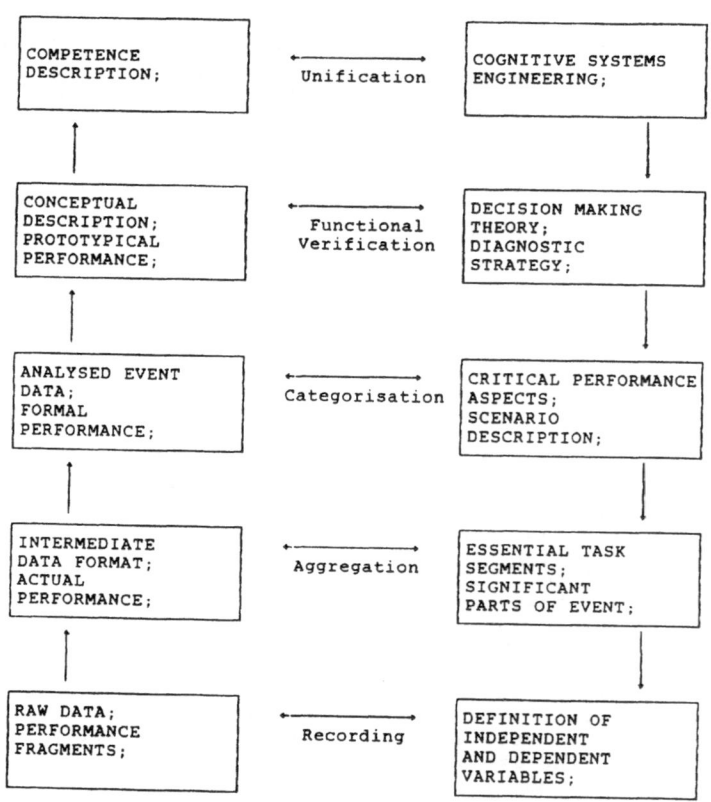

Figure 1: Data-driven and Concept-driven Aspects of Performance Analysis.

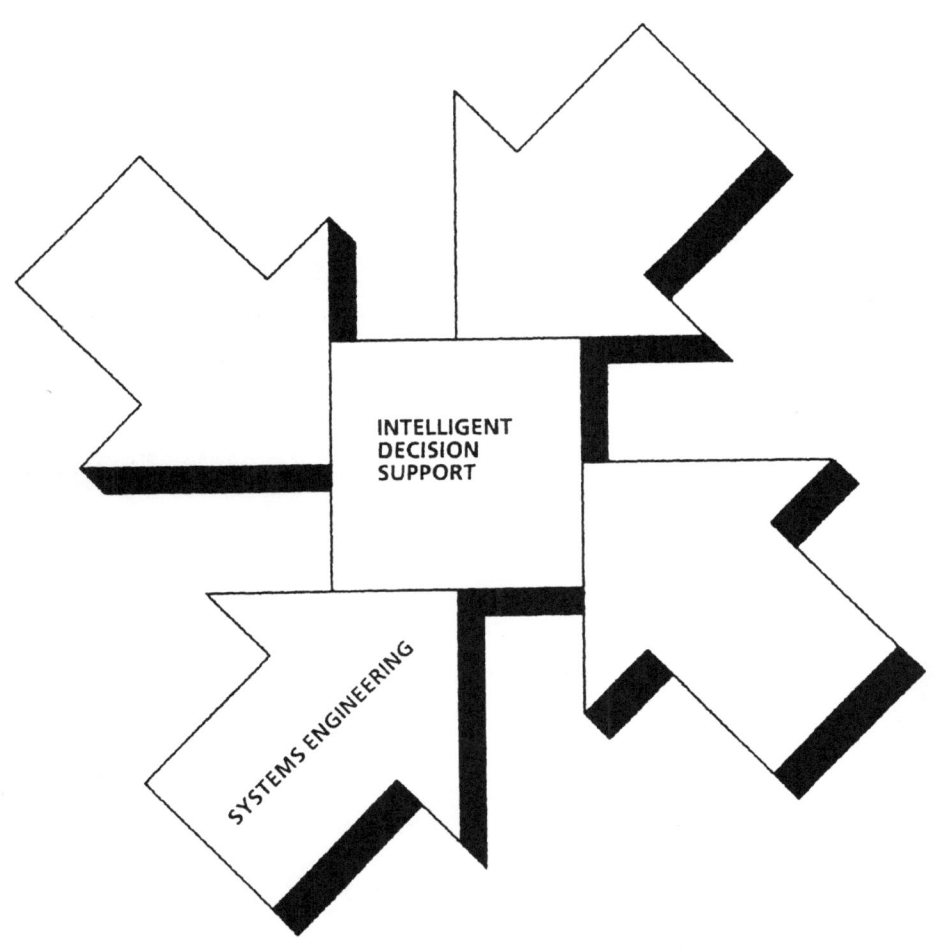

INTELLIGENT
DECISION
SUPPORT

SYSTEMS ENGINEERING

COGNITIVE ACTIVITIES

TECHNICAL ASSISTANCE TO THE OPERATOR IN CASE OF INCIDENT: SOME LINES OF THOUGHT

Veronique de Keyser
I. P. S. E. – Batiment B. 32
Universite de Liege au Sart Tilman
Liege, Belgium

1. INTRODUCTION

If one asks oneself how to assist the human operator faced with an accident, one is led to ask two questions :

1) Has he the necessary knowledge to understand the situation ?

2) Taking into account time pressure, has he the opportunity to use it ?

I shall attempt, based on the scientific literature and taking into account the current technological development of plants, to give some answers to these questions, in order to develop more effective technical assistance.

2. SOME CHARACTERISTICS OF OPERATOR CONTROL AND MONITORING ACTIVITY

Some characteristics of this activity are well known and I briefly recall them because they will be useful in the subsequent discussion. The results have mostly been obtained from field studies.

2.1. Operator activity is based on the control of the system and the detection, diagnosis and correction of anomalies that can occur in it (Faverge & al. 1966 ; Leplat 1968 ; Bainbridge 1981 : Terssac, 1980).

2.2. The control, the detection and some diagnosis strategies are largely based on patterns : these are configurations of indices or of signals which allow the operator, in an economic way from the mental load standpoint, to recognize the state of the process or the possible nature of an incident (De Keyser & Piette, 1970; Goodstein, 1981; Rasmussen, 1974).

NATO ASI Series, Vol. F21
Intelligent Decision Support in Process Environments
Edited by E. Hollnagel et al.
© Springer-Verlag Berlin Heidelberg 1986

2.3. The knowledge which the worker acquires of the system is an
operating knowledge based on his options to manipulate variables
and to check the results of his actions. In this respect, practice
is primary.

2.4. The system which the operator controls is a complex system (Dörner,
1984; Montmollin, De Keyser, 1985) comprising a physico-chemical
process, machines, men and instrumentation. These sub-parts interact
and each has its characteristic malfunctions (De Keyser, 1985).

Accordingly, the operator often works with a wide margin of uncertainty. His objectives are somewhat unclear and sometimes contradic-
tory and he uses a heuristic approach (Iosif & Ene, 1978)·

2.5. The knowledge of the system which the operator acquires is based
on the frequency at which the incidents to be regulated occur; the
worker learns gradually to catalogue the incidents with the con-
comitant pathological manifestations which arise from them. Over
time, experience leads to the ability to recognize rare abnormal
situations (De Keyser, 1985).

2.6. The quest for the information necessary for diagnosis is done
starting from the assumptions made by the operator about the nature
of the anomaly (Faverge & al., 1966; De Kevser & Piette, 1970).

The a priori probability of occurence of an incident guides the
diagnosis process; the operator starts by testing those which are
the most commonplace, the most probable. By proceeding in this way,
there is the greatest chance of success - reinforcement of his
behaviour - and the lowest mental-routine cost (Rasmussen, 1974).

2.7. The operator represents to himself the system through various
operating images (Sperandio, 1980), these images always being
internal representations, structured by and for the action
(Ochanine, 1981). Seeing the heterogenous nature of the possible
sources of information and the forms of intervention, these images
are multiple and can, in particular, be :

- operating images in terms of incidents and of the symptomatology
of incidents. The operator does not hammer out a representation
of the process in terms of physico-chemical laws which govern it,
but of the manifestations which characterize it (Bainbridge, 1981).
The major characteristic of this type of representation is that it
is very effective from the operating standpoint, but it leads to a
piecemeal disjointed knowledge representation. Adding the knowledge
of one incident to knowledge of others does not necessarily result
in a complete understanding of how the system functions. However,
some highly experience operators can find theoretical laws from an
empirical approach (De Keyser & Piette, 1970).

- spatial images of the installations. They can be very different :
static or dynamic, three-dimensional or plane, referring to the
actual installations or the representations offered by the wall
mimic boards. They serve as the basis of diagnosis strategies
of the topographic type (Pailhous, 1970).

- images linked to direct contact with the system : sight, hearing,
smell, etc. They are based on the knowledge of the informal indices
associated with system behaviour. Touch, hearing and recognition
of shapes, for example , favour complex operating images at the
cognitive level and are important from the social standpoint, for
they link the operators by an implicit knowledge which only they
have (De Keyser & Piette, 1970; De Keyser, 1972).

- images of the instrumentation (sensors, software, etc.), their
characteristics and their reliability (De Keyser, 1982).

- images of the social environment and in particular of stresses
linked to this environment (group norms, peculiarities of some
persons, etc.) and of services which the operator can expect
from them (information, assistance, in case of high load or
incident), etc. (De Keyser & Decortis, 1985; De Keyser, 1983).

- images of the general operation of the system and of the movements
of materials, products, men with the constraints of the
neighbouring work stations and departments.

2.8 Operator competence comes not only from the richness of his stock of images , their internal cohesion, but also from the interrelationships which he will be able to establish among them (De Keyser & Piette , 1970).

These interrelationships can be :

- the correspondence between different types of images and the same object , which they represent. This is the case of relationships to be established between a plane spatial representation , a three - dimensional model , and the actual installations. Or between information on analogue indicators and C.R.T. displays.

- the correspondence between these images from the system sub-parts (social environment , machines, instrumentation etc...).

This integration is made through the action wich attemps to harmonize the stresses and the information from highly different sources , but is difficult. The unity of the complex system , its global cohesion is never complete. The degree of integration can vary from one operator to another . One finds operators working with different styles , for example, relying primarily on the instrumentation information , or on the verbal communications , or on informal indices , etc... (De Keyser , 1985).

2.9 There is a continuous attempt by the operator to reduce the uncertainty as to the appearance of the incidents and to grasp the dynamics of the system , to forecast the results of his actions (De Keyser & Piette , 1970 ; Bainbridge , 1981 ; Iosif , 1970). The anticipation is based on series of mechanisms : knowledge of the probability of appearance of dysfunctions, functional links between some parameters , extrapolation of observed trends, etc..

2.10 In the operator activity , there are at least two functions , one of economics and the other of safety. These functions are taken on different ways, depending on the persons and the mental load level required by the work situation (De Jong & Koster , 1974 ; Sperandio , 1980 ; Lees , 1981 .

2.11. The fact of assisting the operator by computer, even leaving him a
      margin of freedom in the selection of his decision, does not
      necessarily lower his mental load. Actually, the psychological
      necessity of mastering the computer can lead the operator to detect
      possible malfunctions by parallel checks, thus increasing the infor-
      mation sources to be monitored (De Keyser, 1982; Ephrath & Young, 1981 ;
      Wickens & Kessel 1981).

2.12. Even in the case of continuous processes monitored from a control room,
      the operator activity cannot be isolated from the one of the team
      surrounding him : roundsmen, supervisors, etc. Recent research shows
      the impact of the team and the organization of work on the system
      regulation ; it introduces the idea of collective control
      (De Keyser, 1983; Fichet-Clairfontaine, 1985; Leplat & Savoyant, 1984;
      Marine & Navarro, 1980).

3. IMPACT OF TECHNOLOGICAL DEVELOPMENT ON OPERATOR KNOWLEDGE

Research recalled in 2. shows a remarkable convergence of results. It
all highlights the intimate knowledge of a complex system by the operator (Montmollin & De Keyser 1985; Montmollin, 1984), which allows him to
diagnose and even to forecast some accidents.
However , recent industrial catastrophies raise questions about
human reliability . It is well known that human reliability can be
reduced under the influence of fatigue, monotony, stress. But, is one
not creating, with a poorly mastered technological development, work
situations where the minimum structuring conditions for this knowledge
are no longer fulfilled ?

Only a few field studies are available to answer these questions : their
results express the emerging trends and must be interpreted with
caution.

3.1. Process stabilization and the changing role of the production
     operator

     Automatization applied in the firm is succeeding in controlling more effectively the production process which becomes, in the sight of the operator, extremely stable. A recent study in five cellulose firms in Norway shows that, in the oldest one, 15 % of all incidents occur daily; in the most recent, this figure is only reached in one month. This study concludes that there is a loss of operator knowledge because incidents have, up to a point, a teaching value (Skorstad, 1985).

When the firm thus succeeds in stabilizing the process, it will tend to restrain, or even to prevent any attempt of the operator to optimize the parameters, for fear that he might upset them. The consequence is that the operator no longer tries to anticipate incidents. Instead of following any process change and to regulate it, he only acts when an incident occurs and acts so that its recovery will be as quick as possible. Thus there is a switch from an anticipation logic to a recovery one. In the latter case, the operator can, in the limiting case, be but a dispatcher of information to the specialized department : maintenance, data processing engineers, supervisors called to intervene. In the steel industry, we have had the occasion to analyse this displacement of competences by observing the same sector at different automation levels. At the highest automation stage, the production operator has only very sketchy operating images of process and installation; he no longer seeks to anticipate risks but, by appropriate strategies, to install prompt recovery procedures. Starting from an initial diagnosis of the anomaly which had developed, he calls on the technicians concerned, and knows where to find them, he makes decisions to regulate the uninterrupted production flow, taking into account the economic and safety constraints. He now excells in a structural regulation (Faverge 1966) , intervening in the whole system - the men, the installations, the adjoining departments, the product and material flows - and not only the process. He was no longer capable of explaining the laws of the process or the intimate topography of the installations but, it should be noted that he had in his control room all the usual means of information : wall-mimic boards, screens, etc.

This result can be explained by the impossibility for the operator to act on the process and to experiment with some actions. Thus he will no longer hammer out the operating images capable of sustaining a diagnosic approach, but this is only partly true. Much research and many observations show that when the operator, because of lack of basic theoretical knowledge or of practical experience, was unable to build himself some strong operating images of the system which he believes he has to master, he delves elsewhere : in his daily life and in his previous experience. Thus, new steel plant operators attempted to explain how the stopper end broke and obstructed the ladle outlet by recalling the image of a bathtub or handbasin emptying. They extrapolated from the suction force which drew their hand to the handbasin outlet, to the force which parted the stopper end in the steel mill. These analogies are not always happy : they can cause errors and unwanted results. Somes technical school students are wrong in interpreting the mechanisms of a decompression device, because they imagine it similar to the one of a bicycle brake (Rabardel, 1982). Workers of a fertilizer plant mistakenly stopped an incident recovery operation, because they believed it would cause a decomposition of the product into a toxic gas (Fichet-Clairfontaine, 1985). Some operator behaviour, apparently aberrant, can be explained, when one understands the erroneous representation of the system on which it is based. These images are there even if distorted or if they contain fanciful cause and effect connections due to a lack of experience and feedback; the person calls them to mind to solve a problem. And it is thought that, with appropriate training, these images could be improved (Cavozzi ,1980). But in the case of the highly automated processes which we have studied, the operator does not appear to be concerned to develop such representations. The feeling that automation is reliable or that in the event of a problem specialists will take over appears to prevail. The knowledge which the operator acquires complies with the necessity of his job. He will not make a huge investment in observation, checking, judging, establishing relationships, gathering of data without being certain of its usefulness. The operator does not invest psychologically in a role which escapes him.

Does this mean that he becomes an unimportant button-pusher ? Not at all : the structural competence which he develops is fundamental in the systems which are becoming increasingly integrated. And today one discovers how much the cohesion of teams and how the information circulating among them can influence the collective control of automated systems. But in this case, the displays, means of communication, screen images which are supplied to him, must take into account this new facet of his activity to a greater extent than they do today.

Thus two scenarios are appearing on the horizon and their risks must be carefully weighed. The first concerns the stable processes, where the danger of an accident is small. Automation can be continuously very advanced and the operator will not be expected to be the one on whom the production parameter control rests, for one must be aware that whatever information is brought to him on the subject, the knowledge which he will obtain from it will be very slight. This possibility can be accepted if the risk is well calculated and if the operator's structural regulation is developed whith ad hoc technical assistance at the same time.

But if the process is unstable, its components fluctuating, or if there is a high potential risk, then the operator must retain control, together with that of the installations. Therefore the orientation must be towards other forms of organization : a flexible automation which still allows a range of experimentation, and a personnel mobility, which maintains strong operating images (De Keyser, 1985).

The information aids to be offered to the operator, his training, the design of his work station, the organization of work, are not the same in both cases : they stem from the prior analysis of the implicit or explicit role he is asked to play in the control and supervision of the process.

3.2. Different information sources and characteristics

Different sources of information are available in control rooms : C.R.T., mimic boards, verbal communications etc. All of them have their characteristics which can entail variations in the cognitive strategies of the operator. The wall mimic boards are a continuous background

allowing the operator to visualize economically the whole process :
fild studies show that they favour the gradual acquisition of relationships of cause and effects between the variables, and facilitate the
anticipation of accidents. They are rarely examined on a on-off basis;
more often the operator takes them in at a single glace and judges a
state through patterns. From this standpoint, the analogue indicators,
with pointer variations, and trend curves, provide an economic spatial
identification. Actually, from the mimic boards, the operator recreates
significatant pieces of information - his operating image.

The cathode screens allow designers not only to present the information
in different way, but also to process it. However, this information is
limited spatially and delivered sequentially. If one recalls that one
of the problems met by the operator in the development of his operating
images, is to make a coherent whole of it, to assemble the puzzle pieces,
to find relationships between the apparently isolated phenomena, one
understands that this method of presentation, a priori hardly favours
this establishment of relationships. One needs to be already initiated
to make sense of it. At present, and in most control rooms which we have
been able to study, images on screens are largely underemployed, even
when the operators have participated in their development. Table 1 shows
the results of an investigation undertaken in continuous casting three
months after the start of the installation. The operator is in the
control room where he has wall-mimic boards, intercoms, seven cathode
screens. Without leaving his work station, he can see through the
window the installation and the giant display of the main process parameters outside the booth. The observation of information sources
consulted has been made at different instants of the casting sequence,
the crucial instant being the start of casting (*). The consultation
frequencies have been weighted by the time which these different moments
lasted, to obtain comparable indices. Fifteen casting operations without
incident have been followed. The results clearly show that during the
critical phase - the start of casting - the consultation of information
sources is the greatest. But the screens and the mimic boards, themselves

---

(*) The sequence is the number of ladles which can be poured without
interrupting the process.

90 % redundant, are little used compared to verbal communication and direct observation of the installations which dispense uncoded information - faster and more certain according to the operators. It should be noted that these trends stand out even more when an incident occurs : at that instant the operator only trusted the uncoded sources and the giant display of some of the most important parameters. This is the way in which he believes he can intervene the fastest. -see fig. 1

Verbal communications provide a cohesion in the team (De Keyser & Decortis, 1985). They are a transmission channel for the collective knowledge and are judged by the operator, sometimes rightly and some- times wrongly, as fast and reliable. Moreover, they attract in the flood of the functional, the jokes, the ironic allusions, the games that break the nervous tension and cause an emotional discharge. To think of reducing them or eliminating them - or even to wish at all costs to separate the functional aspects of work from the relational and social aspects - is a mechanistic view, which does not take account of how the workers live their work, intimately mixing the affective and the cognitive (Sainsaulieu, 1977). Therefore this aspect must not be limited but the mechanism, which can entail perverse effects on team reliability, should be better understood : causal attributions in collective discussions of incidents which block the analysis and entail some errors (De Keyser, 1984), collective risk-taking etc. Social regulations and their interrelationships with the cognitive dimension of the work of the operator have as yet been little explored in the field, in a work situation. However, the analysis of incidents and accidents shows that here there is a fruitful research path in the area of human reliability.

Often, the same information support systems are used by a number of persons, having different tasks, objectives and training and consequen- tly they have dissimilar process operating images. To develop in a design office graphic images for screen without knowing the relevant information for the operator, the weight which he gives to some alarms, the indices which for him are antecedent signs of an anomaly, is an aberration committed daily. However, even when the operators are involved in the design of these images, the contradictions which can

appear between the desiderata of the production and maintenance personnel can be a source of tensions. And more often the selection of
maintenance is seen to prevail for the design of information supports
(Cockerill-Sambre, 1984). This is to implicitly recognize the role
that it plays in productivity and reliability.

These characteristics of information support allow one to understand
at what point it is beneficial for information sources to be redundant
rather than mutually exclusive. Redundancy allows the operator to check
an item of data but also to adopt a flexible strategy opting for a
certain support rather than another depending on the circumstances.
Operator technical aids will therefore have to be understood inside an
information system - and not in an isolated way, like some technological gadgets.

4. TIME REGULATION AND OPERATOR BEHAVIOUR

In the foregoing, we have attempted to answer the first of our questions
- cf. § 1 : "Does the operator have the knowledge necessary to
understand the situation during an incident ?" We may conclude that
if, over a long time, he has acquired, by practice, operating images
allowing him to ensure the reliability of the system, today, there is
a threat over this experience. The increasing stability of installed
automated systems, the increasing complexity of the systems, the privileged role of some specialized departments, such as informatic or
maintenance, reduce the production operator's operating margin. On
the other hand, while the quantity of information which he receives
remains plentiful, the sometimes exclusive resorting to cathode
screens can raise certain problems.

But to suppose that even if he has adequately sound knowledge to solve
these problems "from cold" - in simulations, for example - nothing as
yet says that he might be able to use it under pressure - that is to
say at an instant when his mental load is increasing. It has been
shown that under the influence of increasing constraints, operators
simplify their work strategies and their information gathering to retain
only the essentials. Ordinarily they may handle different criteria,

optimize parameters and refine some operating strategies; if the work
load increases, they will tend towards more stereotypical behaviour
and to retain but one clear objective to be safeguarded. Thus, some air
traffic controllers, in case of dense aircraft traffic, maintain the
safety criteria - to avoid collisions - but drop the idea of optimizing
each individual track of aircraft (Sperandio, 1980). At this instant
the verbal communications become shorter, more functional. Similar
observations are made on railway dispatchers (Lukau, 1985). Generally,
it is seen that the operator reduces his field of view and temporally
his strategies, shift from open-loop towards closed-loop. But by
controlling blow-by-blow he can, in some situations, be drawn into a
vicious circle : a more open, planned strategy allows him to "economize"
the mental load, but he no longer has the option of investing the
necessary time for this back-tracking which an open type strategy
assumes (Turner, 1984). Thus one reaches a spiral where the work constraints increase and as the strategies become less well organized, the
operator can lose control of the system. In a glass fibre plant, one of
our students was able to observe   operators adopting two types of
behaviour : perceptive , or cognitive - see fig. 2 . The operators had to run a group of machines where fibre spools were wound;
their tasks consisted in removing the full spools and to restart the
machines in the event of a break. The cognitive behaviour consisted in
planning interventions on sets of machines, by forecasting the stops of
full spools and the perceptive behaviour of answering, case by case, by
going all over the shop. After the analysis of errors made by these
operators, it was seen that the cognitive ones made significantly less
errors that the perceptive ones. But the observation also shows that,
taking into account the machine characteristics, the temporal constraint is less strong with the cognitive ones. Their open strategy, their
lower number of errors, the communications centred mainly on the work
appear permitted by the work constraints. As for the perceptive ones,
far from reducing their social communications, they maintained them at
a high level; it is permissible to imagine that the jokes which they
made   , served, as we have already observed elsewhere, to remove a
latent tension, sometimes up to taking a fundamental place in the life
of the team, to the detriment of some functional aspects and also of
security (De Keyser, 1983). The adoption of a particular type of beha-

viour, with control organized in varying degrees, thus appeared closely
linked, in this cas, to objective constraints of the work environment.

Many studies today are focused on performance under stress. It appears
difficult to maintain this performance when temporal constraints increase , leaving the control of the system to the operator. Recent research
shows, for example, that training in decision-making processes, wich has
been found to be effective without temporal pressure, does not persist
when the latter increases (Zakay & Wooler , 1984). And in a work situation, the time constraint is very often linked to the seriousness of the
context : if time is short, this generally means that there is a poten -
tial danger.

Many incident and accident reports recall fixations including time
aspects. In some cases , the operator has been seen to let the
system deteriorate for some hours, without even becoming aware of the
urgency or seriousness of the situation : the alarms were deemed to be
barely credible or false. In other cases, an operator would keep on
actuating an inoperative control, hoping to see it works the way in
wich he is hoping, in spite of everything (Daniellou , 1982). Or he
will repeatedly consult the same parameters, disbelieving the values
wich they display, which do not correspond to his analysis of the situation. In all these cases, the operator has an idea in his head, an
assumption from his previous experience, but it does not fit it in with
the reality and he does not perceive the lack of connection. The past,
the history of the installation can be a snare for the operator : if
he does not trust it, he cannot work - if he does not doubt it, he
in at the mercy of an unexpected event. We have recalled elsewhere
(Montmollin & De Keyser , 1985) the usual unreliability of instrumentation and notably of alarms and the part of operator experience
wich is based on the learnin of the system times. The information
wich reaches the operator goes through the prism of this experience :
his estimate of the duration of en operation will outweigh the information of a dial wich is often out of adjustment, a cause of a rare
incident will only be suspected as a last resort.

One cannot ask the operator whenever he overrides an alarm, that is to say that he acknowledges it without taking time over it, to go into long checking procedures : in many installations, this would paralyze the daily work. Moreover, the operator confident of his assumption, will neglect what is presented to him if he does not have the proof of the falsity on his own. Therefore, given the nature of operator logic, when the system appears to drift, he will have to explain his starting assumption and the signs on which he bases himself, to check the relevance of these signs and from there, supply him only with another set of possible assumptions taking into account the context. In this very case, the operator aid would be a diagnostic aid.

5. OPERATOR AIDS AND THE DIFFERENT SITUATIONS

5.1. The aids

The aids which current technology can offer the operator are manifold and take over the variable parts of the information which is to be processed. At one end of the spectrum, this part of the processed information is slight, but the proposed supports are no less valuable. In this category, one may cite all the supports which favour a temporal and geographic enlargement of the operator references : process dynamic images, summarizing mimic boards, chronological scanning of alarms, etc. The operator is referred to a wider context than the one of the present instant. Next one finds all the data checking automatic processes : comparison of some data with threshold values etc. Among these checking procedures, some have a special interest, for they allow to reduce the uncertainty in which the operator usually works : they are the ones envisaged to check sensors, regulations and instrumentation generally. They are not of standard use in all processing industries. Next there are not real aids but rather increasing automation levels. The process

can be increasingly computer controlled; parameter optimization
gradually escapes the operator who only retains the initiative in the
event of incident. The next step consists in using automatic diagnosis
procedures for dysfunctions; we will see that they are built on models
which sometimes differ greatly. Finally, and at the other end, there
are decision aids, which attempt to aid the operator in intervening
and in measuring, through different criteria and taking into account
the situation, the results of the action which he could take. From
simple monitoring to diagnosis and recovery, the way is marked out by
possible technical aids.

5.2. The situations

The situations in which the operator is liable to be in can be
highly different. However, very schematically, two variables may be
distinguished : one constraint level, varying in degree (situation
complexity, temporal pressure, etc.) and an operator response   which
varies in degree of adaptation to this situation. The level of
pressure - and therefore the mental load - of the situation stems from
the interaction of these two variables. The interaction of these two
variables will leave visible traces on the system : rebalancing, or
on the contrary, degradation. This is a gross over-simplification ;
however, it allows four types of situation to be distinguished and
these can be of interest for our subject - see fig. 3

Situation 1 : this is a normal situation. The system parameters
fluctuate, but within an average bracket. If some incidents occur,
they are seen sufficiently early so that they do not entail damaging
consequences. The situation is not necessarily simple; its degree of
complexity will depend on the stability of the process, the number of
variables controlled by the operator, the functional links between
the variables etc. But the operator controls it by the play of anticipation and the optimization of the variables.

Situation 2 : it is a subnormal situation, where the working stresses
increase for the operator, entailing strategy changes. These constraints can be linked to a greater temporal pressure, or to a complex
problem to be solved - or even both simultaneously. The operator

strategy changes, he selects the processed information and only attempts
to attain the fundamental objective of the system - for example the
safety objective.

Situation 3 : it is an abnormal situation, where the constraints
weighing on the operator no longer allow him to maintain the system in
a satisfactory state from the safety standpoint. This can be the outcome of a situation 2, or even the sudden result of an incident.
Operator control tends to become disorganized and closed loop blow-
by-blow strategies replace open loop ones.

Situation 4 : this is a situation which drifts towards the abnormal,
without the operator appearing to be fully aware, through his
behaviour, of this drift. He repeats unsuccessfully inoperative
behaviours,without modifying them, and consults the same information
sources without looking for others.

These situations can be characterized objectively in different
industrial context.

5.3. The situations and the aids

If one attempts to project the aids on the situations, one obtains,
taking into account what has been discussed in 1, 2, 3, 4, the following propositions :

Situation 1

One lives by one's habits, and it is in normal situations where, the
operator forms his operating images. It is therefore in this situation that one must keep for him :

- an adequate intervention margin with action feedback

- high degree of redundancy of information sources (direct view
  of installations, mimic boards, screens) to allow flexible strategies and instrumentation checks. Verbal information, communication
  within the team must be favoured and stimulated

- operator involvement in electronic data processing. If one wishes
  that the latter should use it and make a living tool of it - just as
  he would mentally restructure the wall mimic board - he must be associated with its design : construction of screen images, data bank modification, option of remaking oneself variable processing programs,
  cause-consequence diagrams, etc.

- an extensive and dynamic view of the system allowing anticipation and
  structure regulation : flow diagrams, real-time data on constraints of
  neighbouring work stations etc.

This assumes flexible automation and that the parts of the system on
which the operator should be able to intervene in case of indicents,
should operate on manual long enough for him to acquire the mastery to
control it. The objective of assistance for the operator here must be to
develop and to integrate to the maximum the operating images, in a way
compatible with the economic and safety objective. A French cement plant
has chosen for example to stop its computer one month per year so that
the operators may maintain their competence. In this situation, the
functions of monitoring, diagnosis and of recovery rest in the operator's
hands.

Situation 2

It is often difficult, under industrial conditions, to impart urgency
and complexity when work constraints increase : often these constraints
arise from an incident where both aspects are linked but for greater
clarity we shall attempt to distinguish them artificially.

Temporal pressure is one of the constraints which the operator finds
most difficult. Two ways are offered by technology to help him in
facing them :

- within the framework of flexible automation, to have the computer take
  over a set of associated functions, to filter some variables and
  nonetheless leave the process fine-control to the operator

- to remove the process fine-control from the operator by installing
  automation to a maximum level. To focus instead the operator attention
  on what is not of immediate urgency, but which might be revealed as
  an incident or accident factor : the surrounding system. That is to
  say, to highly emphasize the structural regulation aspects by forcing
  him to take some information about what he would have neglected by
  closely following the urgency of the moment.

From the reliability standpoint, if the process can mastered by
automation, the second way appears better. For, in a critical situation,
it is often the interference of events, at first sight innocuous, with
the initial development of the process, which reveals itself as
pathogenic. It moreover offers the benefit of utilizing the operator
competence optimally, by releasing him from the temporal constraint
and putting him on to the control of a complex system. At that instant,
the operator must be capable of using all the supports giving him an
extended view of the system - hence the importance of having understood
them well beforehand in situation 1.

The foregoing approach may be envisaged if the constraints have
increased without there being a major disturbance : for example, at the
end of an increase of production, when the information to be processed
per unit of time is higher, etc. The objective of operator assistance is
then to release him from the temporal constraint and to direct him to
the tasks at which he excells. His control function is displaced from
the process to the system and in particular to the work team, but things
change if a serious incident occurs.

In this case, and always on the assumption of situation 2, that is to say
within a tolerable parameter drify bracket, the operator is forced
to intervene, ordinarily and taking into account the indications
reaching him, he already has one or several assumptions on the origin
of the incident. He then needs :

- the means of checking very quickly the validity of these initial
  assumptions and, if necessary, the instrumentation

Actually, the initial assumptions have an effect of masking the whole situation; as long as the operator is not sure of how sound they are, he will hang on to them.

If checking contradicts these assumptions, it is useful, at the operator's request, to help him in widening his horizon. Two great families of aids stand out : those which are based on a modelling of the process in pathological terms and those which are oriented towards normal operation of the process. We shall examine them briefly, including their implications for the operator.

In the first family, the underlying logic is the fault-tree (Fussel, 1976). It is possible to collect the fault-trees concerning the incidence in a process, but one is soon up against the problem of the malfunction propagation in a dynamic system. Actually, as the fault-trees are static, it is a matter of establishing from them a dynamic synthesis, which corresponds to the functional relationships existing in the system. Experiments of this type with fault-trees (Apostolakis, Salem & Wu, 1978; Powers & Tomkins, 1974; Hollo & Taylor, 1976; Martin-Solis, Andow & Lees, 1977), with cause/consequence diagrams (Nielsen, 1974; Lind, 1981), with decision tables (Berenblut & Whitehouse, 1977), are found. In petrochemicals, a fairly complex example has been developed (Lihou, 1981) with an automatic recognition of the process state on the basis of symptom patterns : each abnormal state of the process is characterized by parameter drift patterns. A symptom matrix is stored in the computer memory : this is the correlation of all the incidents with the respective parameters of the system. On-line, the computer generates on the process development basis a state matrix of the system : the comparison between the state matrix and the symptom matrix deduces the presence of any anomaly, which is communicated to the operator on the basis of a glossary. On the same lines, recent work at the University of Technology of Valenciennes uses fuzzy models to propose to the operator possible causes of disturbances from symptoms which he enters in the computer and which he expresses in highly subjectively, qualitatively, trusting his sensors experience.

One may ask whether this modelling is accessible to the operator and what are its   benefits or dangers. It is accessible, without any doubt : mapping symptoms to faults is one of the strategies which the operator uses spontaneously (Rasmussen 1981; De Keyser & Piette, 1970; De Keyser, 1972). Concerning the actual construction of a fault-tree, we were able to have workers make them, after three days of training, during participatory safety training. A priori, there should not be a logic comprehension problem, but what can be feared is that when continously used :

- it reinforces a piecemeal, disjointed knowledge of the process
- it eliminates anticipation,
- it forces the operator to algorithmic procedures while in many cases, he takes some heuristic short cuts,
- it releases the operator from seeking out himself the incident symptoms and organizing them coherently

With the exception of the case where the automatic diagnosis is so reliable that the circumstances for operator reasoning can be omitted, the foregoing procedure should only be applied in particular circumstances - night-work, process complex state, emergency, etc.

In the second family of diagnostic aids, one finds less work, at least in building technical aids; the principles, in contrast, have already been tested in training (Cavozzi, 1980; Cuny, 1980). The idea is to release the operator from the concrete representation of the installations, his hands-on experience, to lead him to reason starting from the system laws. This method of reflection should be found to more rewarding when traditional experience fails in the case of rare incidents or those never encountered.

Some experiences of this type have been undertaken at Risö, Roskilde with flow diagrams. The system is summarized by the diagrams, starting from data processing by means of the laws of conservation of mass and energy. What serves as support for the representation is only the network of the observable subsystems and their interrelationships. Thus the operator works with processed variables and in a probabilistic universe. The process does not allow the making of a causal diagnosis, but rather to guide detection and identification of the anomaly (Goodstein, 1981) based on variations of flow and of changes in the mass and energy balances (Lind, 1981).

What the operator uses here is a combination of functional and topographic strategies. One may think that this way of approaching problems does not exceed the limits of operator experience and that it has a strong generalization power. However, one must not underestimate the resistance that it will cause with the operators used to trust their senses only : actually, it is a break with the knowledge acquired day-by-day. Therefore there will be a whole learning and appropriation process to be set into notion outside the critical situation. However, we should recall that initiatives very close to these have been undertaken in technical training, for the familiarization of diagrams. They have shown the marked superiority of diagrams that code the states and functions of the system, rather than reproduce the installations mimetically, in the discovery of solutions to never-encountered breakdowns (Cuny, 1980).

The objective of diagnosis aids is therefore to allow the operator to discover the cause of incidents quickly. If the problem of diagnosis is urgent and anticipation has not been possible, a symptomatological method by anomaly pattern will succeed in quickly detecting an already well known cause, but if the problem is complex or atypical one can reasonably believe that a functional method will be more effective.

Situation 3

Situation 3 is characterized by a system whose parameters are beyond the limits considered as permissible from the safety standpoint. Either a serious incident has occurred suddenly or a situation has deteriorated without the operator being able to control it. The objective of the aids here is to prevent the operator from panicking, to shield him as far as possible from temporal pressure and to relieve him from his responsibility by framing and guiding his action. On this line of thought, it may be useful to dissociate at this stage, the function of diagnosis from the one of recovery.

Actually, the seriousness of the situation assumes a whole series of recovery measures to prevent damage; these measures do not require knowledge of the first cause of the incident, but rather to appraise, taking into account the system state, the consequences of some action. Provided that automation equipment will have been installed to allow the operator to withdraw somewhat, some tools to aid in decision-making, for example multicriteria ones, can be useful. Procedures to be followed must be fully defined, not only for the operator, but the various team members.

The responsibility for causal diagnosis should be removed from the operator in this situation : this does not mean that he or his team should not collaborate by contributing some information to it, by proposing assumptions, but their task here would be directed fundamentally to recovery. Team of specialists - engineers, experts, etc. - would have to take over with some supports in the form of diagnosis aids designed in terms of their theoretical knowledge and their operating images and not those of the operator. Accordingly, there would be the issues of how to develop the supports required by an engineer or an expert to solve a serious problem - with possibly access to data banks outside the plant.

Situation 4

Situation 4 is a subnormal one, even clearly abnormal, but it does not produce the required answer from the operator. Sometimes he just fails to answer repeated and alarming signals, or sometimes he persists in a disastrous behaviour at odds with common sense. These fixation phenomena are pathogenic, but experience teaches us that the operator persists in his behaviour because he has a certain analysis of the problem with masks the whole of the phenomenon from him. Operator aids should be based not on the system state, but on its degradation trend (increase or persistence of alarms, for example); if this trend appears, one could force the operator to re-formulate his assumptions and to give him the means of checking it quickly. Fast instrumentation checking would be useful because if the operator doubts the reliability of one information source, then he will neglect it and resorts to other information sources.

6. CONCLUSIONS

In this overview of operator knowledge and of the technical aids which can be provided, we have but traced the outline. However, we have attempted to stress both the operationality and flimsyness of the operator cognitive dimension which is modelled by the industrial context, the automation level, and the supports provided. The aids provided will also contribute to change it : consequently, it is important to know their benefits and their limits. Some lines of thought will be found in this paper.

source of information \ state of the process	preparation phase	startup phase	normal operation	end of casting
mimic boards	4	22	3,3	2
screens	5,6	23	5,4	4
installations	1,1	73	5,3	2,9
intercom communications	33,6	66	27	25
giant display	0	55	5,5	4,5

Fig.1 Inquiry index (*) of the information carriers during the various
states of a sequence of continuous casting (estimate on 30 sequences)

(*) this index takes into account the frequency of the enquiries made
of the source , weighted by the time of the phase concerned.

(from De Keyser , 1984)

criteria performance	perceptive operators	cognitive operators
Nbr errors	93	65
stoppages	122	70
breakages	99	55

Fig.2
Some differences between perceptive and cognitive operators in a
supervisory task in textile industry   (2 weeks).

(from Grosjean , 1986)

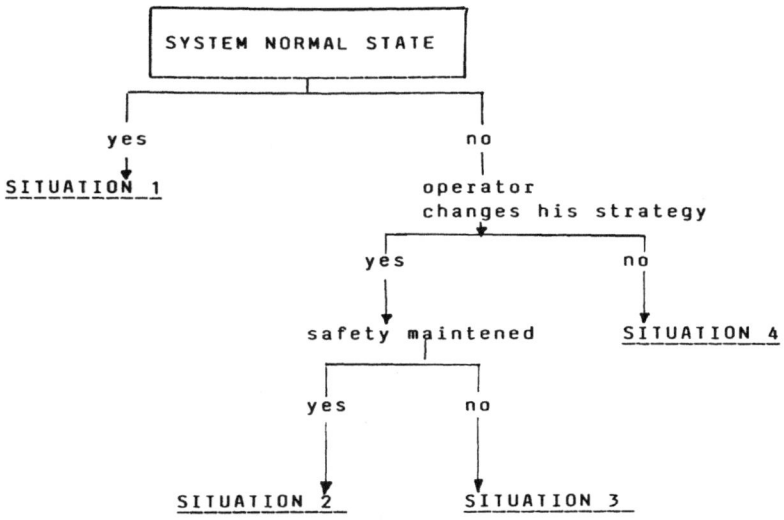

<u>Fig</u>. 3 The different situations

RECURRENT ERRORS IN PROCESS ENVIRONMENTS: SOME IMPLICATIONS FOR THE DESIGN OF INTELLIGENT DECISION SUPPORT SYSTEMS

James Reason
Department of Psychology
University of Manchester
Manchester, United Kingdom

AIMS AND SCOPE

This paper has two related aims. First, to summarise what is known about the forms and origins of the more predictable varieties of human error. Second, to use this cognitive analysis as a basis for suggesting ways in which Intelligent Decision Support Systems (IDSS) might be employed in the process plant environment. Although the error data are drawn mainly from nuclear power plant (NPP) operations, it is assumed that their implications will generalize to other high-risk process environments.

Central to the arguments presented here is the belief that systematic forms of human error have their origins in fundamentally useful processes. However, the centralized operation of high-risk, complex, and incompletely understood process systems can, on occasions, transform these normally adaptive properties of human cognition into dangerous liabilities. The technology now exists to provide IDSS to compensate for some of these situationally imposed limitations. A necessary step along this path is to identify which aspects of mental function could most profit from such 'cognitive prostheses'. It is in this area that the limited but relatively well-documented history of operator errors in NPP emergencies provides a valuable guide.

THE NATURE OF HUMAN ERROR IN NPP EMERGENCIES

Brief descriptions of the 11 well-documented operator errors are given in Table 1. They are grouped into three clusters according to apparent similarities in their form.

NATO ASI Series, Vol. F21
Intelligent Decision Support in Process Environments
Edited by E. Hollnagel et al.
© Springer-Verlag Berlin Heidelberg 1986

Table 1: Brief details of 11 NPP operator errors

==
   PLANT   *NATURE OF OPERATOR'S ERRORS*
==
OYSTER CREEK   Closed four instead of two valves. Shut off
                water to core.

NORTH ANNA   Failed to close main steam injection valve.

NORTH ANNA   Failed to realign pump orifices after safety
========================injection.=======================
THREE MILE   Did not recognise that relief valve was stuck
ISLAND   open. Panel light indicated OFF — taken as
--------------- valve shut.

OYSTER CREEK   Took water level in annulus to indicate level
==============in shroud.=================================
THREE MILE   After reactor trip, crew cut back (diminished)
ISLAND   safety injection to reactor coolant system.

OYSTER CREEK   Cycled valve — wrong strategy in circumstances.

GINNA   Did not isolate faulty generator prior to
--------------- cooldown start.

GINNA   Misinterpreted instructions. Locked valves shut.

GINNA   Cycled valve — wrong strategy in circumstances.

GINNA   Delayed stopping safety injection.
■===============Exacerbated problem.====================

Primary sources: Kemeny et al (1979); Pew, Miller & Feeher
(1981); Woods (1982).

The first three errors show the characteristics of slips
(Norman, 1981; Reason & Mycielska, 1982) in that the
intentions were appropriate enough, but the actions were not
carried out as planned. The next two appear to involve the
application of inappropriate diagnostic rules of the kind —
IF (situation X prevails) THEN (system state Y exists). In
both cases, these rules had proved valid in the past, but
yielded the wrong answers in these particular emergency
conditions. The last group of errors are all characterised
by the selection of an inappropriate strategy or plan.

The first and last of these three groups correspond to
Norman's (1983) succinct distinction between slips and

mistakes: "If the intention is not appropriate, this is a
mistake. If the action is not what was intended, this is a
slip." However, the middle group is anomalous. These two
errors conform to Norman's description of mistakes, yet they
also show clear slip-like features. One way of resolving this
problem is by using Rasmussen's (1980) three performance
levels — skill-based (SB), rule-based (RB) and knowledge-
based (KB) — as the basis of a tripartite error
classification. This is shown in Table 2.

Table 2: Level, control mode and error type

LEVEL	CONTROL	ERROR TYPES
SKILL-BASED	Low-level automatic processors, e.g. schemata	SLIPS AND LAPSES
RULE-BASED	Largely automatic processors, e.g. rules/productions	RULE-BASED MISTAKES
KNOWLEDGE-BASED	Attentional: slow serial, laborious resource-limited.	KNOWLEDGE-BASED MISTAKES

The possible cognitive origins of these three error types are
indicated by the Generic Error-Modelling Systems (GEMS). The
purpose of GEMS is to provide a largely context-free conceptual
framework in which to locate the principal roots of human error.

AN OVERVIEW OF GEMS

GEMS is a simplified composite of two sets of error theories:
those of Norman (1981) and Reason & Mycielska (1982) relating
to slips and lapses; and the GPS (Newell & Simon, 1972)
tradition of problem-solving theorizing which has been applied
to operator failures in high-risk technologies by Rasmussen
(1981) and by Rouse (1981). The operations of GEMS divide
conveniently into two areas: those which *precede* the detection
of a problem (SB level), and those which *follow* it (RB and KB
levels). Errors (slips) occurring prior to problem-detection

are primarily associated with *monitoring failures*, while those
which appear in the subsequent phases (mistakes) are subsumed
under the general heading of *problem-solving failures*. The
basic features of GEMS are outlined in Figure 1.

FIGURE 1 NEAR HERE

1. *Monitoring failures*

In its simplest form, routine action (i.e. well-practised
tasks carried out in familiar surroundings by skilled
operators) comprises segments of preprogrammed behavioural
sequences interspersed with attentional checks upon progress,
carried out either consciously or preconsciously. These
checks involve bringing the higher levels of the cognitive
system momentarily into the control loop and seek to establish
(a) whether the actions are running according to plan, and
(b) at a more complex level, whether the plan is still adequate
to achieve the goal. Diary studies of naturally-occurring
action slips (see Reason & Mycielska, 1982) clearly indicate
the contingency of slips upon failures of attentional checking.

2. *Problem-solving failures*

In purely psychological terms, a problem can be defined as
some externally or internally produced change which requires a
deviation from currently intended behaviour. Departures from
plan can range from relatively minor contingencies, rapidly
dealt with by preprogrammed corrective procedures, to entirely
novel circumstances or goals, requiring new plans and
strategies to be derived from 'first principles', with heavy
demands being made upon the attentional control mode.

The problem-solving elements of GEMS employ a slightly
modified version of Hunt and Rouse's (1984) 'fuzzy set'
variant of Rasmussen's (1980) 'step-ladder' model, and are
based upon a recurrent theme in the psychological literature,
namely that "...humans, if given a choice, would prefer to act
as context specific pattern recognizers rather than attempting
to calculate or optimize" (Rouse, 1981). Thus, the sequence

of operations within GEMS indicates that human problem-solvers always confront an unplanned-for change by *first* deciding whether or not they have encountered this particular pattern of local signs and symptoms before. If the pattern is recognised, they then determine whether or not they have some corrective rule readily available. Only when this low-effort pattern-matching and rule-applying procedure fails will they move to the more laborious mode of making diagnostic inferences from knowledge-based mental models of the system in question, and, from these, go on to formulate and try out remedial possibilities.

Human beings are furious pattern matchers (see Norman — this volume). They are strongly disposed to exploit the parallel and automatic operations of specialized, low-level processors: *schemata* (Bartlett, 1932), *frames* (knowledge structures for representing familiar scenes — see Minsky, 1975), and *scripts* (comparable structures for representing stereotyped episodes or scenarios — see Schank & Abelson, 1977). Knowledge structures are capable of 'filling in' the gaps in missing data on the basis of 'default values'. These default settings are provided by the memory system's remarkable ability to encode frequency-of-occurrence information without intentional guidance or apparent mental effort (see Hasher & Zacks, 1984). Higher level manipulations, however, require the slow, laborious, resource-limited involvement of the attentional control mode.

The distinctions between these three basic error types are summarised in Table 3. A more detailed account of the underlying arguments is given elsewhere (Reason, 1985).

Table 3: Form and process distinctions between the three error types

FACTORS:-	SKILL-BASED SLIPS	RULE-BASED MISTAKES	KNOWLEDGE-BASED MISTAKES
ACTIVITY	Routine actions	Problem-solving	
CONTROL	Mainly automatic processors operating in parallel (Schemata)	(Rules)	Resource-limited serial processes
FOCUS OF ATTENTION	On something other than present task	Directed at problem-related issues	
FORMS	Largely predictable "Strong-but-wrong error forms (Actions)	(Rules)	Variable
DETECTION	Usually fairly rapid	Hard, and often only achieved with help from others	

Errors at the three levels of performance are further distinguished by the nature of the psychological and situational factors which combine to shape characteristic error forms (see Table 4).

Table 4: Differential error-shaping factors

PERFORMANCE LEVEL	ERROR-SHAPING FACTORS
SKILL-BASED I	1. Recency and frequency of previous use 2. Environmental control signals 3. Shared schema properties 4. Concurrent plans
RULE-BASED II	1. Mind set (It's-always-worked-before") 2. Availability ("First-come-best-preferred") 3. Matching bias ("like-relates-to-like") 4. Over-confidence ("I'm-sure-I'm-right") 5. Over-simplification (e.g. "halo effect")
KNOWLEDGE-BASED III	1. Selectivity (bounded rationality) 2. Working memory overload (bounded rationality) 3. Out-of-sight-out-of-mind (bounded rationality) 4. Thematic "vagabonding" and "encysting" 5. Memory-cueing/reasoning-by-analogy 6. Incomplete/incorrect mental model

What can be said about the relative predictability of errors at each of the three Rasmussen levels? Since both the SB and RB levels are characteristic of skilled performance, it is reasonable to assume that the quantity of the errors in both cases will be small, but that when errors do occur they are likely to take 'strong-but-wrong' forms (i.e. will be recognisable as well-used parts of the person's repertoire in that context).

At the KB level, however, it is only possible to state that a relatively large number of errors are likely, and that their forms will be highly variable. While it is often possible in retrospect to identify the influence of the primary error-shaping factors — bounded rationality, contextual cueing, shifting foci of concern, confirmation bias, inadequate or incomplete knowledge — it is largely impossible to predict the precise forms taken by KB mistakes in advance. In this sense, mistakes at the KB level have hit-and-miss qualities not dissimilar to the errors of beginners. No matter how expert operators are in coping with familiar problems, their performance will approximate to that of novices once their repertoire of pre-established routines has been exhausted by a novel or mysterious emergency. The primary differences between the novice and the expert are to be found at the SB and RB levels of GEMS. Indeed, it could be argued that expertise means never having to resort to the knowledge-based level of problem-solving.

IMPLICATIONS OF GEMS FOR DECISION AIDS

The global message of GEMS and of the large research literature it seeks to encapsulate can be expressed very simply: Aside from their proneness to 'strong-but-wrong' errors in conditions of change or overload, human beings function extremely well at the SB and RB levels of performance. This is the domain of expertise. They are far less effective, however, when forced to operate slowly and laboriously at the severely resource-limited KB level, and especially so under the kinds of stresses present in NPP emergencies.

There is nothing new here. In 1964, Shepard commented upon "...the obvious disparity between the effortless speed and surety of most perceptual decisions and the painful hesitation and doubt characteristic of subsequent higher level decisions." The twenty intervening years of NPP operations has merely provided a wealth of costly real-life evidence to confirm Shepard's laboratory-based observation.

The crux of the problem is this: The experience of NPP emergencies shows that most serious incidents very quickly throw operators and their advisers back onto their limited knowledge-based resources. Most serious emergencies are novel and unique events — usually as a consequence of complex and, at the time, mysterious interactions between operator and component failures. Rule-based solutions are quickly exhausted — the combined result of inadequate training, limited experience and deficient operating procedures. It is not that people can't operate at the KB level — given time in the tranquillityof a 'think tank', they will come up with all manner of elegant solutions — they are simply not at their best when forced to reason analytically in situations like those prevailing in the TMI or Ginna control room. Nevertheless, at the present time, there are no intelligent machines that could do any better. Short of the Luddite solution, an obvious way forward is to provide suitable cognitive prostheses for the human operators.

WHAT KINDS OF IDSS ARE NEEDED,

The answer to this question is relatively straightforward, at least in general terms. IDSS should capitalise upon cognitive strengths and compensate for (situationally determined) cognitive limitations. Paradoxically, systematic errors provide valuable clues to cognitive strengths, since they are inextricably bound up with those things at which the cognitive system excels relative to other information-processing devices. Below are presented some of the more conspicuous strengths of the cognitive system together with some of their implications for the deployment of IDSS.

1. *Delegation of control*

One of the most dramatic achievements of the human cognitive apparatus is its ability to delegate control to low-level processors, thus liberating the resource-limited higher levels for longer term concerns (e.g. planning, error recovery, monitoring present progress in relation to future goals). Automatization is evident in all aspects of cognitive function.

At both the SB and RB levels of performance, the cost of 'regional' autonomy is the tendency to produce 'strong-but-wrong' error forms. The penalty of control by a vast community of 'local experts' is that they are only 'tuned' to very restricted aspects of the world, and have a strong inclination to trot out their expertise whenever some part of their triggering conditions are present in the environment. Most of the time, this is both economical and effective, but there are certain conditions — notably changes in either the goals or the context of these familiar routines — when this disposition to make strong associate substitutions (i.e. default values) leads to predictable error forms. The more often a particular routine (schema, script, frame, prototype or rule) has been employed successfully in the past, the more likely it is to appear as a 'strong-but-wrong' error in situations involving both change and attentional capture (see Norman, 1981; Reason & Mycielska, 1982).

Situations that should invoke exceptions to the rule do not necessarily declare themselves in an obvious fashion, particularly in the NPP context. It is likely that such deviations from typicality will contain three kinds of indicators: (a) *signs*: conditions which satisfy the (situation) component of the most frequently applicable rule; (b) *countersigns*: conditions which suggest that some other less familiar rule is appropriate; and (c) *nonsigns*: conditions which do not relate to any particular rule, but which contribute to 'noise' within the pattern recognition system. The point to stress here is that contra-indications may well occur in the company of an abundance of both signs and nonsigns.

One important application of IDSS would be in alerting
operators to the often very subtle cues (countersigns) which
distinguish familiar or expected circumstances where 'strong
rules' may be applied from those in which they cannot. It
would require 'intelligence' because it could only make these
discriminations on the basis of a record of previous events
within that particular system. In other words, it would have
to log past instrument indications (probably in cognate
clusters), compute the statistically typical patterns
associated with given system states (both normal and abnormal),
and to signal significant departures from them. Such notions
have already been discussed by Rasmussen (1980).

Would such 'change detectors' constitute a necessary and
sufficient condition for the prevention of 'strong-but-wrong'
errors? The experience of incidents like the Oyster Creek
loss-of-feedwater emergency (in which the triple-low level
alarm persisted for some 30 minutes while the crew continued
in their error of taking annulus level as an indicator of
shroud level) suggests not. But even if they are not
necessary and sufficient proofs against such errors, they are
clearly necessary.

2. *Selectivity and 'bounded rationality'*

The 'sharp end' of the cognitive system — working memory,
consciousness, and the limited attentional resource — can
only apply itself to very restricted regions of a 'problem
space' at any one time. As Simon (1957) pointed out: "The
capacity of the human mind for formulating and solving
complex problems is very small compared with the size of the
problems..." This ability to think only in a highly selective
manner is a remarkable achievement without which human life
would be impossible. But in the time-scale and informational
loading of NPP emergencies, it is a serious liability.

The resource limitations of the attentional control mode come
into prominence at the knowledge-based level of processing.
KB mistakes due to 'bounded rationality' will be characterised
by over-simplification. The search for objectives will be

restricted, and those chosen will be short-term rather than ideal. Similarly, fewer ways of finding problem solutions are likely to be considered than are potentially available, and — in NPP operations (see Woods, Wise & Hanes, 1982) — choosing between even these small number of alternative strategies poses a serious difficulty for operators. Categorizations of problem-related factors will be based upon global rather than subtle discriminations, and the possibility of complex interactions between them will be either overlooked or underestimated. Attempts at 'thinking through' the consequences of remedial actions will be partial rather than complete, since this activity is extremely demanding of working memory capacity.

At least two aspects of cognition at the KB level offer themselves as suitable cases for treatment. One obvious need is to augment the very limited capacity of working memory. This part of the memory system serves two closely related functions: (a) it acts as a working data base within which logical and transformational operations may be performed; and (b) it keeps track of progress by relating current data to plans stored in long-term memory.

Another need is for IDSS to direct this limited 'working data base' to relevant and logically important regions of the problem space, as opposed to those which are merely psychologically salient. There is now a wealth of both laboratory (see Evans, 1983) and NPP data (see Pew, Miller & Feeher, 1981) to show that an important source of knowledge-based mistakes lies in the selective processing of task information. Errors will arise if attention is given to the wrong features or not given to the right features.

Those IDSS that guide the problem-solver's attention to the critical aspects of the situation will, at the same time, restrict the load imposed upon working memory by irrelevant data. But they will not necessarily help in keeping track of an individual's path through the problem space. Reasoners interpret data by constructing an integrated mental model of them (Johnson-Laird, 1983). In orer to check whether an inference is valid it is necessary to search for different mental models of the situation that will explain the available data. This activity of accessing and testing several possible models places a heavy burden upon working memory. As new hypotheses are considered, earlier conclusions are likely to 'spill out' of working memory and be forgotten. The evidence from a variety of reasoning studies indicates that working memory operates on a 'first-in-first-out' basis. Thus, as Doerner (1984) has described, problem solvers can behave as 'thematic vagabonds', flitting quickly from idea to idea, treating each one superficially, and often returning to previously rejected hypotheses as if anew.

At the very least, NPP operators tackling emergencies at the knowledge-based level should be encouraged to externalise their reasoning, rather in the style of bomb disposal experts (not altogether a remote analogy). This will require specific training, as it does in the case of subjects in verbal protocol studies. Externalisation of thought processes should be coupled with 'shadowing' by technical advisers, preferably on a paired basis. (This element of moment-to-moment 'accountability' may also have the added bonus of minimising the yet unexplored consequences of the Ringelmann effect, or 'social loafing', within the control room — i.e. the greater number of persons working on a task, the less effort each one invests (Latane, Williams & Harkins, 1979)). Certainly, the intermediate decisions of the Shift Supervisor could be routinely voiced aloud and monitored by specifically trained 'trouble-shooters', located outside the control room, but with access to system state information and 'real-time' modelling facilities. There is no reason to suppose that IDSS should not be other people as well as clever machines.

3. *Retrieval, metaknowledge, and availability*

Another remarkable feature of human cognition is an 'operating system' which, for most of the time, is capable of locating and retrieving items from an enormous knowledge base with extraordinary speed and accuracy. No existing machine can compete with this facility.

People's personal knowledge (metaknowledge) of the workings of their memory retrieval system becomes embodied in a few intuitive rules of thumb or heuristics. The usually unconscious application of these heuristics directly colours the interpretation placed upon certain pieces of evidence in situations where information is incomplete or uncertain (Tversky & Kahneman, 1983). Perhaps the most potent and pervasive of these is the *availability heuristic* which can be summarised as follows: "Things that come readily to mind are likely to be more frequent, more probable, more important, more useful, and better understood than less readily available items." Or, it can be expressed more simply as: "first-come-best-preferred". The order in which information is retrieved ought not, of itself, to affect its probative value or evidential worth — these should be determined by other, unrelated criteria. But, in the cognitive system, it does.

Planning and problem-solving are activities which involve the generation from memory of possible goals and courses of action. There is now a wealth of experimental evidence to show that when people are required to generate exemplars of a particular semantic category, they do so — on average — in an order which is highly correlated with the dominance of items, i.e. the proportion of the group citing a particular item (Bousfield & Barclay, 1950; Battig & Montague, 1969). In a recent series of studies (Reason, 1984), subjects were asked (a) to generate items according to various category descriptors, and (b) to make metacognitive ratings of frequency and recency of encounter, feeling of knowing and the affective tone for each item. A highly significant concordance was found between serial order of item generation, dominance (reflecting 'salience in the world' for the group as a whole), and the metacognitive ratings (reflecting 'salience in the head' for the individual). A further study demonstrated that even in categories with zero dominance (i.e. where generated items are unique to individuals), correlations in excess of 0.95 were found between the mean metacognitive ratings and order of output.

These data indicate that people possess reasonably accurate metaknowledge regarding the relative salience of items within a given semantic category. One possibility is that, having located a given knowledge domain, subjects make a crude 'frequency scan' of the category, and then use this to guide subsequent memory search and knowledge retrieval. Although they are not necessarily *aware* of using this metaknowledge in their memoric activities (Nisbett & Ross, 1980), they can produce rapid 1-7 ratings in answer to questions like "How frequently have you encountered any reference to Indira Gandhi?" with a high degree of confidence when directly asked.

Metacognitive ratings appear to offer a means of mapping the frequency or salience 'contours' of an individual's knowledge base. As such, they can provide valuable inputs to decision aids designed to counter the availability bias (the tendency to prefer the possibilities that spring soonest to mind).

One such decision aid (WISE FRIEND) is currently under investigation in our laboratory. The prototype version is very simple. It first invites the user to state a goal. The user is then asked to list at least seven ways of achieving this goal in the order in which they come to mind. He or she is then asked to rate each of these items (in random order) according to (a) its frequency of encounter, and (b) preference. If either the order/preference or the frequency/preference correlation exceeds 0.5, the user is informed, and it is suggested that they might wish to reconsider their decision in the light of this knowledge — though the original choice may have been perfectly acceptable. WISE FRIEND, therefore, is a device which knows nothing about the decision area in question (though an intelligent domain-specific version could be provided with such information), it merely alerts the user to what may or may not have been an excessive yielding to the availability bias. The final decision must rest with the user, but it will be made in the full knowledge of the effects of this normally covert heuristic. A secondary benefit is that it requires some externalization of the decision process.

SUMMARY

A generic error-modelling system (GEMS) is outlined which differentiates three basic error types: skill-based slips (and lapses), rule-based mistakes and knowledge-based mistakes. This tripartite classification is an extension to the simpler slips/mistakes distinction made earlier, (Norman, 1983; Reason & Mycielska, 1982), and is based upon a variety of discriminating factors. Different dimensions yield different dividing lines: it is only on the basis of the total pattern of distinctions that a case for three basic error types can be made.

This GEMS-based error analysis identifies a number of possible applications for intelligent decision aids in process plant environments: (a) To enhance the often very subtle cues that distinguish situations in which readily available routines may be applied from those in which they cannot. (b) To augment working memory (bounded rationality). (c) To direct limited attentional resources toward critical aspects of the 'problem space'. (d) To counteract the effects of availability bias.

270

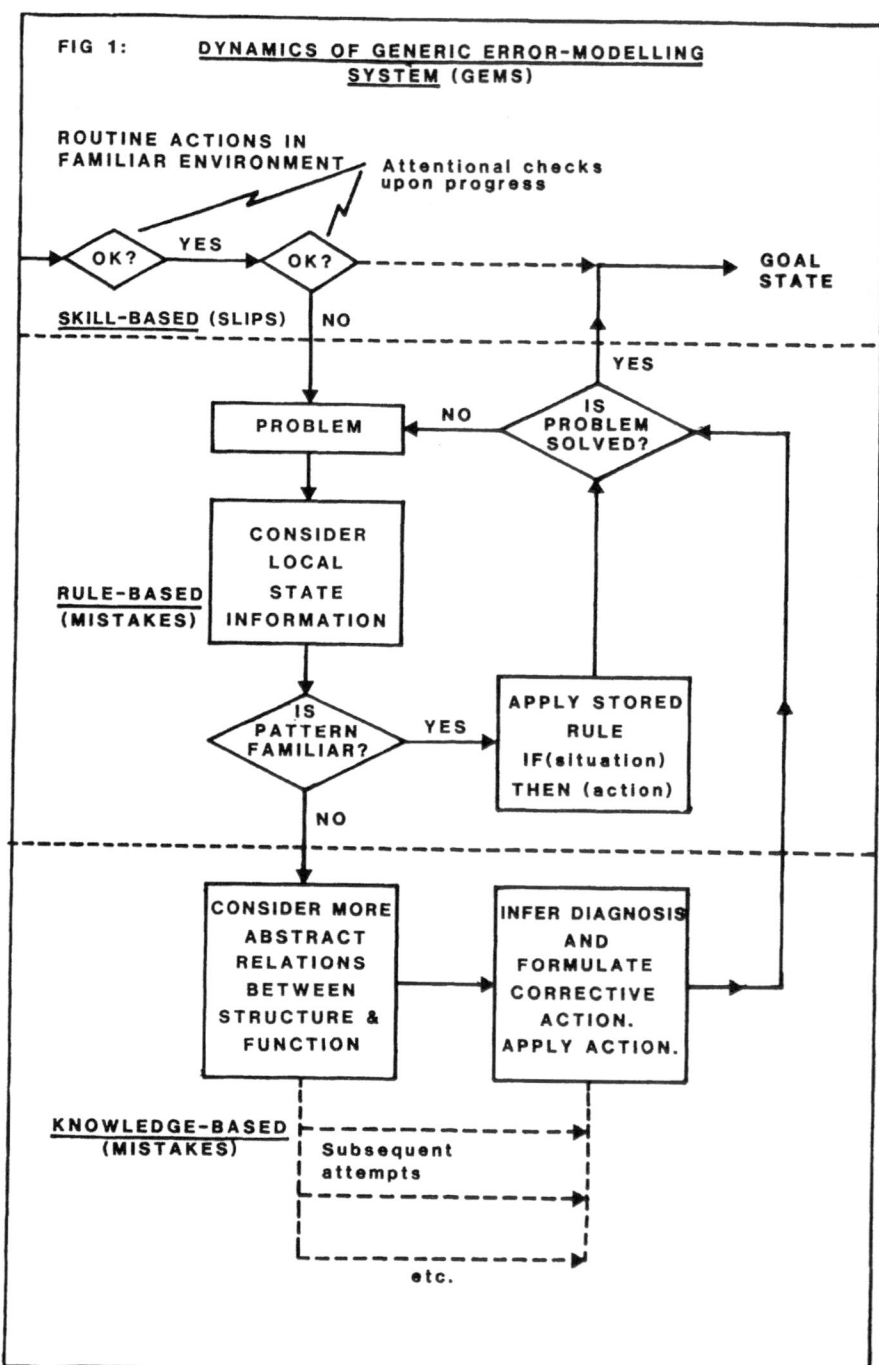

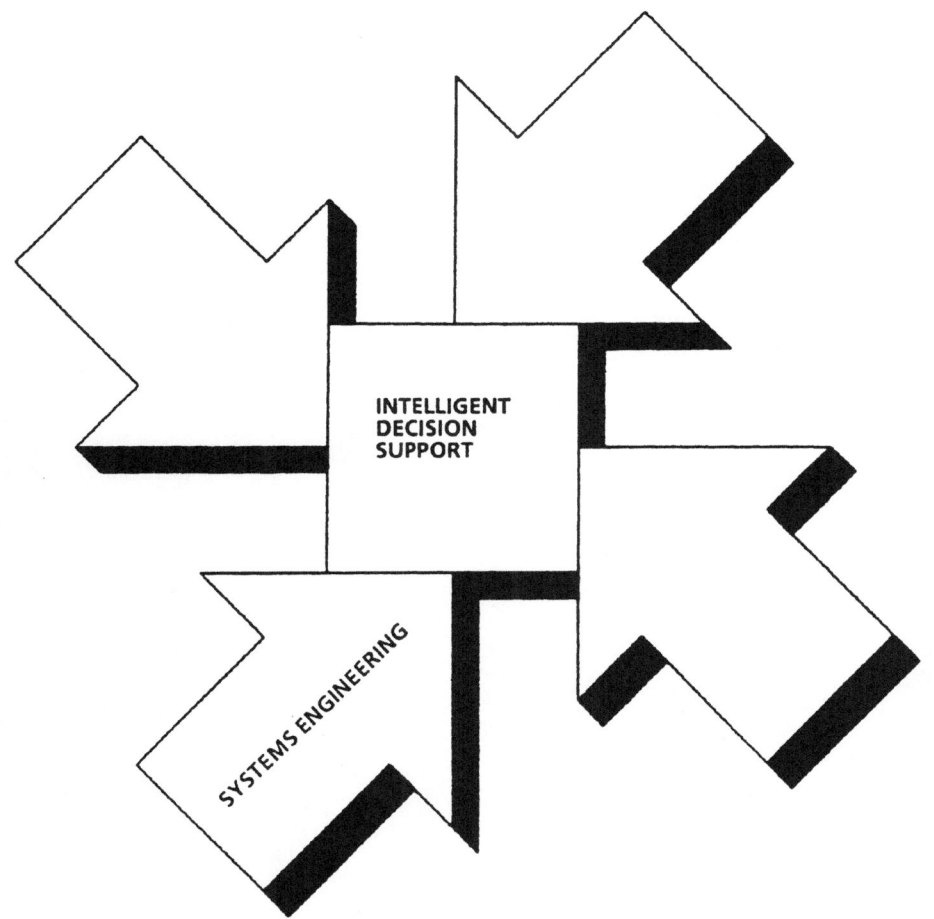

INTELLIGENT
DECISION
SUPPORT

SYSTEMS ENGINEERING

MODELS

MODELLING COGNITIVE ACTIVITIES:
HUMAN LIMITATIONS IN RELATION TO COMPUTER AIDS

Neville Moray
Department of Industrial Engineering
University of Toronto
Toronto, Ontario, Canada

INTRODUCTION

In the design of a human machine interface for process control the designer must couple three elements as closely as possible, namely the properties of the environment, the properties of the plant, and the properties of the operator. Information and control pass backwards and forwards between these three components, and within each component information and control pass between subcomponents.

Section I: Properties of Systems and Their Parts

Axiom 1

The degree to which events in one component can be mapped onto events in one or more of the other components determines the degree to which we have a safe and efficient system.

NATO ASI Series, Vol. F21
Intelligent Decision Support in Process Environments
Edited by E. Hollnagel et al.
© Springer-Verlag Berlin Heidelberg 1986

Axiom 2

The purpose of artificial intelligence is to improve this mapping, compensating for properties at each interface which reduce the power of the mapping.

A useful heuristic name for this mapping is that used by Green (1980), "cognitive impedance matching". It is used by analogy with electrical impedance matching, which allows the maximum power transmission across an interface. Here we are concerned with the maximisation of transmission of "communication and control".

Before considering specifically what properties of a human may reduce the effectiveness of cognitive impedance matching, let us consider some ways of representing the matching process, and the role A.I. may play in solving the problem.

FIGURE 1. CONANT'S REPRESENTATION OF A COMPLEX SYSTEM.

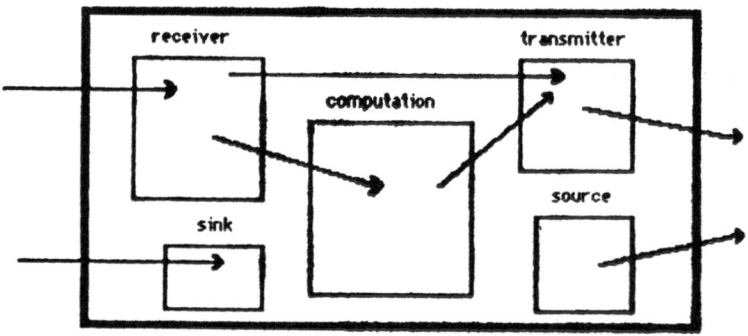

A SYSTEM MAY BE COMPOSED OF A VARIETY OF MORE OR LESS LOOSELY COUPLED SUBSYSTEMS WHICH HAVE A GREAT VARIETY OF NEEDS AND PROPERTIES WITH RESPECT TO CONTROL AND INFORMATION.

Figure 1 shows a generalized model of a system decomposed into its subsystems (Conant, 1976). The feature of interest is the emphasis placed on the fact that subsystems may be sources, sinks or transmitters (computation subsystems). In any system, (we may note particularly in the case of a human,) information may emerge from a subsystem which did not recently enter it, or disappear into a subsystem without emerging. In the former case such information would, in an inanimate system, usually appear as noise, and the latter disappear as equivocation, to use the classical terms (Shannon, and Weaver, 1949). But in the case of humans, a subsystem may behave as a source because it generates goals and subgoals on the basis of information which may have entered the operator either at a time or in a context quite remote from the current state of the system. For example, a goal may be chosen for "social" reasons arising from personal, familial, or management value judgements, and may arise from events which occurred several days or more prior to the time at which the goal expresses itself.

HIERARCHICAL OF COMMUNICATION AND CONTROL

Figure 2 illustrates problems of hierarchical control in supervisory systems.

Axiom 3

Whichever component in a system has the ultimate authority for its actions must have complete knowledge of the properties of the system.

Axiom 4

Whichever component a system has the ultimate authority for its actions must have complete control.

FIGURE 2. VENN DIAGRAMS OF HUMAN-COMPUTER INTERACTION

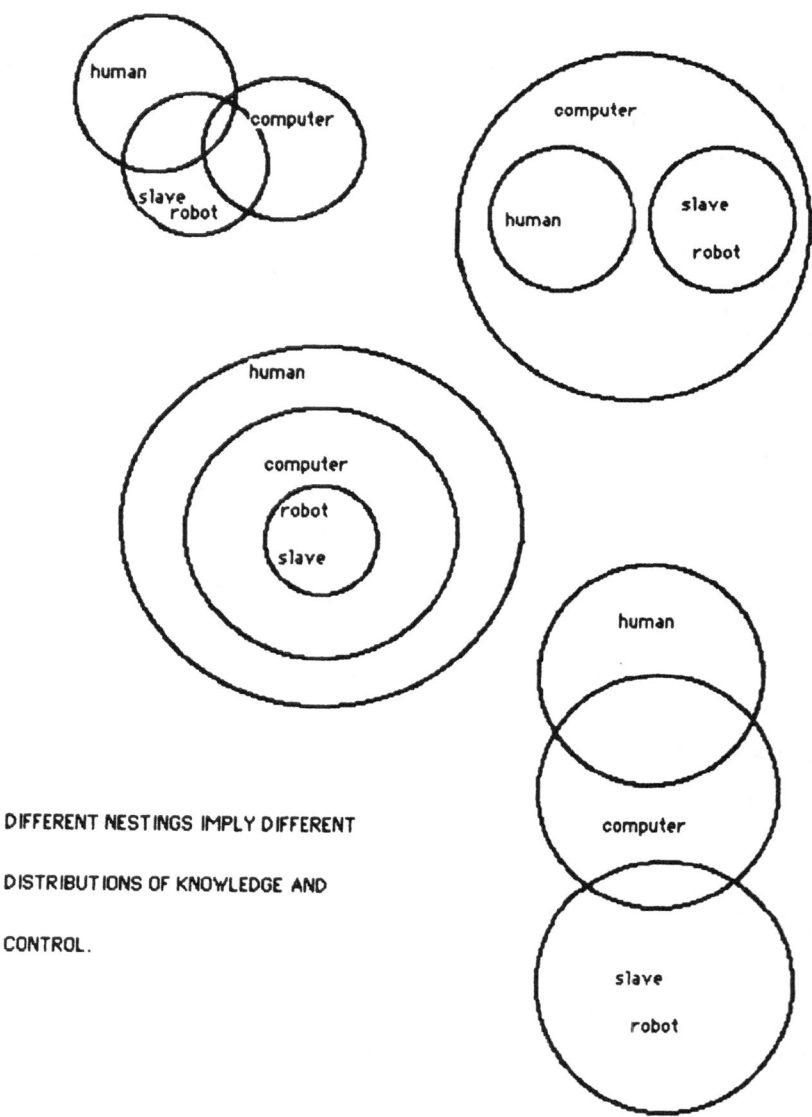

DIFFERENT NESTINGS IMPLY DIFFERENT

DISTRIBUTIONS OF KNOWLEDGE AND

CONTROL.

The circles in Figure 2 are sets of information or control. The question is, which should contain which? The figures are Venn diagrams. Each circle represents a set of knowledge or control abilities. Problems arise when an inner set, which contains only a subset of knowledge or control, may temporarily demand or be required to take control of an outer system. In general, the ignorance of a subset about the properties of its superset may be expected to give rise to difficulties.

Axiom 5

A subsystem can only guarantee in principle to exercise safe and efficient control over subsystems which are completely contained within it.

Figure 3 shows the implications of a failure hierarchically to nest sets of information adequately. (Sheridan and Hennessy, 1984; Moray, 1985).

When we consider the relationships between a system and the subsystems of which it is composed, we can define a role for artificial intelligence in the design, implementation and functioning of human machine interfaces.

Axiom 6

It is the role of A.I. to guarantee that a subsystem which is currently in control of the overall activity of the system shall function as a superset containing all subsystems over which it currently is exercising control.

Section II. Rational and Empirical Models of Human Information Processing.

Many models have been proposed for the subsystems involved in human information processing, but there are few models for the human considered as a holistic component. For example, the Theory of

FIGURE 3. IMPLICATIONS OF IMPERFECT NESTING OF SUBSYSTEM PROPERTIES

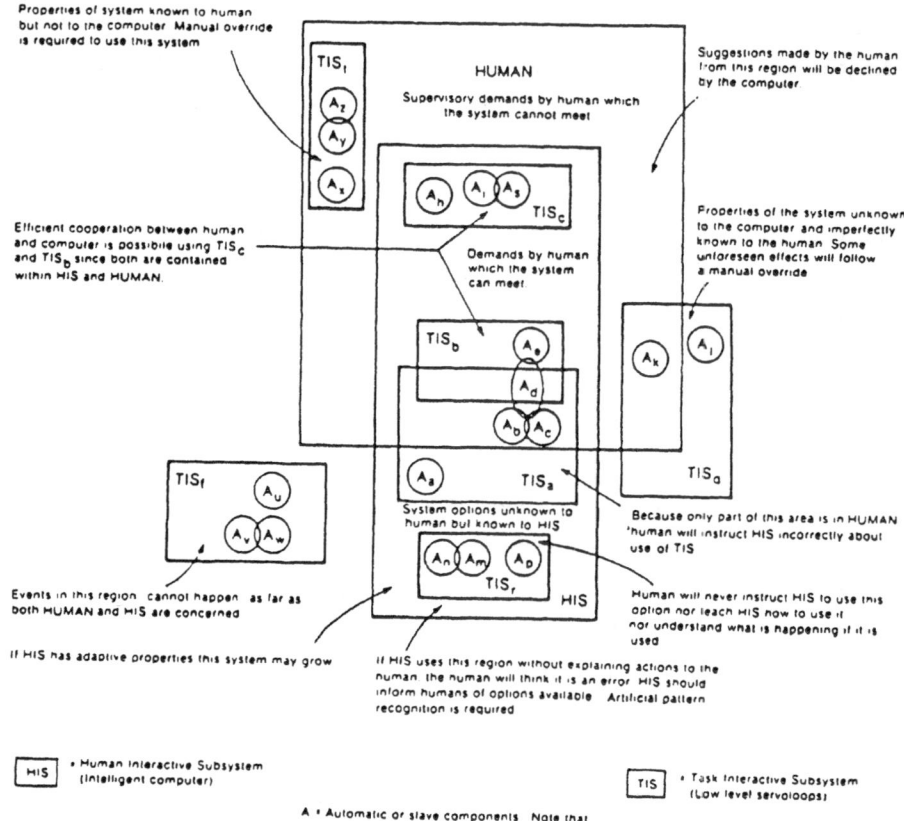

EACH BOX OR CIRCLE IS INFORMATION DEFINING THE PROPERTIES OF THE
RESPECTIVE HIS, TIS, OR A.

Signal Detection is adequate in many cases to describe the detection of
signals, speed-accuracy trade-off, and for modelling certain aspects of
pattern recognition and memory. Classical and optimal control theory
have been used to model manual control at bandwidths in the range
0.5 to 10 radians/sec (McRuer and Krendel, 1974; Kleinman, Baron, and
Levison, 1970). Queueing theory and Information theory have been used
to model visual sampling (Carbonell, 1966; Senders, Elkind, Grignetti, and
Smallwood, 1964; Senders, 1984). Sequential decision theory has been

used to model fault detection, (Curry and Gai, 1976; Wewerinke, 1981). For an overall review of such micro models in the context of monitoring behaviour and supervisory control see Moray, (1985).

On the other hand, there are very few models of humans "as a whole" which are of use in user-machine systems design. (A review will appear in late 1985 or 1986 from the U.S. National Academy of Sciences Press, edited by Baron.) The most impressive attempt to model the operator in a complex system is without a doubt PROCRU, which combines optimal estimation and optimal control theory with an expert rule-based production system. (Baron et al., 1980).

Furthermore, the vast majority of the studies have been laboratory studies of highly simplified systems. The contrast between such systems and those which are the topic of this conference cannot be too strongly emphasised. As I have done elsewhere, I would like to quote Jens Rasmussen's comments on this point:

Laboratory tasks tend to have a well-defined goal or target. Payoff matrices are artificial and have low values. The subject is controlled by the task. Task instructions are specific. Task requirements are stable. Subjects are relatively untrained. By contrast in "real" tasks only a (sometimes vague) overall performance criterion is given and

detailed goal structure must be inferred by the human operator from rather general commands about how to perform the task. The task may vary as the demands of the system vary in real time. Operating conditions and the system itself are liable to change. Costs and benefits may have enormous values. There is a hierarchy of performance goals. The operator is usually highly trained, and largely controls the task, being allowed to use what strategies he will. Risk is incurred in ways which can never be simulated in the l laboratory.

The settings in which we wish to model humans as information processors are those of manual or supervisory process control. The most important characteristics of such systems are the following:

1. They are large - up to 50 degrees of freedom.
2. They are complex - there are many couplings between variables.
3. They are hierarchical or heterarchical - they can often be decomposed into "almost" independent subsystems.
4. They are dynamic - the most important statistics for describing their behaviour are some kind of time series statistics and differential equations.
5. They are interactive - actions by a controller affect the future state of the system.
6. They are normally slow, often with time constants of the order of minutes or hours.
7. Events are often very rapid in fault conditions, with time constants of the order of seconds or fractions of seconds.

The design of a human-machine interface must support the human who is trying to stay in contact with the state of the system. Sheridan has identified five roles for the operator: planning, learning, intervening, monitoring and teaching, (Sheridan, 1985). To design a suitable interface requires us to use artificial intelligence to ensure that the human maintains the appropriate pattern of information intake to keep in touch with all the relevant state variables - at least as many as are required to define the system state.

We could approach this through a formal analytic model of human information processing, but none exists at present of sufficient power. Central to any such model would have to be a model of attention, pattern recognition and memory. That is because a human reads information sequentially from displays, integrates information from a number of sources, and interprets the resulting pattern of "facts", combining "atomic" data into meaningful "molecules" whose structure embody the overall state both of subsystems and the system as a whole. Given a detailed quantitative model we could point to nodes in the system where bottlenecks might occur, or where information processing is particularly inefficient. Intelligent interfaces and computer aids could then be designed to cope with such problems.

Unfortunately, the most sophisticated research on human information processing is all laboratory based and is largely irrelevant. Consider the recent intense interest in questions of controlled vs. automatic information processing, and parallel vs. serial processing. Leaving aside the considerable conceptual confusion due to inadequate definition of terms, the studies are not really useful. It may well be that under intense practice or special conditions of presentation, highly intelligent and highly motivated subjects can seem to process information "automatically" and "in parallel". But in an industrial setting, we are concerned with design for workers of average intelligence and average motivation whose information intake is above all limited by eye movements and body movements. The question for the designer is: if the operator can make no more than 2 eye fixations per second, and takes 2 or 3 seconds to change the display on a VDU, how can he or she be helped to maintain up to date knowledge of a 40-degree-of-freedom system? Changes in response times of 100 msec to tachistoscopic displays have nothing to say on such matters.

Let us then start by listing empirically known limitations on the operator's ability to process information.

1. Eye movements in "real" tasks are limited to about 2 per second.
2. Duration of fixation is increased as the noise/signal ratio of the display increases.
3. For "average" processes it seems to take about 10 seconds to detect a 50% change in mean or variance of a (single) zero-mean Gaussian display.
4. Dynamic memory span for unrelated data is around 3 items.
5. Observers make use of only about 3 items in a time series to

predict the next event or value in the series.

6. Limitations of attention - especially visual scanning.

7. Limitations of static memory span

   -memory span ~ 7 items

   -interference

8. There are a further 27 known sources of inefficiency in human decision making listed by Sage (1981). They are as follows.

9. Anchoring - to reduce information load.

10. Availability - easily available information is overvalued.

11. Base rate - tendency to use relative rather than absolute estimates of probability.

12. Conservatism - unwillingness to update estimates in the light of new evidence.

13. Data presentation modes - e.g. effects of scales in graphs.

14. Data saturation - unwillingness to consider all evidence.

15. Ease of recall.

16. Desire for self-fulfilling prophecies.

17. Expectations bias.

18. Fact-value confusion.

19. Chance-cause confusion.

20. Gambler's fallacy.

21. Habit - tendency to use past successful strategies.

22. Hindsight enhancement - events seem more probable with hindsight.

23. Law of small numbers.

24. Primacy and recency.

25. Local summit - tendency not to explore distant regions in problem space.

26. Data quantity - more is thought to be (necessarily) better.

27. Redundancy enhancement - redundant information is given undue weight.

28. Reference effects - past experience irrelevant in current context is used as if relevant.

29. Regression ignored.

30. Representativeness - small samples thought unduly representative of population.
31. Selective perception.
32. Wishful thinking.
33. Confidence in optimistic outcome.
34. Cognitive 'lockup', or tunnel vission.

APPROACHES TO DESIGN

Axiom 7

A good interface must compensate for items 1 - 34 in the above list.

However, before considering how this may be done, it is also worth considering the recent increase in interests in what has come to be called "ecological" information processing models. One origin of such models is the suggestion by Simon (1981) that before trying to understand and model the deficiencies of human information processing, it is as well to examine the structure of the environment. Although I disagree with the following quotations, it should be prominent in the view of all of us who deal with human factors in "real", not laboratory, environments:

A man, viewed as a behaving system, is quite simple. The apparent complexity of his behaviour over time is largely a reflection of the complexity of the environment in which he finds himself.

(Simon, op. cit. p. 65)

Axiom 8

Simplify the task by simplifying the environment.

FIGURE 4. EXAMPLE OF 6-DEGREE OF FREEDOM DISPLAY

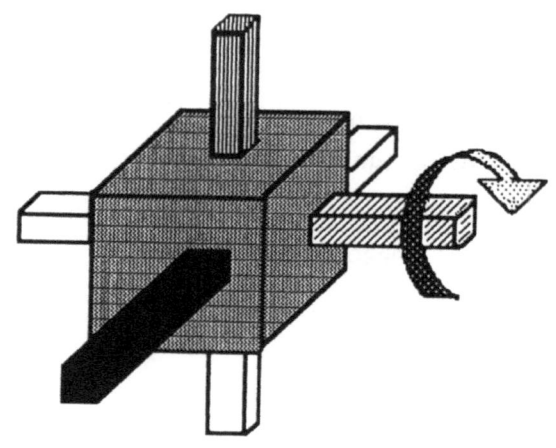

THE DISPLAY SHOWS THE VALUES OF THREE FORCES AND
THREE TORQUES AROUND THE WRIST OF A REMOTE
MANIPULATOR. THE DIRECTION OF EACH FORCE IS SHOWN
BY SHADING ONE END OF THE X, Y, OR Z FORCE AXIS. THE
DIRECTION AND MAGNITUDE OF TORQUE IS SHOWN BY
WRAPPING AN ARROW AROUND THE AXIS ON WHICH IT IS
ACTING. IN THE ORIGINAL COLOUR CODING IS USED TO
INDICATE THE MAGNITUDE OF THE FORCE, GOING THROUGH
THE SPECTRUM TOWARDS RED AS THE FORCE OR TORQUE
APPROACHES A DANGEROUS MAGNITUDE. IF THE DIRECTION
OF THE FORCE CHANGES, THE TORQUE ARROW WILL STAY
AT THE APPROPRIATE END OF THE BAR TO MINIMISE THE
NEED FOR EYE MOVEMENTS. (MORAY AND HO, 1985.)

Normal process control often involves very slow processes with very
long time constants. It is not possible to see trouble developing because
it occurs too slowly over too long a period. On the other hand,
catastrophic faults tend to have very short time constants. It is not
possible to tell what has happened because too many events occur too

rapidly. Each of these facts suggests a requirement to replay a recent sequent of history at an unreal speed: the operator should be able to display the last hour in a few seconds, or the last seconds in a few minutes. The first is "time-lapse photography", the second slow motion. Each moves a time series into the band of the information processing spectrum where our "information receptors" can "see" "visible information", by analogy with mapping infra-red or UV pictures into the visible spectrum. In doing so, the complexity of the environment is reduced. It becomes an environment matched to the abilities of the nervous system.

Another example is the use of "star" diagrams to represent multi-degree of freedom systems, or the colour-analog display we have recently developed for 6-degree of freedom manipulators (Figure 4). In each case dynamic attention is off loaded at the expense of some perceptual complexity and pattern recognition.

SOME METHODOLOGICAL ISSUES

It is appropriate at this point to raise two problems of methodology. The simpler is that of assessing new systems. Operators in industry work for many years with a piece of equipment. Crossman (1959) has published data showing improvement with practice over a period of 7 years. At what point can we tell whether an "unpromising" decision aid could become an excellent one with practice, or that the problem is in the training programme, not in the aid?

Towill (1974) has published several learning-curve time constants for industrial processes. The conclusion seems to be that in order to establish the shape of the learning curve, we should be prepared to practise operators on new equipment for not less than 10 hours, preferably not less than 10 days and preferably for more than 100 hours before collecting summary data. Data should be collected throughout the training so that the parameters of the learning curve can be determined and an estimate made of final performance levels to be expected.

The second methodological issue is of greater importance, since it speaks to the question of the relation between environment complexity, human information processing complexity, and the design of artificial aids. How well do we really know the limits on human performance? Eindhorn and Hogarth (1981) have suggested that some of the "well known" deficiencies in human decision making are artifacts of laboratory settings. To take an example of my own (Moray, 1981), it is usually thought that Item No. 34 in the list of deficiencies, "cognitive tunnel vision", is a defect. It is bad for an operator to concentrate on one subsystem or one hypothesis to the exclusion of others. But as I argued in the paper cited, it is a good and rational strategy, given that the environment has certain properties. In particular, when a complex system can be "nearly decomposed" into independent subsystems, an operator who detects an abnormal value in a subsystem should examine in the first instance those variables which are tightly coupled to the abnormal variable. In a tightly coupled subsystem, it is far more probable that the fault is one of the tightly coupled components than that the fault has propogated from a distant component in a different subsystem. It is highly adaptive to search the "actions" of your "molecule" before examining other "molecules". In real systems "cognitive lockup" is adaptive during fault diagnosis.

On the other hand, it is undersirable that the rest of the system should be completely ignored. An intelligent interface should monitor and model the operator's tactics in a strategic light, and should remind him or her from time to time that parts of the system are being neglected. (We are at present trying to develop such a system).

The conclusion from the above is clearly that the system should display its structure to the operator. Moreover, this should be done in several ways, since as Rasmussen has pointed out there are classes of information (signals, signs, symbols), classes of behaviour (skill-, rule- and knowledge-based), and classes of tactics (mass balance, topographic, functional) each of which may require a different representation of the system.

A further problem arises due to the existence of individual differences in "cognitive style". It is of course characteristic of real systems that the aim of good performance is the attainment of a goal rather than the use of a particular route to that goal or the maintaining of a particular score on a variable during the path to the goal. (It may be just as acceptable to have a large rms error for a short time as to have a small rms error for a long time.) The magnitude of individual differences in tactics, strategy and performance is far greater than is generally acknowledged, although evidence is now accumulating on this topic. Sage's (1981) paper again provides a useful introduction; we have found large individual differences in the perceptual styles of radar operators; and Sanderson, Thornton and Vicente (1985) and Hart and her co-workers (1984) have emphasised their importance in workload. It should also be noted that the current interest in rule-based expert systems as models of the human operator was anticipated by Beishon (1974) who reported cases of process operators who appeared to have up to 200 rules to use. These had been informally learned. How should such indiosyncratic talents be supported?

Axiom 9

A major virtue of artificial aids is to adapt to the information processing requirements of the individual by modelling the operator and changing their own properties to respond to the model so formed.

A WORD OF WARNING

It is a natural assumption that the power of a computer based intelligent interface lies in its adaptability and flexibility. The following story should act as a warning against over-enthusiasm.

We were recently evaluating a new engine room console which uses an all electronic display and control system. It has three screens, and all displays can be sent to any screen. There are about 24 pages of displays. There are two engines, port and starboard. The operator had an overview display on the centre screen, showing a mimic of both power trains and variety of engine parameters. On the starboard screen was a close-up mimic of the starboard combustion chamber sub-system. An alarm occured indicating a fire in the starboard gearbox. The operator put up the <u>starboard</u> gearbox mimic and parameters <u>on the port screen.</u> Simultaneously, the starboard gearbox on the centre screen overview alarmed in red, the rest being green. The operator correctly diagnosed the trouble and started to reduce power to the starboard engine. <u>But he used the port controls which were immediately in front of the port screen.</u> For two minutes he tried to reduce rpm, and then told the simulator operator that the system was not responding. During this time, he repeatedly looked back and forth between the port screen and the starboard side of the centre overview. But so strong was the link between the position of the screens and controls that he failed to notice what he was doing. Had the ship been at sea, it would during this period have executed a very sharp and unexpected turn to port. <u>The classical stimulus response stereotype completely overwhelmed the flexibility of the intelligent interface.</u>

<u>Axiom 10</u>

We do not know how to use the power and flexibility of intelligent interfaces.

PROBLEMS OF RESEARCH

The list of limitations on human decision making is largely derived from laboratory studies. In general the experiments are artificial, limited in time, and over simple. We do not have a large body of information about real-time real-world experiments, and hence, while the evidence in the decision making literature is suggestive that the results can be extrapolated from the laboratory, it is desirable that more realistic studies should be conducted. What should they be like?

Real-world systems are dynamic, complex, real-time, and demand goal-directed behaviour sustained over a considerable period of time and taking place in a meaningful context. This argues for a high degree of realism and involvement by the participants. They should be employed in a task, not paid as subjects. They should be motivated as people usually are, not paid or rewarded as student participants. (Most people work for salaries or wages, not for occasional rewards.) At least 10-20 hours practice is required before data are collected - after all, the world we live in is a stochastic, not a deterministic world, and people must have time to learn the statistics. Tasks should be such that usually several different ways of doing them are equally acceptable: behaviour in the world is usually a matter of reaching a goal, not making a single correct response; and ways of analyzing such variable behaviour must be developed. There should be provision for "informal" information in de Keyser's sense: many tasks provide signals such as noises, vibration, smell, etc., not just visual data on a screen.

When it comes to the analysis of such experiments classical experimental design is not usually appropriate. Analysis of variance based on two or three orthogonal variables will seldom capture the nature of process behaviour. Rather, time series measures, Markov models, etc., are more appropriate at least some variables should be partially correlated, and the operator's responses should affect what happens next. Even then there are severe problems. In process control, and especially in supervisory control, operators make very few actions. To understand what is happening will require not just data-logging, but eye movement recording, protocol analysis, and other methods. The reduction of raw data to summary statistics almost always destroys significant patterns of behaviour, and produces a description which is typical of no single individual. To invent new ways of analyzing and representing such complex behaviour is a major challenge, since otherwise the raw data remains the best model of itself.

A final challenge is to develop a good methodology for investigating fault management and human error. Plant failures are rare, and human errors relatively rare. How many data are required to characterise operator behaviour? How are cognitive errors which never emerge as overt behaviour to be collected, classified characterised and modelled?

Without a rich and realistic experimental environment realistic performance will not be seen. If the work setting is not realistic and as well human factored as a real work place, irrelevant variations will intrude to mask the significant behaviour. And if realistic behaviour is not studied, we will not be able to extrapolate from our investigations to real world settings. This will be particularly true of advanced systems with advanced technology. Research on the topic of this conference will remain difficult, time consuming, and expensive. Without such research the potential for industrial disasters will increase dramatically over the next two decades, for when plants fail they will fail on a catastrophic side, not on a scale local to the operator.

The more intelligent and flexible we make them the more we open the door to as yet unknown types of catastrophic error which will be far harder to control than in conventional systems. The more we avoid such options and flexibility the less point there is in having such systems. The better the system adapts to the information requirements of one operator the harder it will be for another operator to understand what is going on. Caveat creator!

CONCLUSIONS

1. We do not have an adequate empirical data base to feel confident of modelling the human operator's limitations as a controller of complex systems, especially in supervisory control.
2. We do not have an analytic model for that purpose either, especially in view of individual differences.
3. Without such research we do not have a basis of rational design to incorporate the human operator in complex systems.
4. Assessing designs is costly and time consuming.
5. We had better design systems which monitor the operator as well as the system, which warn the operator of unusual patterns of his or her behaviour.
6. We had better go very cautiously in the introduction of powerful artificial aids, since we do not really understand what we are doing.
7. **Axiom ll**

We will see the introduction of intelligent interfaces marked by new and unexpected classes of major accidents.

ACKNOWLEDGEMENTS

This paper was supported by an operating grant from the National Scientific and Engineering Research Council of Canada for work on the human operator in process control.

DECISION DEMANDS AND TASK REQUIREMENTS IN WORK ENVIRONMENTS: WHAT CAN BE LEARNT FROM HUMAN OPERATOR MODELLING

Henk G. Stassen
Man – Machine Systems Group
Laboratory for Measurement and Control
Delft University of Technology
Delft, The Netherlands

0. Summary

Supervisory control can be distinguished from manual control which is a closed loop, skill based action, by four important aspects:

● The control of highly complex, multivariable processes with mostly large to very large time constants.

● The discrete action patterns based on decision making processes.

● The variability in tasks, such as process tuning, start and stop procedures and fault management.

● The often vague information on the ultimate supervisory control perspectives.

Hence, supervisory control tasks are different from those in manual control. Taking into account the three major process control modes which may occur in supervisory control -Normal operation, start and stop, and abnormal operation- it is of interest to classify the different tasks with reference to these control modes, as well as with reference to the three levels of control behavior as introduced by Rasmussen: Skill-based, Rule-based and Knowledge-based behavior. At the Skill-based level only manual control activities play a role; at the Rule-based level activities like process tuning and to a certain extent fault management may occur, whereas at the Knowledge-based level intelligent, cognitive activities such as optimisation, planning and fault management are thought to be.

In understanding at which level of human control the different tasks can be placed best, it is instructive to study where successful human operator models are reported in literature. At the Skill-based level certainly successful control models describing manual control behavior have been reported. At the Rule-based level, less but still to a certain extent, some control and artificial intelligence models are known. At the Knowledge-based level, tasks like fault management can only be modelled, if and only if, these tasks are that well defined that they are far away from real world situations; hence only very few

NATO ASI Series. Vol. F21
Intelligent Decision Support in Process Environments
Edited by E. Hollnagel et al.
© Springer-Verlag Berlin Heidelberg 1986

models have been published. So, the conclusion can be drawn that only in those cases where tasks are well defined, successful human operator models are developed. This important statement elucidates exactly the basic problem:

> The tasks which yield a correct description of human supervisory behavior are to be find at the Skill- and Rule-based level, they are all well defined; those which do not lead to any description of human behavior are Knowledge-based level tasks, and they are not well defined.

In particular, the last class of tasks is dealing with those tasks where one needs the operator for his creativity, knowledge and intelligence.

As a concluding remark the following can be said. The human operator should mainly be the adaptive, creative and intelligent supervisor who does things which can not easily or at all expected. So, those tasks which can be carefully defined, thus where modelling of the human behavior is expected to be successful, can easily be taken away from the operator. Often intelligent MMIf can help the operator to support him to do the job. Whether this is wise or not is dependent on factors such as the need for training, for learning, for building up an Internal Representation of the process, etc. So the final and major question becomes:

> To what extent should one automate, or, to what extent should one build in artificial intelligence in man-machine system interfaces?

1. Introduction

The oldest form of human system interaction is manual control; in this mode the human operator acts as a human controller who responds directly on the information presented to him, taking into account his limitations. This type of human performance is just a skill, it can be learned by time. In general, the tasks to be performed are well and precisely defined.

In the supervision of highly complex and multi-variable processes, the circumstances are different. The supervision of an industrial plant is now-a-days mainly a cognitive and intelligent task which can be distinguished from manual control by at least four important aspects.

- The supervision relates to highly complex, multivariable processes with mostly large to very large time constants.
- The discrete supervisory control observations and actions are mainly based on decision making processes.
- There exists a large variety of tasks to be performed, such as process tu-

ning, starting-up and closing-down and for instance fault management.

● The human supervisor is assumed to perform his tasks with often very vague
information on the ultimate supervisory control goals in mind, such as:
Protect the personal staff from injuries or death, protect the process plant
from damages, and avoid environmental pollution.

From the foregoing, it is clear that supervisory control tasks are completely
different from manual control tasks. The variety in supervisory control tasks
probably can be illustrated best by considering the tasks during the different
control modes which can be recognized: Normal operation, Start and stop procedures, and Abnormal operation. Moreover, it is of interest to try to classify
the tasks with reference to the qualitative Three-level Model, a model where
a distinction is made between a Target Controlled or Oriented Skill Based Behavior, SBB; a Goal Oriented Rule Based Behavior, RBB; and a Goal Oriented
Knowledge Based Behavior, KBB [RASMUSSEN, 1983].
A taxonomy of human operator tasks is given in Table 1. The table shows that

Table 1: A taxonomy of human operator tasks.

Level of Control \ Process Control mode	Normal operation	Start and stop procedures	Abnormal operation
SBB	Manual Control	Manual Control	Manual Control
RBB	Process tuning	Process tuning/ testing	Compensation in Fault Management
KBB	Plant optimisation/ planning	Trajectory planning/diagnosis	Detection and Diagnosis in Fault Management

at the SBB-level only manual control activities play an important role. At
the RBB-level tasks which are to be performed according to well defined strategies and procedures, such as process tuning, start and stop testing procedures and the compensation phase when one is coping with system failures, are
carried out. Finally, at the KBB-level cognitive activities such as optimisation, planning and diagnoses in fault management are thought to be executed. This taxonomy might give a guide to what extent intelligent decision
aids may be helpful in process control, in particular it may be possible to
predict to what extent artificial intelligence concepts may benefit human

operator control behavior. This yields that a detailed description of the
decision demands and task requirements in the work environment needs to be
available.

The many research that is carried out on the topic of human operator modelling
is providing us with some of these data, since the idea behind human operator
modelling is to obtain an abstraction that involves an explicit mathematical
or computer-based formalism that can be used as a design tool in Man-Machine
Interface developments.

2. Human operator modelling: Skill based behavior

In the review paper "Perspectives on human performance modelling", Pew et al.
state:"A model of human performance implies the existence of a model of the
environment or system in which that performance takes place" [PEW, BARON,
1982], or in other words, it states that the human operator is expected to
possess an Internal Representation, IR, of the environment in terms of [VELD-
HUYZEN, STASSEN, 1977]:
● The process to be supervised,
● the disturbances acting on the process, and
● the tasks to be performed.
For consistent operator behavior the existence of such an IR even seems to be
a necessary condition, i.e. the lack of an IR always leads to such a variance
in human operator behavior that modelling will be difficult. However, the reverse can not be argued: The existence of an IR does not imply consistent human
operator behavior [STASSEN, KOK, v.d. VELDT, HESLINGA, 1985].

At the SBB-level a lot of studies has been devoted to manual control problems;
in particular the manual control behavior has been modelled successfully. Well-
known models in frequency and time domain, such as the Describing Function
Model, DFM, [MCRUER, JEX, 1967; RUITENBEEK, 1985] and the Optimal Control
Model, OCM, [KLEINMAN, BARON, LEVISON, 1971] have proven their value in the
design of controls and displays. In general the studies refer to the control
of linear, SISO and fastly responding systems. Extensions to MIMO systems
[VAN LUNTEREN, 1979] also have shown the value of the describing function
method in human behavior modelling. Also many non-linear models have been proposed [STASSEN, VAN DER VELDT, 1982]. An interesting example is that of the
helmsman's behavior in the maneuvring of Very Large Crude Carriers, VLCC's.

Veldhuyzen showed that the decision processes involved in the generation of the rudder calls rather precisely could be indicated in the phase plane of heading error and rate of turn [VELDHUYZEN, 1976]. He also showed that two essential problems in the control of such a slowly responding system did occur:

● The lack of a good IR: That is to say, the understanding of the effects of the very large time constants (τ > 250 sec.) on the VLCC dynamic behavior is very difficult.

● The process of making a decision between a large rudder call for a short time and a small rudder call for a long time is very dependent on the skills of the helmsman. Here it should be mentioned that large rudder calls are much more speeding down the VLCC than small ones, so the well-experienced helmsman and/or navigator knows that by chosing small rudder calls for a long time he will be able to preserve the consumption of energy.

Both the shortcomings in helmsman's control can be overcome by intelligent man-machine system interfaces. The lack of IR can be compensated by the use of a predictive display, which, by the way, is not easy from the technical point of view. The decision process with respect to the energy saving can be enlighted by a better training, hence building up a better IR of the task, or an arti-ficial feedback of the change in velocity of the VLCC. Hence by defining the task variables more precisely, the human operator tends to show a more consis-tent behavior.

The remarkable similarity in all these models is the fact that the tasks to be performed are very well defined, and hence, the human operator was forced to act as a stationary process with low variance. As a result, human behavior could be modelled rather precisely, even as a function of time. In this way valuable tools became available in the design of MMIf, i.e. the controls and displays.

3. Human operator modelling: Rule based behavior

A very essential distinction between SBB and RBB is the existence of explicit decision making. In stationary supervisory control situations the task mainly consists of monitoring and interpreting, and is only interrupted occasionally by moments of actions such as set-point correction and collection of additional information. Difficulties in modelling human supervisory behavior now arise with respect to the decision making processes in particular. Only, where con-trol and observation strategy are basically non-varying, it is possible to achieve a reasonable fit of the data; but even then such a model is burdened

by a large number of parameters, a complicated identification scheme, relatively few measurements and, above all, a lack of a general decision making
mechanism which applies to all situations.

For supervision of stationary processes, the Observer/Controller/Decision
Model, OCDM, has been postulated [KOK, VAN WIJK, 1978; KOK, STASSEN, 1980].
In this model the functions of the observer part and the controller part are
similar as in the earlier mentioned OCM which means that the Internal Model
concept, IM, is fully integrated in the mechanisms. The functions of the decision making part are directly related to the observation actions and the
control actions. The outputs of the observer part (Fig.1) are the estimate
$\hat{x}(t)$ of the system state $x(t)$ and the variance $V_{\tilde{x}}(t)$ of the estimation error.

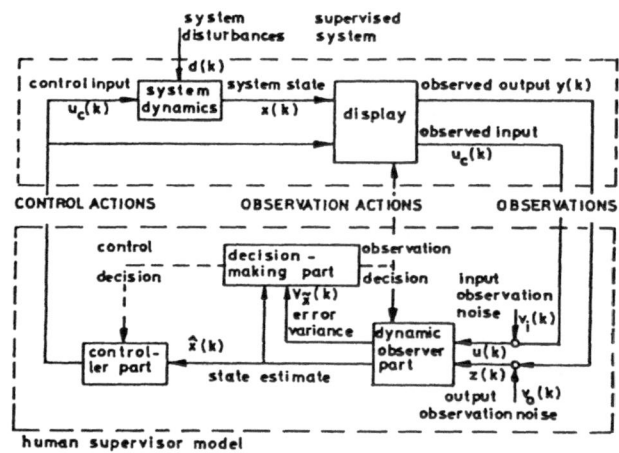

Fig.1: Structure of the Observer/Controller/Decision Model [KOK,VAN WIJK,1978]

Based on this information, the decision making part of the model determines
the instants in time of the sampling requests, the observation actions, thereby keeping the variance of the estimation error at an acceptable level. The
task of the supervisor was defined as the provision of the supervised variables
within a given target area by means of set-point corrections. Since practical
observations have shown that such a task is usually accomplished by discrete
control actions of the supervisor, a model structure was accepted which pre-
serves this character of control. The instants in time of the set-point corrections, the control actions, are determined by the desicion making part of
the model, where the decisions are based on the estimated values of the supervised variables as derived from the state estimate. The amplitudes of the control actions are determined by the controller part of the model in relation

with the running value of the state estimate and, with the overall goal of
the control. To allow for different control strategies depending on the typical control situation at hand, i.e. single control actions or maneuvres, the
possibility to define a sequence of subgoals each of which can be realized by
a constant control input over a certain period of time was introduced. The
overall goal of the control is then achieved by subsequent realization of the
different subgoals. The moments in time of the controller actions and the
observer actions are determined by the decision making part of the model.
Since the supervisory task is given in terms of the supervised variables
$s_i(\kappa)$, i=1.....,d_s, it is assumed that the decision making process evolves
with respect to these variables. By accepting a linear relation $s_i(\kappa)$ =
$h_i^T \times (\kappa)$, and a normally distributed estimate $\hat{x}(\kappa)$ a sufficient statistic
in the pair $\{\hat{s}_i, \sigma_{s_i}^2\}$ is obtained, and so the decision rules can be stated
in these terms. For the sampling requests or the observer actions a reasonable fit of practical data was obtained by a hyperbolic decision rule as
shown in Fig.2. However, these results apply only to simple situations with

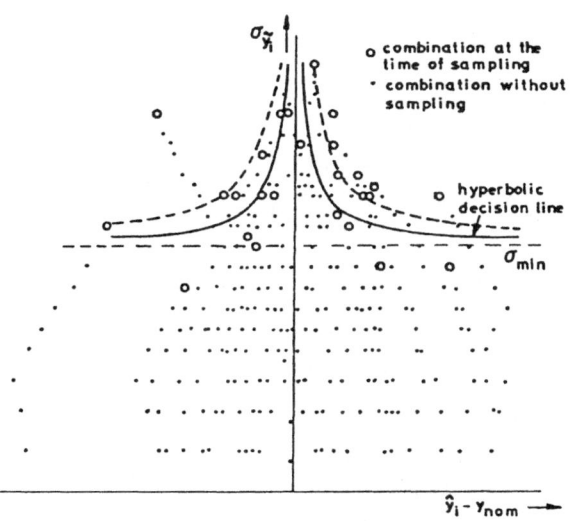

Fig.2: Combinations of $\hat{s}_i$ and σ_{s_i} and the supervisor's sampling
requests [KOK, VAN WIJK, [1]1978]

little or no interaction between the observed outputs; it could not be confirmed in a more realistic set-up (Fig.3), [WHITE, 1983].

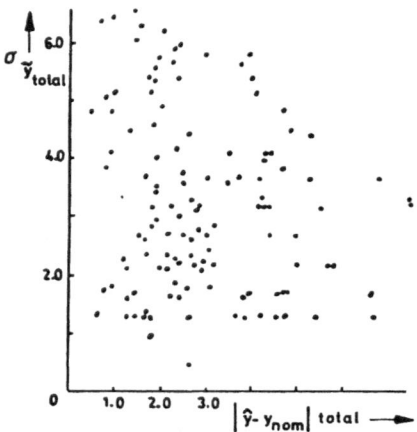

Fig.3: Position of sample points after a weighed summation of the estimates and the estimation uncertainty [WHITE, 1983]

For the set-point corrections or controller actions, the inconsistent behavior of the supervisor is even worse. According to the theory, the optimal detection rules for a control objective to maintain the supervised variables within a symmetrical region $\{-d_i, d_i\}$ around a nominal value s_{nom} takes the

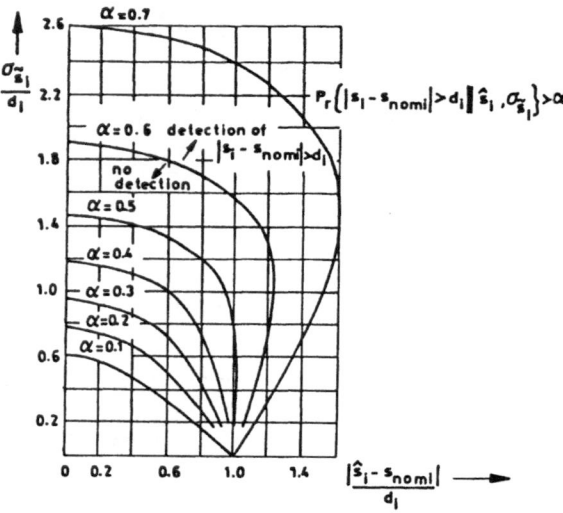

Fig.4: Optimal detection rules for various values of the utilities [KOK, VAN WIJK, 1978]

form of Fig.4, where α_i is the utility parameter of the decisions, given by
the task. However, this mechanism has never been observed in experimental data.
Instead, a crude simplification like $|s_i - s_{nom\ i}| > q_i$, which makes the detection
independent of the accuracy of the estimates, gives essentially no better fit
of the data. No improvement could be obtained by the introduction of a more
sophisticated rule [KOK, VAN WIJK, 1978].

In spite of the limited relevance of the decision making mechanism assumed, the
input-output relation of the OCDM resembles the actual supervisor behavior to
a certain extent in the control of simple systems. However, application of the
model in more complicated situations, such as the supervision of a process, is
not satisfactory in this respect; only occasionally a successful fit as a function of time could be achieved due to the variance of the supervisor's behavior.
So, the attempts to model RBB in supervisory control tasks were only successful in those cases where fast responding and/or simple processes had to be
supervised, and where decision making is rarely involved. As consequence,
many extensions of the control theoretic model were devoted to the decision
making process. To mention a few: Blaauw introduced the Supervisory Driver
Model describing the automobile driver as a combined observer/predictor, controller and decision maker [BLAAUW, 1984]; Pattipati et al. developed the
Dynamic Decision Model for predicting human sequencing [PATTIPATI, EPHRATH,
KLEINMAN, 1980]; and Muralidharan et al. reported the DEMON, a decision-moni-
toring and control model for analysing the en route control of multiple remote-
ly piloted vehicles [MURALIDHARAN, BARON, 1979]. However, these models as well
as the single decision models [GAI, CURRY, 1976] have not incorporated a number of very significant features common to many supervisory control problems
[PEW, BARON, 1982]. For example, these models do not consider the detection
of events not explicitly related to the system state variables, they do not
model multi-operator situations and the effects of communication among operators, and they do not include procedural activities of the operators. Inter-
estingly, just these neglected features are the prime concern of psychologically oriented models. It is therefore that Baron et al. developed PROCU, a
Procedure Oriented Crew Model that has been developed to analyse flight crew
procedures in a commercial ILS approach-to-landing [BARON, ZACHARIAS, MURALID-
HARAN, LANCRAFT, 1981]. PROCU includes a system model and a model for each
crew member, where the crew is assumed to be composed of a pilot flying, a
pilot not flying and a second officer. The second officer model does not include any information processing or decision making components, it is modelled
as a purely deterministic program. Both the pilots are represented by complex
supervisory control models which have the same general form, but differ in

detail. It is assumed that the pilots have a set of tasks or procedures (check lists) to perform in some optimising fashion [PEW, BARON, 1982].

The preceding section illustrates that modelling at the RBB-level is rather difficult, in particular due to the variety in human decision making behavior. In fact one can argue that all tasks where decision making is a prime aspect, yield KBB. Only in those cases where by means of checklists, handbooks and procedures the tasks to be performed are structured, modelling seems to be possible. Hence, in attempting to model RBB, one learns to what extent the human supervisor is able to perform certain tasks, and how he can up date his IR of his environment; one also learns the weak aspects in supervisor's performance. In this way modelling can lead to some insight at which aspects and to what extent intelligent MMIf's such as predictive displays [VAN DER VELDT, 1984; 1985] and rule based decision support systems, or in general human operator support systems can be applied fruitfully.

4. Human operator modelling: Knowledge based behavior

In those circumstances where tasks are poorly defined human supervisory behavior can not be modelled in detail [STASSEN, KOK, VAN DER VELDT, HESLINGA, 1985]. Hence, at the level of KBB no general model exists; apparently it is just the human operator who is needed for his creativity and intelligence in order to solve unexpected and unforeseen problems. As soon as this behavior could be modelled one could leave those tasks to the computer, and as a consequence it is no longer human intelligence but it is called Artificial Intelligence, AI. The existing models on SBB and RBB have one thing in common, they describe the operator's behavior as a function of time, where mainly the decision making processes lead to modelling problems. Such models on KBB are not reported, but at least two alternative approaches can be mentioned: (i) Measurement of global human performance and (ii) the normative model approach. Instead of developing specific models of operator behavior describing his actions as a function of time, it is possible to adopt a more global, quantitative modelling approach in order to investigate human performance, human limitations and human capabilities. As an example an ongoing study on predictive behavior of human supervisors in the control of complex, multivariable processes can be mentioned [VAN DER VELDT, 1984; 1985], where the essential question is why the human operator is able to achieve under certain conditions a better performance when predictions -future values of process varaibles- are explicitly shown. In fact, the underlying philosophy is to understand in which manner the operator's behavior is imposed by the environment, such as task and system to be controlled. The first experiments indicate that the positive effect of predictive information on human performance increases with system complexety; furthermore, it appears that showing prediction can almost completely compensate for the negative influence on human supervisory behavior of processes with time delays.

The second way out is a normative approach as elucidated in a normative fault management theory [MENDEL, THIJS, 1983]. This theory which is based on Savage's decision theory [SAVAGE, 1971], and which contains elements of the utility theory and subjective probability theory, describes the decisions a human operator should be made. Studies at Delft University of Technology have indicated that the theory can be used for accurate definitions of notions as rationality, knowledge, optimal observation and optimal acting [THIJS, STASSEN, KOK, VAN DER VELDT, 1984].

Both the methods just-mentioned are, at the one hand, basically different, but, at the other, they have a common element. The first method leads to the provision of an intelligent MMIf to the human operator in order to enlighten his task, whereas the second method yields the training of human supervisors to come to a more consistent fault management behavior. However, although changing the environment and the human behavior are of a totally different order, the final goal which has to be achieved is similar, i.e. the shifting from KBB to SBB or RBB by defining the task variables -task, system and disturbances- in a better way. It is obvious that modelling techniques can help in detecting such a shift.

As elucidated in the introduction, fault management is a task which typically leads to KBB, and as such the modelling attempts should be discussed. Literature shows that the often poor definition of the task makes the decision making processes very difficult to structure and to understand; hence it yields that direct, precise modelling of fault management behavior, and thus of KBB is not possible. Instead, one has to choose between global performance modelling and normative prescription. The claim that artificially well-defined tasks to train operators in fault management will yield KBB, so that operators achieve knowledge about fault management tasks can hardly be defended [ROUSE, 1979]. Instead, these tasks will evoque RBB, as modelling efforts have proven! Furthermore, to what extent such a training might be fruitful for real world fault management problems is still not precisely known.

Just to close the gap between the laboratory and the real world, Shirley et al. undertake a study to use an expert system for fault detection and analysis on a chemical process. The expert system is a computer program which uses knowledge about some field to solve problems in that field, problems that are difficult enough to require an expert, or at least a significant amount of expertise. This is accomplished by encoding book knowledge and rules of thumb into a knowledge base, building an inference engine to use that knowledge base, adding an explanation capability, surrounding it all with an easy and intuitive man-machine interface, and using the system to solve or assist in fault management problems [SHIRLEY, FORTIN, 1985]. The expert system has been named Falcon Analyses Consultant, FALCON.

The use of expert systems in process control can be considered as an effort to model the knowledge of the expert-operator in such a way that other operators can use it. In fact, one makes this knowledge accessible to other operators with the consequence that part of the overall task of the human supervisor can degrade to well-defined or at least well-structured tasks. In this way one can appeal to the operator's RBB. Although an expert system is completely dependent on the quality of the knowledge base and the inference engine, one might expect that they will become interesting, new developments.

Finally an example not related at all to process control, but certainly of interest: The use of intelligent predictive decision consulting aids in the field of rehabilitation of quadriplegics. The treatment of quadriplegics is a highly complex, multidisciplinary decision making process, where medical, paramedical, psycho-social and technical disciplines are involved. A treatment team planns a particular treatment on the basis of many observations of the state of the patient taking into account the goals to be achieved. This treatment planning is a mixture of using knowledge, estimating the effects of a possible treatment, estimating the future state of the patient and making final decisions. In trying to understand this planning process, an attempt to model it was made. To a certain extent the results were indicating a possible solution. Somewhere some kind of standard tratment was given to the patients, although the results obtained varied in terms of treatment quality and length of duration of the clinical treatment. Three major short comings were recognized: (i) The, during the treatment time varying set of goals which was not similar for the different team members, (ii) the IR of the team members of the rehabilitation process itself, and (iii) the prediction of the future state of the patient. Furthermore, it was strongly believed by all team members that the final decision making process was such an intelligent activity, where only KBB was involved, that AI concepts could not be used.

Hence, on the basis of this modelling approach [STASSEN,VAN LUNTEREN, HOOGEN-DOORN, ET AL., 1980], an intelligent treatment planning support system could be developed. In fact, the three shortcomings were compensated by the following support systems: (i) A program to verify the opinions on the goals to be achieved by the team members individually, (ii) a treatment optimisation program based on a model of the rehabilitation process, and finally (iii) a prognosis model that for a reduced state vector of the patient predicts future states (Fig.5). Implementation of these programs into a treatment planning support system turned out to be successful. Again, a part of the tasks where KBB was involved was shifted to rule based activities, leaving the final medical decision making processes to those who are responsible. Similar support systems for process control can be developed along the same way.

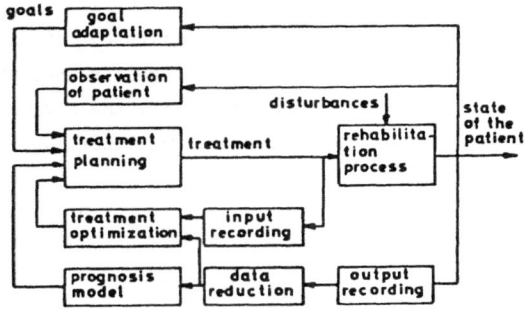

Fig.5: Diagram of the rehabilitation process of a patient with a lesion of the spinal cord

5. Concluding remarks

To summarize very shortly: Modelling is useful and worthwhile for SBB; it is occasionally possible for RBB, in particular if no decision making is involved; and it is not possible to model KBB otherwise than in a global or normative way. In fact, the final conclusion is close to one of the conclusions of a workshop on Research and Modelling of Supervisory Control Behavior [SHERIDAN, HENNESSY, 1984]: "No single or simple model of supervisory control is appropriate at this time. Various models have emerged and are useful as paradigms for analysis and experiments. More sophisticated models will be in demand to guide research and design in the future. Supervisory control, while reducing the human operator's participation as a manual controller, depends on human decision-making skills and is vulnerable to error in that decision making".

As a final remark one might add to this statement. Intelligent human operator support systems can take over parts of the difficult job of the human supervisor; it reduces often KB-activities to SB- or RB-activities. How intelligent support systems should be, is not to estimate at this moment, but the variety of tasks in the different process control modes might lead to conflicting demands and requirements.

6. Acknowledgements

Part of the research reported in this paper is carried out at the Delft University of Technology. This research was sponsored by a number of organisations and industries, i.e. the Dutch Organisation for the Advancement of Pure Research (ZWO), the TNO Institute for Mechanical Constructions, and the Foxboro Company, Mass., USA.

MODELLING HUMANS AND MACHINES

Giuseppe Mancini
Commission of the European Communities
Joint Research Centre – Ispra Establishment
Ispra (Varese), Italy

INTRODUCTION

The study of Human-Machine Interaction is approached here with the aim of
discussing the needs of an analyst for the correct design of a complex system. The two terms which we have introduced above merit some further definitions:

- system is in our terminology the set of all those features which are
  needed for the performance during the time of a certain function. These
  features include machines, humans, interfaces, computers, procedures,
  etc.;
- complexity is here referred to the many dependencies which intentionally
  or not are built within the various items of the system.

Reference, although sometimes not explicitly, will be made in the course
of the paper to the case of nuclear reactors in which a great number of
functional systems are engineered (more than 100); within each of these
systems variables are tightly coupled to each other, often through feedback
loops. Because of these couplings, many correlations exist between the
values of variables, not merely of statistical nature, but also in a causal
sense, so that it is particularly difficult to analyse the propagation in
time and "space" of disturbances generated in some value of a variable
(analysis of transients, incidents, accidents). Variables are a function
of time and are dependent at any moment on the present status and on the
previous actions of the various system features, either human or machine.

It is indeed important to approach the system design from this global viewpoint if we wish to properly take into account all the interactions between the human and the machine in the control of a process and/or in the
management of transients and accidents which may occur.

The design of the system will, therefore, include also the definition of:
- the automatic control system tasks;

NATO ASI Series, Vol. F21
Intelligent Decision Support in Process Environments
Edited by E. Hollnagel et al.
© Springer-Verlag Berlin Heidelberg 1986

- the interfaces between the control system and the operator in terms of panels, displays, operational procedures, etc.;
- the role of computers either as a help in the control of machines and as a more "intelligent" help in the strategic setting of the process variables;
- the role of the operator from monitoring to supervisory control tasks;
- the optimal allocation of duties between operator and intelligent computers during accident conditions, etc.

In the following we will first describe the methodology (DYLAM) which has been set up at the JRC-Ispra for the analysis of the dynamics of complex systems in which machine and operator tasks are modelled either in nominal conditions or in failure states. We will then focus on the aspects of operator models, discussing their application in DYLAM. Two examples of the methodology in respect to operator behaviour will also be shortly summarized.

THE SYSTEM MODEL

The DYLAM methodology, which is used as the approach for system modelling, originates as an improvement to existing system reliability (such as fault trees, event trees) techniques, in order to achieve a more complete and synthetic way of system representation which could account, at the same time, for physics of the process and system performance (Reina, Squellati, 1978; Amendola, Reina, 1981; Amendola, Reina, 1984). Indeed, either fault trees or event tree methods do not satisfactorily take into account the dynamic aspects of the random interaction between the physics of the transients, represented by system variables such as pressure, flows, temperatures, and the logic of the systems, represented by the state of its components. The current probabilistic techniques separate the system reliability analysis from the dynamic development of the accident: the accident in reality develops according to the values of the system variables which trigger the intervention of protection and control systems and/or operators, and to the occurrence of logical events such as lack or delays of intervention of demanded systems, failure or delays of operators in recognition of events, misdiagnosis, partial component failures, etc., which occur at random time or are linked by complex causal relationships. An adequate methodology must account for the possibility that a given initiating event

a. <u>INPUT-OUTPUT SCHEME</u>

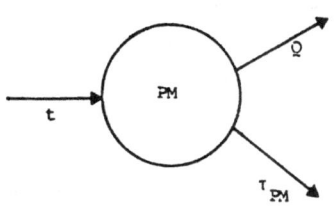

t = transient current time

$Q(t)$ = primary sodium flow-rate

τ_{PM} = pump coast down decay constant

b. <u>NOMINAL BEHAVIOUR</u>

$Q = Q_o$ = const. = 6400 kg/s

c. <u>FMEA</u>

Failure Modes	Event Variables
- Pump coast down with 3 different values of τ_{PM}	SPM: 0 Nominal Condition ... $\tau_{PM} = \infty$ 1 $\tau_{PM} = \tau_1 = 5.71$ s 2 $\tau_{PM} = \tau_2 = 7.145$ 3 $\tau_{PM} = \tau_3 = 8.70$ s

Component States	Event Variables SPM	Effects τ_{PM}
S1	0	∞
S2	1	τ_1
S3	2	τ_2
S4	3	τ_3

d. <u>PARAMETRIC OPERATORS</u>

$OPM = OPM (SPM; \infty, \tau_1, \tau_2, \tau_3)$

with the meaning "choose the ith value of the parameter τ_{PM} according to the integer value i of the event variable SPM"

e. <u>FINAL LOGICAL-ANALYTICAL MODEL</u>

$\tau_{PM} = OMP$

$Q = Q_o e^{-1/\tau_{PM}}$

Fig. 1 - Pump model.

will trigger in its temporal progression new logical events which, in turn, may act on the physical evolution of the accident in a continuous dynamic interactive process. Moreover, such methodology must take into account the statistical spread of physical parameters relevant to the transient beha- viour. It is clear that by providing this framework, the analysis of the system may accommodate a dynamic modelling of the operator.

The system is described by the set of the models of its components. The models are obtained by a quantitative failure mode and effect analysis of the components resulting in analytical relationships characteristic of each component state (nominal, failed, degraded). The set of analytical relation- ships, characteristic of a component, is controlled by parametric logical operators which select the appropriate relation for the component behaviour under each state; the component states are introduced by means of discrete event variables. Relations constituted by non-linear functions are possible, such as for instance step functions, simulating on-off response of compo- nents. In Fig. 1 the model of a pump is reported as an example.

The system behaviour is thus synthetically described by a set of parametric equations which contains both logical and physical information. The appro- priate relations are selected through the parametric operators according to a generation procedure of event sequences. Such a procedure is guided by some rules in order to restrict the investigation of the system beha- viour to a prefixed level of resolution (for instance by rules concerning the probability of events to be scrutinized or the maximum number of events in the incident sequence). The physical consequences of a generated event sequence are obtained by the numerical solution of the equation set: in this way all the system variables are evaluated during the transient time while simulating incident occurrences by introducing the possible failure events. These variables constitute usually the input for the human operator actions, thus various operation tasks (with related failures) could be equally modelled and inserted in the incident simulation, as it will be shown in the next paragraph. Eventually, the probability associated with the various sequences is calculated.

The solution of the equation set will be bounded by the definition of the "TOP" condition which the analyst wishes to investigate; in the case of DYLAM, this TOP condition can also be described by a numerically quantified value of a system variable and not only by a logical on—off event as in the usual fault tree or event tree approach (for instance primary coolant temperature greater than ... after ... seconds from incident start). In classical fault tree and event tree approaches, indeed, quantification is obtained only by separate calculations of the consequences corresponding to a particular logical TOP event; furthermore, different fault trees must be constructed each time a different TOP event is to be analysed.

By the DYLAM methodology one is able, by changing only the computer instructions related to the TOP conditions, to analyse a large variety of incidental situations either from a logical point of view or by weighting the sequence importance with the corresponding probability. In fact, DYLAM, being born as a substitute and an improvement of existing reliability techniques, is showing its versatility especially as the driver of simulation programs of system behaviour; DYLAM constitutes, indeed, a conceptual framework for representing systems which consist of discrete state components, continuous system variables, operational task elements and interaction between them; it also provides the analytical tools for the analysis of such a system.

By exploiting these possibilities, the development of a System Response Analyser (SRA) is under way at the JRC for the simulation of accidental event sequences in Pressurized Water Reactors (Cacciabue, Amendola, Mancini, 1985); for this application the complex physics of PWR accidents is being simulated by implementing in DYLAM a fast version of a reactor transient computer code (ALMOD) (Cacciabue, Lisanti, Tozzi, 1984).

As a final remark on the DYLAM methodology, we would like to stress two aspects which are of particular importance when dealing with operator modelling:
- the system description is based on the functional description of its components with no limitation as to the level of their detail, thus allowing more precise description of the part of the system under study;
- there is ample possibility of structuring dependencies either between the random variables controlling the state of the components or between these variables and the system variables (pressure, temperature, flow, etc.).

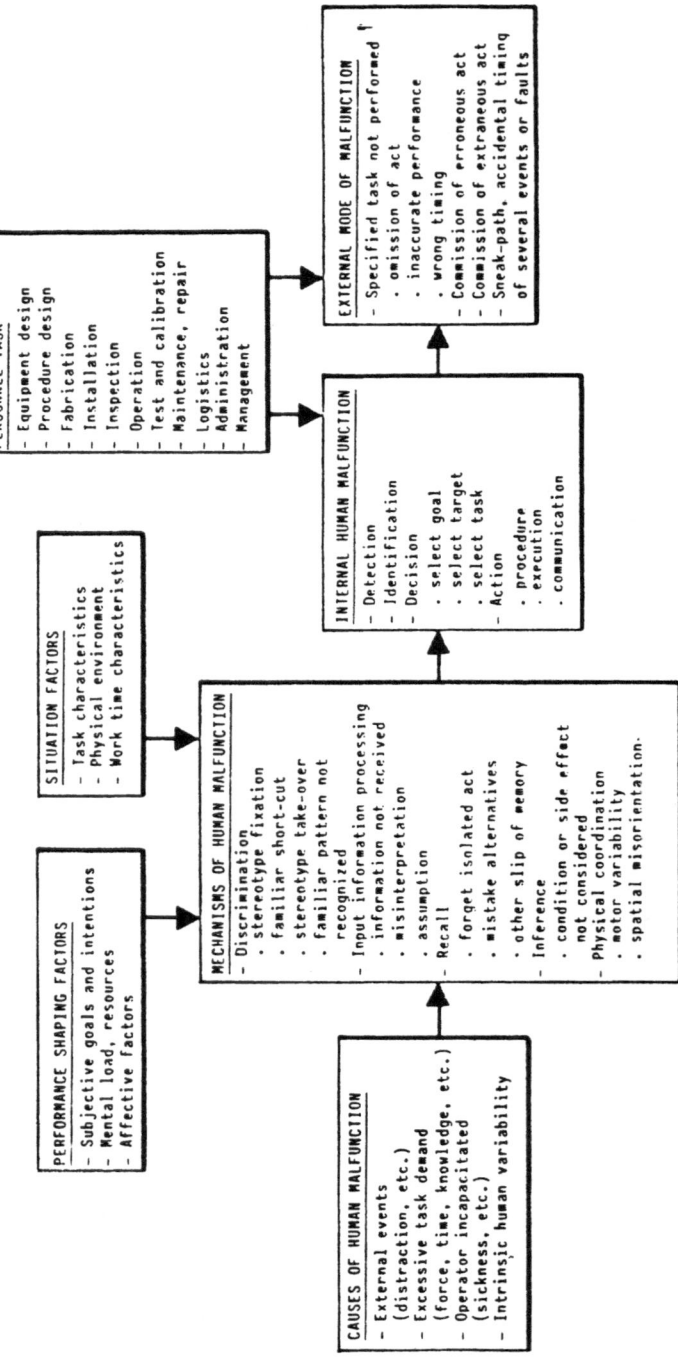

Fig. 2 - Multifacetted taxonomy for description and analysis of events involving human malfunction.

HUMAN MODELLING

As we have seen in the previous paragraph, in order to properly account for
the dynamic interaction of human and machine, each component of the system
is described through models derived, in case of machines from failure mode
and effect analysis. When considering human behaviour, this analysis has to
be directed to the task the human is called to perform: we are concerned
here mainly with tasks of operation of process plants. In Fig. 2 part of
the taxonomy related to human malfunction as developed by Rasmussen,
Mancini et al. (1981) has been reported; it is evident that for simple situations and specially when attempting to estimate reliability of specific
tasks, a failure mode and effect analysis based on the "external mode of
malfunction" classification (errors of omission, commission, etc.) may
satisfy the analyst's needs: simple models can thus be constructed similarly to what is done in PRA (Probabilistic Risk Assessment), where use of
THERP (Swain, Guttmann, 1983) methodology if often made. This simple model
approach reveals its inadequacy anyway in many real situations such as:
- errors dependent on some deliberate decision based on a specific understanding of the plant and of its state, and therefore on a prethought
  strategy;
- extraneous interference acts due to misinterpretation of protective
  automatic device functioning (deactivation of alarms, safety systems, etc.);
- more generally, all those situations in which human decision making is
  needed leading, therefore, to a multiplicity of possible outcomes.

In fact, in these cases, errors are no more stochastic in nature but they
are the result of causal processes of decision making; it is important to
try and model such processes to properly account for all the dependencies
in the system analysis.

Operator Task Model

A first step in this direction is represented by the modellization of human
behaviour no longer based on the "external mode of malfunction" attributes,
but on the "internal human malfunction" ones (Fig. 2). This classification
attempts, indeed, to identify internal, mental data processing elements
or decision functions used to characterize steps in decision sequences.
Modelling in terms of identification, decision and execution can be done
at several levels of detail of the task description. In Table 1, for example,

an overall operator parametric model is presented, developed by Amendola, Mancini et al. (1982) and Cacciabue, Cojazzi (1985) for the study with the DYLAM methodology of the Auxiliary Feedwater System (AFS) of a real reactor. The operator is modelled, in each of the tasks as identified in the "internal human malfunction" classification, according to a certain number of "states", with a similar procedure as followed when constructing the pump model of Fig. 1.

TABLE 1 - Operator model

Component operator	State	Meaning	Effects
Failure detection	0	nominal	detection after 30' = τ_n
	1	delay	detection after $\Delta t_d + \tau_n$
	2	fails	no detection at all
Failure diagnosis	0	nominal	identify SG1
	1	erroneous	identify SG2
Compensation	0	nominal	regulation according to procedures: - to isolate identified SG; - 50 t/h to 3 sane SGs or 100 t/h to 2 sane SGs; - max mass flow = 100 t/h; - do not allow for flows < 0, check at $t_c = \tau_c$
Recovery	0	nominal	recovery after $\Delta t_{rec} = \tau_r$ from erroneous diagnosis

The incident studied consisted of the transient following a Steam Generator (SG) feedline break at the junction of the AFS and the feedwater flow control system. The objective of the analysis was the identification of the conditions leading to the failure of the system in satisfactorily performing its duty, that is: providing sufficient mass flow to the sane steam generators after the break of the feedwater line of SG1. In order to ensure sufficient cooling the assumed minimum flow conditions were: 50 t/h to three good SGs or 100 t/h to two good SGs.

The operator model of Table 1 represents a "high" level description of the operator task, that is: few details of the task description have been accounted for and only global decision elements have been considered. The failure detection is assumed to be accomplished after 30' from incident initiation or after a variable time delay which becomes infinity for failure to detect

the incident. The diagnosis also presents two possible states: a nominal
one which implies the correct identification of the defective SG, and an
erroneous one by which SG2 is isolated instead of SG1. No time delay is
assumed between failure detection and diagnosis. Compensation and recovery
are assumed, in this case, to be always performed correctly; anyway, time
delays for both of these steps have not been fixed in order to allow for
parametrical studies.

Although the above described model is rather simple and approximate, its
use in the DYLAM analysis framework has permitted several interesting
results to be obtained. In Fig. 3 the relation found between operator
time of intervention and delay of regulation for ensuring sufficient feedwater all along the transient is shown.

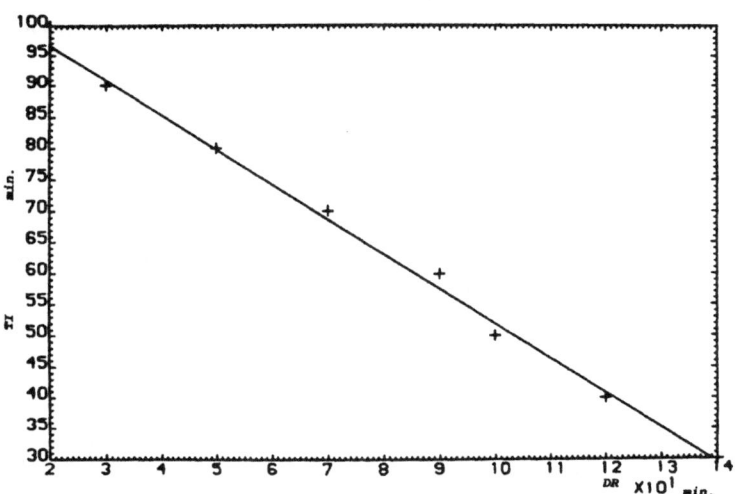

Fig. 3 - Critical operator time of intervention (TI) vs delay of regulation (DR).

Operator Cognitive Models

The model described above attempts to characterize some of the decision
steps of the operator, but it does not give the complete answer to the
need of identifying the thought processes of humans: indeed, how it is possible to quantify the delay times in detection and diagnosis, which is the
probability of a correct diagnosis, which other diagnoses could be possible,

which wrong compensation actions could be started, how recovery could be attempted when not necessary, etc. In order to approach these questions, one has to model the "mechanism of human malfunctions" (Fig. 2), that is: one has to model the cognitive processes of humans.

Very likely, one single model will not alone cover the complexity of the various cognitive processes. A first step in this direction has, anyway, been attempted at the JRC-Ispra by implementation of a fuzzy logic model of diagnosis (Amendola, Reina, Ciceri, 1985): in such a model, decisions have been simulated which are caused by verisimilitudes in actual situations; these verisimilitudes are reflected in proper membership functions. This is the typical case where decisions are taken in a state of imprecision.

In particular, it has been supposed that diagnosis of an event monitored by sensors of related variables, is performed based on the judgement of the operator on:
- the "high" level of the sensor indications;
- their "correctness" and "regularity";
- the "coherence" of the different signals with respect to each other and with respect to expected scenarios.

For instance, the judgement "high" (or "low") is simulated by means of a membership function $S(x,a,b)$ proposed by Zadeh (1974):

$$S(x,a,b) = \begin{cases} 0 & \text{if } x < a \\ 2\left(\dfrac{x-a}{b-a}\right)^2 & \text{if } a \leq x \leq \dfrac{a+b}{2} \\ 1-2\left(\dfrac{x-b}{b-a}\right)^2 & \text{if } \dfrac{a+b}{2} \leq x \leq b \\ 1 & \text{if } b \leq x \end{cases}$$

Another membership function, R, is introduced to simulate the opinion about the "regularity" of the sensor being monitored: an oscillating index makes the operator suspicious about the quality of the sensor. A combination according to fuzzy logic rules of S and R results in the overall judgement about the credibility of "high" or "low" value for variables considered separately from each other.

The "coherence between" the indications of two sensors is simulated through
a compound coherence index defined by

$$\text{Max} \; [\text{Min}(S_1, 1-S_2), \; \text{Min}(1-S_1, S_2)]$$

where S_1 and S_2 are the two membership functions expressing the judgements
about the two sensors. According to the resulting value of the index it
is then established whether ambiguity exists in the interpretation of the
two indications.

The model has been used in the simulation performed with DYLAM of a Loss
of Coolant Accident in a PWR; in particular, the diagnosis in object referred to the identification of the break location in the reactor.

Development in Cognitive Models

Other models are, at the moment, being considered for future inclusion in
the DYLAM frame: they will be shortly reviewed in the following.

Filtering models

In the monitoring behaviour field, normally dominated by vision, simple
models derived by Shannon's Mathematical Theory of Communication and based
on data on eye movements and fixations (Senders, 1983), while deserving
the task of designing appropriate displays and of optimizing the information content of the displays, represent an unnecessary complication if
introduced in simulation frames like DYLAM; indeed, they do not entail any
decision process from the operator and they cannot, therefore, constitute
valid means for describing possible dependencies in human behaviour.
Detection of failure models through the use of the Optimal Control Theory
(OCT) seem, on the contrary, to be more apt to be used in our simulation
environment. Wewerinke in his model (1981) assumes, for instance, that, to
detect a failure occurrence, the operator updates continuously by means of
a Kalman filter his estimation of the mean and variance of system variables.
These best estimates are used to compute the likelihood ratio between the
probabilities of the estimates given that the process has not changed, and
given that the failure has occurred. Such ratio will drive the decision
on the failure occurrence.

Two important aspects can be found in such a model which deserve attention:
- the operator acts in a closed feedback loop as an information processing element, filtering noises;
- the operator is supposed to have an "internal model" of the process (in this case of the system variables being monitored), enabling him to predict future system responses and compare them with observations.

These characteristics will allow not only the modelling of "machine" failure detection but also the testing of the correctness of the operator "internal model", offering thus the possibility of modelling human error recovery and learning.

The assumption of the existence, or better of the progressively building up of an "internal model" would also lead to the recognition that a practised operator would act, in respect of monitoring tasks, as an unconcious actor, that is: in a skilled-based mode of operation, once he has rationally decided a strategy of overall behaviour.

Other models of monitoring behaviour which have also been proposed in the literature, will not be mentioned here as they are more normative in nature and they do not fit, therefore, with the goals of our simulation framework.

Very much linked to the problem of monitoring behaviour is the problem of diagnosis of a failure or of a system state which presupposes the detection of a signal and its interpretation. Models based on fuzzy set theory such as the one previously described, seem to approach in a satisfactory way the situation. Possible alternative models have been suggested, based on pattern matching (Gonzales, Houngton, 1977).

When shifting towards the "superior" operator task of Supervisory Control, one has to acknowledge the increasing role of decision making in the models: indeed, Knowledge Based Behaviour becomes a predominant mode of operation at least in deciding which Rule Based Behaviour to invoke (i.e. rules previously taught, rules contained in procedure manuals, heuristic guidelines). Contemporaneously the closed loop simulation of the Optimal Control Theory type is replaced by the open loop of the Optimal Estimation Theory in which the output from the Kalman filter is not used to optimally change, in the feedback loop, the state variables, but it is processed for generating a decision.

Artificial Intelligence models

Again it must be underlined that the "internal" model of the operator is acquiring an ever increasing importance. With the aim to capture this model, attempts have been initiated to exploit the task and protocol analysis used by cognitive psychologists and implemented in Artificial Intelligence (AI) formalism: operator rules, whether taught or described in procedure manuals, or under the form of heuristics, have been transformed up to now into production rules of AI type computerized models (Baron, Muralidharan, Lancraft, 1980). These models may constitute an alternative to the operator model as presented in Table 1 and could be easily inserted in the DYLAM frame if advantages are proved (i.e. new procedure oriented models, easier programming and updating, easier error simulation). On the other hand, it is hoped that the elicitation of the operator knowledge may eventually lead to the set-up of a more "deep" representation of the thought processes of the operator: the "internal" operator model. Such a model would then serve as a common reference which will be used either in premodelling analysis or in parallel to the DYLAM execution for feeding each singular operator decision model with background information. Only by implementing such operator model dependencies between the possible operator decisions and behaviours will be correctly accounted for.

Obviously, if one wants the model to become manageable, situation-independent, that is: able to account for situations not previously established, and true to the operator, one would have to represent the knowledge in some optimal way which takes into account the peculiarities of the "operator" knowledge acquisition.

Indeed, the operator builds progressively specific bodies of knowledge by experience and mainly through analogical reasoning (de Montmollin, De Keyser, 1985). The operator knowledge is thus structured in parts which may justapose each other, often heterogeneous, corresponding to specific situations linked together by analogies. It is important to recognize that, on the contrary, the knowledge of an "expert" (i.e. designer) is completely different: it has been acquired during many years of University, organized in some nested and hierarchical structure, with strong power of generalization. (This knowledge representation is, for instance, optimally used in "expert systems" for large data bases interrogation (Bastin, Capobianchi, Mancini et al., 1985).)

We have, therefore, to be careful not to construct an operator model based on "expert" knowledge. "Expert systems" which are at the moment thought for aiding the operator, do not need to represent the operator knowledge but, on the contrary, they have to complement it in those aspects where it is more fragile (i.e. when solving new problems by analogy). Operator models, on the other hand, have to represent the knowledge structure and modes of reasoning of the operator, even and specially if these are fragile and erroneous.

Although AI is a very promising and exciting field, many improvements have still to be achieved before a correct application becomes possible; we will mention below only those which are more relevant to our needs (Feehrer, Baron, 1985):
- deep conceptual representation for plants, functions and processes related to the particular activity;
- integration of mathematical and quantitative knowledge with qualitative knowledge (i.e. qualitative modelling (Forbus, 1982));
- more flexible control structures than simple forward and backward chaining;
- techniques for dealing with incomplete, inconsistent and uncertain knowledge;
- techniques for representing and reasoning with temporal knowledge (Volta, 1985).

Eventually, it is important to underline that, despite all these efforts and improvements, we will be able perhaps to model modes of reasoning pertaining to the operational problem-solving area such as deduction, induction, analogy, etc.; what will remain outside the capability of models will still be the representation of cognitive strategies or, more generally, of the problem space.

Learning

Before leaving the subject of human models, it is, maybe, worthwhile to outline some aspects inherent to the learning process or, more in particular, to the possibility of recovery from error. Rarely in the literature models of recovery are explicitly mentioned as they are more or less connected with the usual models of detection, diagnosis and action. Indeed, the representation of an "internal model" of the operator not only brings the possibility of detecting and diagnosing external failures which do not comply with the model (OCT model of Wewerinke), but also gives the opportunity to test the correctness of the plant/function/process internal representation against some adverse or new evidence. Thus the construction or the retrieval of another internal model could be necessary, triggering therefore a process of learning.

It is important to note that this process may have consequences in anterior times as well as in posterior times (Aristotle IVCbC): in anterior times because, the previous ignorance being recognized, all anterior facts, actions and decisions of the operator will be re-assessed by the operator himself according to the evidence of the new model; in posterior times because the new knowledge and the actions and decisions taken accordingly, will allow the operator to predict future outcomes. This notion of mechanicistic "reversible" time regulates the cognitive processes of the operator and continuously interacts, although remaining separate, with the notion of "irreversible" statistical real time. These two time scales, one subjective and the other objective, will drive the modelling of the operator and of the physics, respectively: at various "real" times, possibility will be given to run the "human" models over time lapses covering anterior as well as posterior times.

CONCLUSIONS

After having reviewed the necessity of modelling human and machine and their interaction together in order to optimally design a sound "interface", the frame of such analysis has been described in some detail. In this respect the DYLAM approach has been mentioned with its enhanced capabilities to treat contemporaneously discrete and continuous variables (state variables and system variables such as pressure, temperature, etc.) in a dynamic simulation mode. The possibilities to model multistate components and dependencies among components have also been outlined.

Human models have been reviewed in relation to their possible use within the DYLAM frame: from an overall parametric operator model to a fuzzy set logic diagnosis model (both already implemented), to applications of the Optimal Control Theory models for monitoring, diagnosis and supervisory control tasks. It is evident that the common denominator of all these approaches is constituted by the description of the "internal model" which the operator has of the plant/systems/functions. This is indeed the necessary model which will govern the simulation of diagnosis, prognosis and control of the event.

Particular interest has, therefore, been dedicated to the early tentatives of using Aritificial Intelligence formalism and techniques to encode the cognitive processes of the operator with all the limitations due either to the still immature AI techniques and to the problem of modelling the true operator knowledge structure. Eventually, some thoughts have been expressed on the characteristics of time in the operator learning processes or, generally, in the human cognitive processes.

One area which has not been addressed at all in the paper is that of data: indeed, data are essential for every step of the modelling scenario we have depicted, whether they refer to machine behaviour, human behaviour, or their interaction. We would, herewith, only underline the importance of testing the results of the described system analysis with historical data. Indeed, models are always a simplification of nature and especially when describing complex systems, the structure of dependencies may be overlooked and oversimplified by the analyst.

Some experiments have been run at JRC-Ispra in which accident sequences generated through the DYLAM methodology have been compared with the past history of accidents and failures in nuclear power plants stored in our event data banks of the ERDS (European Reliability Data System). Although no generated sequence occurred in the exact mode as in the past, several occurrences could be singled out which covered most of the steps of some original sequences, permitting, therefore, to give a relative weight to these. But, the analysis of the retrieved sequences outlined also some mechanisms of failure and some human behaviours which were not properly modelled or accounted for in the simulation scenario: this feedback of the past experience on the model is of utmost importance if we wish to build models which can be continuously updated and used; conversely, the use of the models is needed to structure and unify the vast amount of information available.

ACKNOWLEDGEMENTS

This paper contains some of the results of several years of research performed at the CEC JRC-Ispra in the Division of System Engineering and Reliability. In particular, the author is greatly indebted to Mr. A. Amendola for the development of the DYLAM approach, and for the elaboration of the fuzzy set model; to Mr. P. Cacciabue for the application of the DYLAM to the modelling of the Auxiliary Feedwater System. The author wishes to thank also Mr. N. Moray for having provided an excellent survey of the existing human models from which several information and suggestions have been taken.

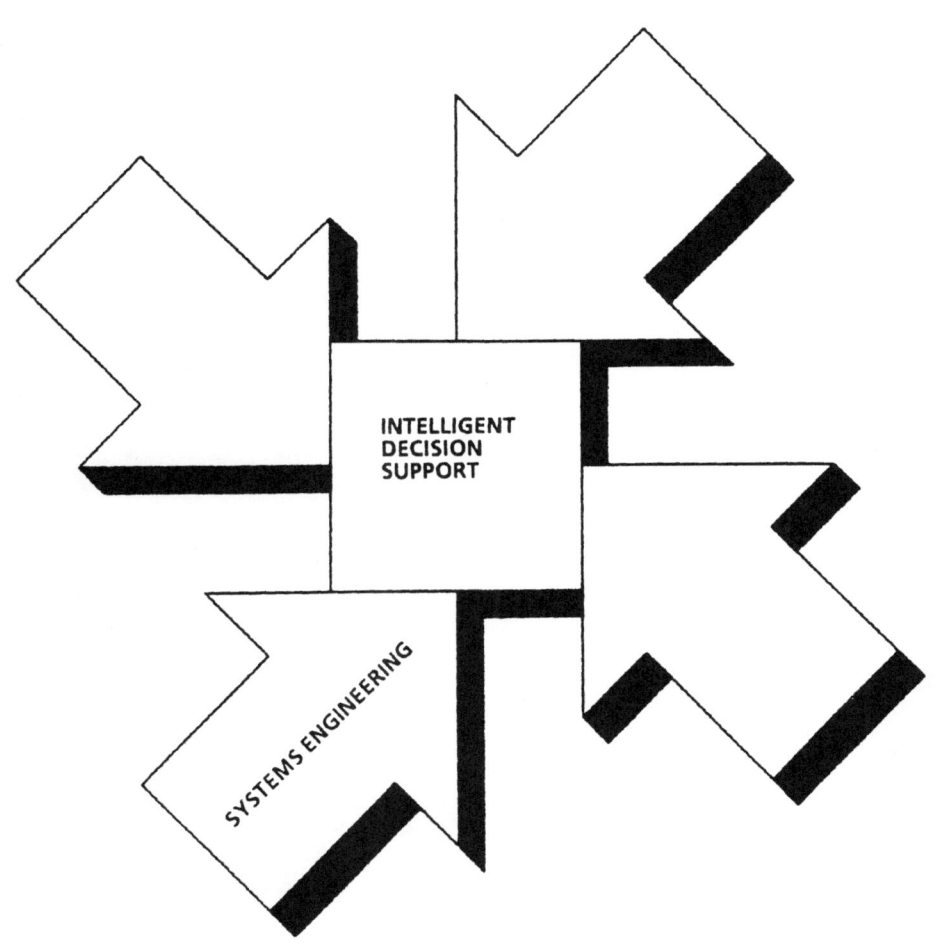

INTELLIGENT DECISION SUPPORT

SYSTEMS ENGINEERING

SYSTEM DESIGN & APPLICATIONS

ARCHITECTURE OF MAN – MACHINE DECISION MAKING SYSTEMS

Gunnar Johannsen
Laboratory for Man – Machine Systems (FB 15)
University of Kassel (GhK)
Kassel, West Germany

INTRODUCTION

The main reason why humans are involved in controlling and supervising process environments of all kinds is the fact that humans are
and have to be ultimately responsible for decision making in critical situations. This has to be considered in the design of technological systems and of their organizational environment. The
pure automation approach fails, at least in complex systems, because the interface to the human can only inappropriately be
added afterwards. The allocations of tasks and responsibilities
between different technological, human, and organizational system
components strongly influences the behaviour of the overall system under normal and emergency situations. This is particularly
important with respect to the decision making capabilities when
time pressure, uncertainties or risk have to be mastered.

The recent advances of computer technology have largely increased the number of alternatives for designing man-machine decision making systems. Computerized decision aids are advisable
in many process environments. Looking into real technological
systems shows, however, that not much has yet been done to develop and implement operational decision support systems. On
the other hand, many disciplines like management sciences, systems and control engineering, human factors and man-machine systems, cognitive sciences, as well as artificial intelligence
have produced a wealth of research ideas and solutions. Several
of these have led to interesting architectures of man-machine
decision making systems. Some of them will be discussed and
evaluated in this paper, particularly with emphasizing their

NATO ASI Series, Vol. F21
Intelligent Decision Support in Process Environments
Edited by E. Hollnagel et al.
© Springer-Verlag Berlin Heidelberg 1986

potential for operational use.

Decision making tasks in process control and supervision can
thus be accomplished in different ways by implementing architectures with different organizational structures, degrees of
decision support, levels of interaction, and adaptability. Architecture will be defined here along three dimensions, namely,
with respect to
(1) a hierarchy of decision making situations,
(2) the stages of decision making tasks, and
(3) the principles of human-process interactions in decision making.
These three dimensions will be discussed seperately in subsequent sections of the paper.

HIERARCHY OF DECISION MAKING SITUATIONS

Decision making is required on all levels of social life and
in all work situations. This is independent of whether many or
few people are involved and whether humans interact with technological systems or just with each other. The decision making
situations range from policy making through urban and environmental planning, economic management, process plant and factory
supervision to process and vehicle control. These application
areas are only examples for the broad spectrum.

The continuum of decision making situations can be structured
into a hierarchy of (1) organizational, (2) macro-operational,
and (3) micro-operational situations. The time horizon, the number of people involved, and the degree of complexity decrease
from the first to the latter whereas the interaction with technological systems becomes tighter and the level of detail in
the decision process increases in the same direction from the
more general to the specific situations.

The last two levels of the hierarchy are comparable with the
two principal activities of problem solving and controlling

which were integrated into a qualitative model described by
Johannsen, Rijnsdorp, and Sage (1983). The organizational level
of the hierarchy will also be dealt with in the following because
some issues are directly related to process environments.

Organizational situations prevail in such fields as policy, trans-
portation and environmental systems, business administration, and
health care. Management information systems (MIS) and decision
support systems (DSS) have been developed for aiding the humans
in these applications. Sprague and Carlson (1982) assume that the
upcoming improvements in computer technology will make DSS feasible in real world applications and that DSS should be a major
new business area for computer technology.

So far, this area has been a very active research field. Sage
(1981) gave an extensive overview on behavioural and organizational considerations with citing 415 references. Cognitive styles
of decision makers need to be incorporated in the design of DSS.
Individual human information processing in decision situations
and biases in the acquisition, analysis, and interpretation of
information are discussed. A hierarchical structure of decision
rules for individual decision situations as well as contingency
task structural models are explained. The latter describe the
decision making process in terms of contingency elements associated with the environment and the decision maker's prior experiences. Different relevant features of task evaluation, information processing preference, decision rule selection, and
degree of stress are considered. Decision making frameworks, organizational settings, and information processing in group and
organizational decision situations are further factors of the
decision making process. The needed information has to be effectively summarized and structured in order to arrive at appropriate systems designs for aiding groups in decision making.

Laboratory studies were performed by Belardo, Karwan, and Wallace
(1984) in order to investigate the possibilities of DSS in emergency management of rare catastrophic events such as the Three
Mile Island incident. Each of the four groups participating in

the experiment was composed of eight subjects who had to play different roles in a team of emergency preparedness officers. The results indicate a clear support for a computerized DSS which was designed exclusively to enhance data display.

Another type of decision aids using preprocessors was proposed by Chyen and Levis (1985) for command, control, and communication (C^3) systems. The processor is located between the information source and the decision maker. The workload of the individual decision maker was reduced and the quality of the organization's decision making was improved.

An individualized message handling system was proposed by Madni, Samet, and Freedy (1982) for C^3 applications. It is based on an adaptive information selector. The adaptability with respect to each user considers individual processing characteristics and immediate decision making needs. This architecture partially belongs also to the next level of the hierarchy.

An overall aspect of all the described organizational situations seems to be the necessity to provide the appropriate information in an effective manner for the human or the group of people involved in the decision making process.

The macro-operational situations are characterized by the need for decision making during such tasks as goal-setting, fault management, and planning in systems operations and maintenance of man-machine systems. One or a few humans are interacting with a technological system. Although organizational aspects may be involved, the operational demands are always predominant. Examples are power plants and aircraft in which such situations as start-up, shut-down, abnormality, and emergency have to be accomplished by the human supervisor. Some of the characteristics of supervisory control were described in the book edited by Sheridan and Johannsen (1976).

The online planning behaviour of aircraft pilots in normal, abnormal, and emergency situations was studied by Johannsen and

Rouse (1983). It includes decision making processes which are related to the different flight phases and subtasks selected for mental engagement in a time-driven and event-driven manner. Decision aids for the process of online planning have not yet been proposed. More insight into the human planning behaviour has first to be gained experimentally.

The information seeking process of human operators is better understood also on this level of the hierarchy. An onboard computer-based information system for aircraft was proposed by Rouse, Rouse, and Hammer (1982). The current and anticipated decision making situations in flight management are supported by means of a computer aid which keeps track of the current flight procedure including inferencing the status of procedural steps and branching among procedures. The results of an experimental evaluation of a prototype system are presented. It is shown that the frequency of human errors can substantially be lessened by such a computer-based information system.

In a recent study in which the author participated (Alty, Elzer, Holst, Johannsen, and Savory, 1985), the state-of-the-art and the potentials for human operator support in industrial supervision and control systems were investigated by a literature and user survey. Some operators complained that, sometimes, more functions are automatically locked than necessary for safety reasons (e.g., during start-up in a gas power plant), hampering an otherwise possible problem solution based on the operator's system knowledge. The literature review showed a wide range of new techniques that could be used to produce a major advance in graphics and knowledge based decision support for supervision and control. Some investigations aim at modelling fault management and, more generally, problem solving behaviour. They have to be intensified and combined with results and new efforts in specifically designed expert systems and graphical presentation techniques.

The micro-operational situations involve decision making as an ingredient of control processes, either manual or supervisory control, in man-machine systems. In manual control, micro-decisions have to be made by the human operator all the time, mostly subconsciously, in order to arrive at control actions which are appropriate in timing and amplitude. With slowly responding systems like most industrial plants and large ships, more time is available for this decision process and it will be handled more consciously resulting in a supervisory control mode. An example is given by Veldhuyzen and Stassen (1977) for large ships.

A dynamic decision model for predicting human task-selection performance in a multitask supervisory control environment was discussed by Pattipati, Kleinman, and Ephrath (1983). The model is a derivative of the optimal control modelling technology. It combines control, estimation, and semi-Markov decision-process theories. The results from general multi-task experiments were used for its validation. The multi-task situations included tasks of different value, time requirement, and deadline, all competing for the human's attention.

Another modelling approach describes the dynamic allocation of decision tasks between human and computer in automated production systems (Millot and Willaeys, 1985). The goal is to reinsert the human operator into the control and supervisory loop in order to achieve appropriate levels of workload under normal as well as abnormal conditions.

Generally, the models of this level of the hierarchy are suitable as a fundamental tool for the design of dynamic decision aids which can adapt to different task situations. However, this design process has often not yet been tackled.

Further aspects of supervisory control tasks and related human operator models are included in Stassen's paper of this Advanced Study Institute (Stassen, 1985).

STAGES OF DECISION MAKING TASKS

Complete decision making tasks can be structured into three
stages. This is true for all levels of the situational hierarchy
just described. The three stages are:
    (1) situation assessment,
    (2) planning and commitment, and
    (3) execution and monitoring,
as described by Rouse and Rouse, 1983.

Situation assessment deals with alternative information sources
and, afterwards, with alternative explanations of a particular
situation, its determining factors, and its variations. Both, information sources and explanations, have successively to be generated (or identified), evaluated, and selected. Information may
be available about past, present, and future states of the overall process environment. Some human operators would like to be
better supported with predicted information about systems states
which presents expectancies of the near future (Alty et al., 1985).
A certain situation is assumed to have been appropriately assessed
when the selections among information sources and among the explanations are terminated.

The next stage of decision making tasks, i.e., planning and commitment, concerns generating, evaluating, and selecting among alternative courses of action. Planning is mainly a future-oriented
activity resulting in mentally produced, anticipated sequences of
actions. These plans may be retrieved or inferred from stored procedures and experience with analogous situations or may be newly
developed by creativity. The consequences of alternative plans
have to be imagined and evaluated leading to the selection of the
most suitable plan.

Finally, execution and monitoring includes implementing the plan,
observing the consequences, evaluating the deviations from expectations, and selecting between acceptance and rejection. In
case of unacceptable deviations, an iteration of the whole decision making process may be necessary starting again with iden-

tifying information sources or with generating explanations for
a better situation assessment. Returning to one of these two
phases may have also been required if no explanation of the present situation or no alternative course of action seem to be
appropriate.

As Rouse and Rouse (1983) point out, five generic subtasks are
involved in all three stages of decision making, namely, generation, evaluation, selection, input, and output. The first three
have repeatedly been mentioned during the above description of
the decision making stages. The latter two, input and output,
are common to all decision making tasks and stages, at least
when decision aids are implemented. They involve information displays, input devices, and dialogue structures.

Opportunities for aiding on all five subtask levels are also
listed and briefly explained by Rouse and Rouse (1983). Further
ideas for operator support were presented, e.g., by Alty et al.
(1985).

PRINCIPLES OF HUMAN - PROCESS INTERACTIONS IN DECISION MAKING

A single human decision maker or a team of decision makers can
either perform all the above subtasks or decision making stages
themselves or can be aided by a decision support system or expert system in some or all subtasks or can be replaced by automatic decision makers. Thus, one or several human decision makers
might form a team together with one or more computerized decision aids. Different organizational structures and interactions
between the team members are possible. Hierarchical and decentralized aspects, overlapping coordination, and leader-follower
relationship can be considered.

Decentralized as well as hierarchical aspects occur in supervision and control of any large-scale systems (Athans, 1978;
Johannsen, 1981). Decentralized systems are complex systems
which can be devided into subsystems with a certain degree of

autonomy. One has to distinguish between local and functional decentralization. Examples for local decentralization, i.e. often spatial distribution, are given in chemical plants, inter-connected power systems, and C^3 systems. The functional decentralization allows a separation into subsystems with different dominant time scales.

There are different methods available for designing control laws for large-scale systems. Among several others, game theoretical approaches have been proposed which assume that a separate decision maker with its own cost functional exists for each subsystem.

The decision makers are sometimes in a goal conflict with each other if couplings between individual subsystems are correspondingly strong. Especially for such decision-making situations, the game theoretical approach seems to be very appropriate. A coordinator or leader on a higher level in the decision structure seeks for a global optimum. Thus, a hierarchical organization of decision making exists. The lower-level decision makers pursue their individual goals beeing influenced indirectly, however, by the actions of the coordinator. This basic leader-follower strategy is also called the Stackelberg strategy (Zheng, Başar, and Cruz, 1984) which was originally adopted from mathematical economics.

The game theoretical approaches have not yet been applied with systems of cooperating human and automatic decision makers although no principal structural differences exist. In fact, many man-machine systems supervised by humans exhibit a hierarchical and decentralized structure with leader-follower relationships (Johannsen and Borys, 1985; Sheridan, 1984; Sheridan, 1986).

Figure 1 shows the architecture of a hierarchical, decentralized dynamic system. Sheridan (1984) introduced the terms human-interactive computer and task-interactive computer which correspond to the higher-level coordinator and the lower-level decision makers, respectively. The lower-level task-interactive com-

336

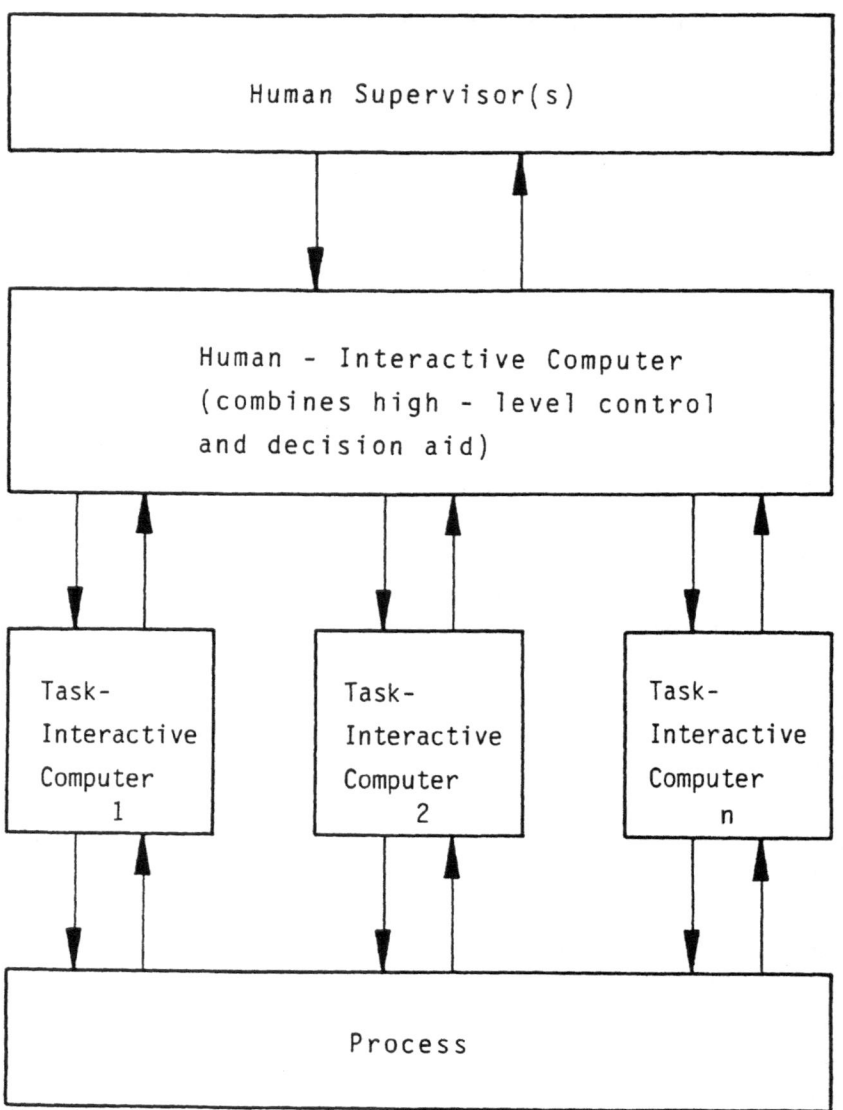

Fig. 1: Architecture of Hierarchical, Decentralized
        Dynamic System
        (after Sheridan, 1984, 1986)

puters control, again, the appropriate number of subprocesses or tasks. The advantages of this architecture are that almost all dynamic systems can be described and, if necessary, either the level of task-interactive computers or human-interactive computers or both can be omitted or expanded. Also, the number of human supervisors may be greater than one.

The three hierarchical levels of human supervisor(s), human-interactive computer, and task-interactive computers emphasize the above organizational, macro-operational, and micro-operational situations, respectively, to a certain extent.

Each decision support or expert system can work off-line to or online with the controlled process. It may be identical with or part of the human-interactive computer as shown in Figure 1. Alty et al. (1985) distinguished between the advisor model and the intelligent front end which produce different levels of interaction. The first of these two types of decision aids is a separate part of the human-interactive computer which is used by the human supervisor in a consultative mode. Using the intelligent front end, the human supervisor interfaces directly only to this decision aid, which in turn interfaces to the high-level control of the human-interactive computer. Therefore, particular real-time capabilities of such an expert system are required (see also Alty et al., 1985). The different architectures are dependent on the goals and structure of the controlled process and the chosen trade-off between speed and accuracy in decision aiding.

The internal architecture of any decision support or expert system consists of a dialog component, a knowledge base, and an inference engine. The knowledge base can be divided into a data subsystem and a model subsystem as usually done in decision support systems. The models in the knowledge base include models of the technological system (and the environment) as well as user models. The inference engine is more explicit in expert systems but is also represented in the management software of decision support systems (Wang and Courtney, 1984). The dialog component is designed as the user-system interface. Graphical and natural language dialogs may be implemented. The paper by Guida (1985) of this Advanced Study Institute explains the expert systems approach in more detail.

CONCLUSIONS

The three dimensions of decision making which have been explained in the preceding sections of this paper determine different architectures of man-machine decision making systems. The combination of these dimensions leads to a multitude of possible solutions adapted to any particular application. The attempt in the preceding sections to classify this multitude involves some course descriptions. The levels of the situational hierarchy are not as clear-cut as one may assume. It presents only a structure for the whole continuum of decision making situations.

The stages of the decision making tasks are perceived in a similar way by many researchers. The third dimension dealing with the principles of human-process interactions shows still progressive changes depending on new technologies and new application demands.

The architecture of any combination of the three dimensions can be designed on the basis of the structure shown in Figure 1. The principles of human-process interactions will then be determined by choosing the number of task-interactive computers, the allocation of control and decision making tasks between the three hierarchical levels of computers and humans, as well as the types of decision aid and information presentation.

The selection among these various design alternatives depends not only on the type of process or application but also on the two dimensions of the situational hierarchy and the decision making stages. It is very important, for example, whether all three stages of a decision making task or just the situation assessment have to be performed. This will have an influence on the required hardware and software as well.

In several power and industrial processes, a certain plant operator may be involved in micro-operational situations with a subsystem of the plant and, more seldomly together with other people,
also with organizational situations concerning the whole plant.
Nontrivial design decisions are questions like whether two different interfaces are needed for these two situations or which information presentation technique one should use.

In many application areas, the new technologies can be introduced
for operational use. They contribute to more effective systems as
well as higher job satisfaction of the human decision makers. As
general design rules are missing, an overview of the available
research as given in this paper can he a useful guide into the
many possible architectures for the practitioner.

DESIGNING AN INTELLIGENT INFORMATION SYSTEM INTERFACE

Nicholas J. Belkin
School of Communication, Information & Library Studies
Rutgers University
New Brunswik, New Jersey (USA)

INTRODUCTION

An information system, in our sense, arises when a person, with some goals and intentions, recognizes that her/his state of knowledge is inadequate for attaining the goal. To resolve this anomaly, the person goes to some information provision mechanism (IPM). The IPM, in general, consists of a knowledge resource and some intermediary access mechanism. Thus, the information system consists of a user, who instigates the system, a knowledge resource (KR), which contains the texts which might be relevant, and an intermediary mechanism (IM), which mediates between the user and the KR. Some typical systems of this type are in student advisory services, social security benefits offices and bibliographic retrieval systems. The system as described is certainly a joint cognitive system (Woods, this vol.; Hollnagel, this vol.), and the IPM has the basic characteristics of a classic decision support system.

At the present time, almost all information systems of this type must have a human intermediary in order to perform at all adequately. One goal of our research group is to design automated IMs which would accomplish the functions that good human intermediaries would in the same situations. Any such IM would perforce be an intelligent interface between user and KR.

DISTRIBUTED EXPERT PROBLEM TREATMENT

Our basic approach has been to focus on the functions which need to be performed in user-intermediary interaction, and on why they are performed. We have suggested one such specification (Belkin, Seeger & Wersig, 1983), which we have subsequently developed and used as a basis for a number of detailed investigations. It appears that these functions (figure 1) must be performed by the intermediary in some way (usually in cooperation with the user)

NATO ASI Series. Vol. F21
Intelligent Decision Support in Process Environments
Edited by E. Hollnagel et al.
© Springer-Verlag Berlin Heidelberg 1986

in order for any reasonable response to be achieved.

NAME OF FUNCTION	DESCRIPTION
Problem State (PS)	Determine position of user in problem treatment process, e.g. formulating problem, problem well-specified, etc.
Problem Mode (PM)	Determine appropriate mechanism capability, e.g. document retrieval
User Model (UM)	Generate description of user type, goals, knowledge, etc, e.g. graduate student, dissertation, etc.
Problem Description (PD)	Generate description of problem type, topic, structure, subject, etc.
Dialogue Mode (DM)	Determine appropriate dialogue type and level for situation, e.g. menu
Retrieval Strategy (RS)	Choose and apply appropriate retrieval strategies to knowledge resource
Response Generator (RG)	Determine propositional structure of response to user
Explanation (EX)	Describe mechanism operation, restrictions, etc., to user as appropriate
Input Analyst (IA)	Convert input from user into structures usable by functional experts
Output Generator (OG)	Convert propositional response to form appropriate to user, dialogue mode, etc.

Figure 1. The functions of an intelligent interface for document retrieval. From Brooks, Daniels & Belkin, 1985.

Each of these functions can be considered as a discrete 'expert' whose role is to accomplish that particular function. Together, these experts cooperate to achieve a single goal: helping the user in the problem treatment task by providing an appropriate response. Our problem then is to specify how these experts do, or should, interact to achieve their goal.

In our initial work, we noted that there is no obvious single order in which these functions are, or should be performed, but rather that any sequence is circumstance driven. We also saw that each of the experts depended upon results from several of the others in order to accomplish its function. These conditions of high interactivity and situation dependence, together with the necessity for function distribution, led us to consider 'distributed problem solving' approaches to complex problems in artificial intelligence (see, e.g. Chandrasekaran, 1981) as a basis for a

generalized IPM. Our eventual model, on which our subsequent
research has been based, is a distributed expert system, in which
the experts are the functions we have specified, organized to
communicate with one another until they have agreed upon how to
achieve their common goal, a response appropriate to helping the
user in problem treatment.

ACHIEVING AN INTELLIGENT INTERFACE

In order to achieve our research goal, it is necessary to specify
at least the following: 1. the 'ideal' functions of the IM; 2.
the knowledge necessary for performing the functions; 3. the
dialogue structure in which the functions are performed; 4. representational schemes for knowledge and messages; 5. retrieval
strategies; 6. general system architecture. To do these, we have
used a mix of methods, including: 1. functional analysis of
human-human information interaction; 2. interviews with participants in information interaction; 3. system simulation; 4. classification of problem and knowledge types; 5. retrieval 'testing'. We have concentrated most of our efforts on the first
method, because: our problem situation is such that we can readily
observe the human intermediary in action; the situation is so
complex that it requires much exploratory data analysis; and,
we wanted data as unconstrained as possible by pre-conceived ideas
about 'ideal' or 'correct' interaction.

Our method has been to record real-life information interactions
(primarily in document retrieval situations), and then to analyze
the transcripts utterance by utterance in at least one, and usually all of the following ways. By assigning the utterance to
one or more of the pre-specified functional categories; by understanding the specific purposes of each utterance; by determining the knowledge required to perform each utterance; and,
by grouping utterances into functional and focus-based sequences.
We have done this for about 20 dialogues, ranging from 10 to 60
minutes. This, together with our other methods, has allowed us
to approach most of the significant issues in designing an intelligent information system interface.

RESULTS AND DISCUSSION

We have now achieved the following results. We have identified a minimum set of functions for an intelligent intermediary in at least the document retireval system, and perhaps for information systems in general (figure 1). We have nearly completed speci- fication of two of the functions sufficiently for implementing them in a computer intermediary for document retrieval. The UM (Daniels, 1985) consists of about 50 frames, representing the user's status; goals; level of knowledge; previous information system experience; and general background, with default values and rules for filling slots. The PD (Brooks, 1985) is represented as several layers of partitioned semantic networks, with a limited set of mode and relation types. We have suggested a structure for document retrieval interaction, based on its functional pro- blem structure, which could guide and interpret a human-computer dialogue (Daniels, Brooks & Belkin, 1985). We have, through simu- lation, developed an architecture for the overall system (a func- tionally distributed system with blackboard communication and distributed control), and a well-defined set of message types (Belkin, Hennings & Seeger, 1984). And, we have begun to relate explicit retrieval strategies to structural representations of the user's problem (Belkin & Hapeshi, in prep.). Although our results were obtained primarily within document retrieval, they seem also to apply in the other sorts of advisory interaction we have studied. That these all share a number of characteristics with other classes of decision support systems (see Woods, this vol.), indicates that the idea of structuring human-computer in- teraction about human model-building functions applies generally.

CONCLUSION

Much intellectual system specification remains, and all of imple- mentation is to come (but in collaboration - see Croft, 1985). But we have accomplished enough to justify our approach to the overall problem. That is, that in order to design an intelligent information system interface, it is at least necessary to under- stand what happens in human-human information interaction, and why it happens, and to extract from this the functions that are required for any good information interaction. And to do this, functional analysis of real interactions is a useful and appropriate technique. It also appears that our function specification applies to the design of intelligent decision support systems in general, and that the study of human-human interaction is relevant to the human-computer interface in most such systems.

SKILLS, DISPLAYS AND DECISION SUPPORT

Penelope M. Sanderson
Department of Mechanical and Industrial Engineering
University of Illinois at Urbana – Champaign
Urbana, Illinois, USA

I have two concerns in this paper. The first is to raise the point that some decision support displays might have a syntax of their own which obscures the very data they are supposed to enhance, just as a misleading analogy may lead one to overlook an important aspect of a problem. In other words, enhanced displays may sometimes sow the seeds of their own destruction. The second concern is to suggest that supposed expertise is sometimes heavily dependent upon certain display formats. We may be in danger of making overly general claims for the depth of operators' knowledge, when the truth lies ultimately at a far simpler, more perceptual level. The answers may lie in a detailed examination of the problem itself and its context, making an 'ecological' approach to decision support necessary (Gibson, 1969; Neisser, 1976).

The syntax of integrated display formats.

Carswell and Wickens (1984) have looked at the role of displays in supporting failure detection in a simple simulated process control task. (I use their data here because of its richness and availability, not in a spirit of criticism.) Their underlying process consisted of two sine-wave inputs which were combined mathematically to make an output [$O = a1(I1) + a2(I2)$ or $O = a1(I1) \times a2(I2)$]. Carswell and Wickens were interested in seeing whether an 'integrated' triangular display would support superior failure detection performance than a 'separated' bar graph display. In the triangular display, the two inputs formed the base of a triangle and the weighted sum or product of their values formed the height. This display had surface similarities to Polar Star displays used for safety parameter display systems (Woods, Wise and

NATO ASI Series, Vol. F21
Intelligent Decision Support in Process Environments
Edited by E. Hollnagel et al.
© Springer-Verlag Berlin Heidelberg 1986

Hanes, 1981). In the bar graph display, the two inputs and the single output were presented as three separate bars. Carswell and Wickens' results showed that under a variety of conditions the triangle supported failure detection better than bar graphs.

It is tempting to conclude that the triangle provides a better mental model or that its Gestalt properties allow finer detection of failures, and to leave it there. However there are curious aspects of the data which demand a more detailed approach. For example, Carswell and Wickens show that when the two inputs are weighted evenly [$O = .5(a1) + .5(a2)$] and failure consists of a steady decrease in these weights to .4 and beyond (making the triangle flatter than normally possible), failure detection latency averages around 4 seconds. However when the two inputs are weighted unevenly and failure consists of a steady _increase_ in these weights (making the triangle taller than normally possible), failure detection latency averages around 8 seconds. Why should there be such a big difference when the rate of change is the same?

The recommendation of an ecological approach would be to search for answers in an objective analysis of the display itself. In order to explain where detection is aided, and even why the triangle gives superior performance, it is necessary to know a lot about the geometry of the displays being used. Only then is it appropriate to consider how the human might interpret or represent the display. John Flach (University of Illinois) and I have recently been working on this approach and are about to embark on a series of experiments to test the predictions such an ecological analysis would make.

In the case presented above, let us assume for convenience that one of the inputs (I1 or I2) can never be more than five times as great as the other (largest ratio 5:1). Let us now assume that they are weighted equally to create the output ($a1 = a2 = .5$, so that $O = .5(I1) + .5(I2)$). The issue is to find some aspect of the display which might explain operator behavior. When failure consists of a1 and a2 increasing to .7, there is a less dramatic relative change in the numerical value of the output than when failure consists of a1 and a2 decreasing to .4. So for the triangle, as a1 and a2 increase, additional units of height make smaller and smaller relative changes to existing height, whereas the opposite holds as a1 and a2 decrease. Moreover, when weights increase from .5 to .7, the topmost angle of the triangle will decrease, on average, 20 degrees. However in the case of a decrease of weights from .5 to .3, the topmost angle of the triangle will increase, on average, as much as 35 degrees. No wonder, then, that failures involving weight decreases can be detected so much faster than weight increases: there is more in the stimulus to indicate change (Flach, personal communication).

Further detailed analyses provide predictions for dynamic situations. For instance, if a1 = a2 and (I1 + I2) is increasing, then O should also be increasing. On the bar graphs an inference has to be made about whether the sum of I1 and I2 would visually expand or contract, and this has to be compared to O. However with the triangular stimulus only two dimensions have to be considered, as I1 and I2 have already been added. If the sign of the change of (I1 + I2) differs from the sign of the change in O, a failure must be present. Carswell (personal communication) reports that once subjects realise this, their performance shows a dramatic, qualitatively different improvement.

Finally, if I1 and I2 are equal and a failure consists in changes in a1 and a2 (which leave them still adding to 1.0), then the result is mathematically indistinguishable from what would happen if the weights were the same. Subjects cannot be expected to detect failures when I1 and I2 are in this region, and the data bear this out.

Thus, there are clearly process states which are technically failures, but for which neither triangle nor bar graph displays will provide appropriate support. There are other situations where one display will 'afford' the viewer some vital information better than the other. The moral is that if we look at the logical structure of a process and look at how this logical structure maps onto the syntax of a display, it should be possible to identify an upper bound to a subject's range of effectiveness. Both limitations and strengths should be identifiable. It is clear from the above analysis that even such an integrated display as Carswell and Wickens' (1985) triangle has hidden 'pathogens' (Reason, personal

communication) built into its very structure, and if it were to be
used as a decision aid its success might be very uneven. It
certainly yields a higher <u>average</u> level of detection
performance, but this may be only because the particular format
makes certain failures very obvious, while still leaving room for
others to go completely undetected.

Dependence on display formats.

In designing decision support one wants to help operators
focus on all the relevant data and only the relevant data and to
entertain all the plausible conclusions and only the plausible
conclusions. One way of doing this is to rely on the frequently
noted fact that the context, or presentational format, of a problem
can affect the way it is solved (Payne, 1982; Bar-Hillel & Falk,
1982; Kotovsky, Hayes & Simon, 1985; Tversky and Kahneman, 1981).
An interesting related observation is the fact that as decision
makers become more expert, they focus on increasingly superficial
aspects of a problem or display as they perform (Wood, Shotter and
Godden, 1974). What implications does this have for the so-called
'deep knowledge' an expert is meant to have of his or her domain?

As an example, take the following experiment (Sanderson,
1985). Subjects were shown two logic gates A and B, as in Figure 1.
They had to work out the legal connections between them (1. "A can
lead to B only", 2. "B can lead to A only", 3. "A can lead to B and
B can lead to A", 4. "Neither can lead to the other"). The gates
could be AND gates or OR gates and could have a value of 0 or 1,
making 16 conditions in all. Subjects learned the rules of binary
logic and were then given a series of problems. At first they were
slow and inaccurate and claimed to be doing a great deal of logical
processing, reasoning with the rules of logic to decide if A led to
B and then if B led to A. However, after about an hour subjects
became much faster and more accurate. At the same time, they found
it difficult to describe their strategies, making such comments as
"I don't seem to have to work it out all the way any more", "It's
automatic: I just look at it and I know the answer" "I get half-way
through and the answer pops ito my head" and so on.

```
(other gates)              (other gates)
       ¦                          ¦
       ¦                          ¦
  A (AND) = 0    <- ? ->   B (OR) = 1
```
FIGURE 1: NETWORK PROBLEM LAYOUT.

Upon looking closely at the pattern of results it became
clear what subjects were doing. They learned very fast that if
there were two Os or two 1s, the answer alternative was 3. If the
binary values differed, however, a truth table 'lookup' was
necessary to determine if A led to B. After this, subjects could
merely use surface characteristics of the task to determine whether
B led to A. For instance, if the 'lookup' showed that A led to B,
the subject would then note whether the gates were of the same type.
If yes, the subject could safely conclude that B could not lead to A
and that the correct answer alternative was 1. Similar perceptual
strategies were used for all conditions.

In this case, subjects had discovered perceptual aspects of
the task which were isomorphic with its logical structure and which
could be exploited in the interests of efficiency. The new strategy
was logically consistent with, but psychologically distinct from,
the one which it might be supposed they were using. It is
interesting to speculate, then, how this seemingly expert
performance at inferring network links might generalize to other
situations. It is possible, for instance, that if after much
practice subjects were asked to state whether B led to A, subjects
might now take longer to do this than to do the whole task of which
it is supposedly (but now not actually) an element, a truly
paradoxical result. Something similar to this is seen in a variety
of process control data. Moray (1985), for instance, has shown that
well-trained (simulation) process control operators can appear to
have a profound knowledge of a process under normal operating
conditions. However, once a fault is introduced, they behave as if
they have no 'mental model' at all. The truth may well be that
apparent expertise was being supported by superficial aspects of the
display only.

Conclusion.

A great deal of ingenuity is going into the design of
integrated and enhanced displays as decision support devices.
However it is possible that not enough is known about the way the
logic or geometry of a particular display might interact with the
deep structure of the task it is meant to support. When perfomance
is acceptable but uneven, it might be that a display is misleading
in the way an inappropriate analogy can be misleading (Gentner &
Stevens, 1983). Certain critical events may be hidden. On the
other hand, when performance is highly expert, it might be because a
display supports a variety of shallow heuristics whose effectiveness
may not generalise. Apparent expertise may reside in surface
features of the context, rather than in appreciation of task deep
structure and a knowledge of fundamentals. It is only necessary for
the situation to change a little bit, or for a display to be changed
or unavailable, to see whether operators really have the sort of
knowledge their normal performance suggests. Because in each case
the problem lies partly in the structure of the environment, it is
possible to predict it and test for it.

AUTOMATED FAULT DIAGNOSIS

Maurice F. White
Department of Marine Engineering
University of Trondheim
Trondheim, Norway

ABSTRACT

Machinery condition monitoring and fault diagnosis using a
fault matrix and process deviation approach is described. In
particular the application of computer aided diagnosis using
an expert system based on artificial intelligence is considered. A prototype fault finding system has been developed
and tested. Some examples from this system and some more
recent results from an ongoing research programme are outlined.

1. INTRODUCTION

Machinery management systems based on knowledge of the mechanical or thermodynamic condition of individual components of
the machine can be used as an aid for effective planning of
maintenance activities before serious faults occur. They can
also be used for quality control of mechanical plant at the
installation stage or after overhauling or maintenance work
has been carried out. The third, and in many cases most important, application is for energy economising and optimal
safe operation of rotating machinery and plant.

There are many parameters which can indicate the condition of
a machine. These can include process parameters such as performance, pressure and temperature; utility system parameters
such as lubricating oil temperature and pressures; operating
conditions such as operating hours and number of starts or
stops; and mechanical parameters such as vibration or wear,
and other factors indicated by measurement or by visual

NATO ASI Series, Vol. F21
Intelligent Decision Support in Process Environments
Edited by E. Hollnagel et al.
© Springer-Verlag Berlin Heidelberg 1986

inspection. The relation between these parameters and their sensitivity for indicating machinery condition is a complicated function.

The best course of action can be based on many things:
- machinery management objectives
- machine design
- commissioning experience
- operational procedures
- maintenance planning routines
- measurement of performance data
- calculation of condition parameters
- evaluation of performance using
  preset criteria
- fault diagnosis matrices
- machinery management strategy

It is clear that there is a large flow of information within a machinery management system. Condition monitoring and fault diagnosis is based on measurement and interpretation of data.

Machinery management is governed by a set of rules which are triggered by the information obtained from the process or by plant operator experience. The total system can be broadly divided into two main parts, as shown in Figure 1:

a) Diagnosis and evaluation of the fault
b) Strategy about the best course of action

By using a computer a large amount of information can be stored in a data base so that it is easily accessible. This includes design criteria, measured data, fault diagnosis matrices etc.

Such systems for monitoring and diagnosis process deviations and faults in machinery using the fault matrix approach have been described by Brembo (1977) and Fagerland (1978). White (1984) has applied Artifical Intelligence (AI) and Expert Systems (ES) to this area.

There is every indication that machinery condition monitoring
and fault diagnosis is a problem that can be solved in the
future by expert computing.

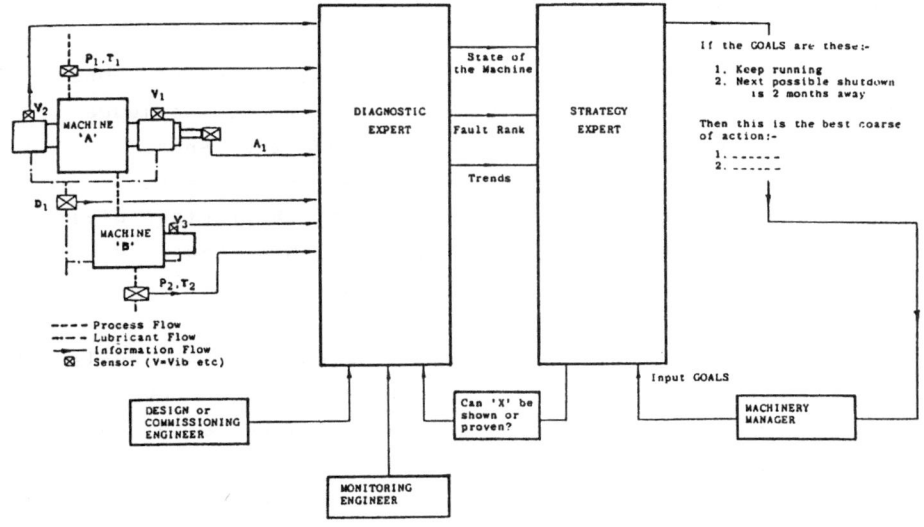

Fig. 1. The flow of information in a machinery
         management system.

2. FAULT DIAGNOSIS METHODS

In a conventional alarm system a machine failure is indicated
by a response in a single performance parameter, and an alarm
is given when the value of the parameter exceeds a predetermined
limit. In a complex machine a failure in one item very often
will produce changes in the values of more than one single
performance parameter. It is therefore necessary to detect
abnormal deviations in several measured parameters, and afterwards recognize a pattern of the abnormal parameters in order
to identify that a fault is present and identify the real
cause of the problem. For those parts of the machine or process which are too complicated to describe mathematically, it
is possible to apply pattern recognition techniques to evaluate a selection of abnormal conditions.

3. THE USE OF EXPERT SYSTEMS FOR FAULT DIAGNOSIS

A prototype expert system for condition monitoring and fault diagnosis of diesel engines is described in some detail by White (1985).

Expert systems are not conventional computer programs. One major difference is the separation of the expert knowledge (the rules forming a "knowledge base") from the general reasoning mechanism (the rule interpreter or "inference engine"). This partitioning, together with the further division or encoding of general knowledge into many separate rules, offers several important advantages.

Ordinary computer programs organize knowledge in two levels; data and program. Most expert computer systems, however, organize knowledge on three levels: data, knowledge base, and control. On the data level is declarative knowledge about the particular problem being solved and the current state of affairs in the attempt to solve the problem. On the knowledge-base level is knowledge specific to the particular kind of problem that the system is set up to solve. The control structure is a computer program that makes decisions about how to use the specific problem-solving knowledge.

How to represent the knowledge is still a matter of debate. One popular approach is to use IF-THEN rules. These rules say that if a certain kind of situation arises, a certain kind of action can be taken. Most existing rule-based systems contain hundreds of rules. In any system the rules become connected to each other to form rule networks. Once assembled, such networks can represent a substantial body of knowledge.

A possible "strategy" to choose a rule and apply it, is to select a goal to be achieved and scan the rules to find those whose consequent actions can achieve the goal. Each such rule is tried in turn. If the antecedents (IF part) for a rule match existing facts in the database, the rule is applied and the problem solved. If an unmatched antecedent is encountered,

arranging conditions to match that antecedent becomes a new
subgoal and the same procedure is applied again recursively.
If there are no rules to establish the new subgoal, the
program asks the user for the necessary facts and enters them
in the database. Since the behaviour of the system is in
direct response to the goals that the system is trying to
achieve, this is known as a "goal-driven" control strategy.
It is also known as "backward-chaining".

3.1 The Knowledge Base Structure

The prototype fault finding system is based on a combination
of fault tree analysis and pattern recognition. It has the
capability to link separate knowledge bases (KB). This enables
the KB chosen by the user (primary KB) to access rules in
another KB (secondary) so rules don't have to be repeated. An
important feature is that in this case (when information from
another KB is used) only the goals in the primary KB will be
attempted, those found in the secondary KB are not investigated
unless required by rules in the primary KB.

This structure gives more control to the user who can specify
the type of goals he wishes to achieve according to his needs.
For example a fault may not be present, but he can find an
out-of-normal-range parameter or verify it is in normal con-
diton. Additionally it provides a faster compilation and a
better structure to store the knowledge in a systematic way.
It is also much more easy to find a rule than in a large system.

The "structured" knowledge base can easily be enlarged. For
example, with a few more simple rules a decision adviser could
be built to prescribe actions when a fault is identified, or
when no fault is identified but some parameters are outside
normal range. This is termed the "strategy expert" in Figure
1. Note that a large part of the knowledge is already built
into the system and that if these extra rules are implemented
in a separate file that calls rules from existing files, none
of the diagnosis capabilities is lost and a very well informed
decision adviser is gained with little effort.

3.2 The Parameters' Condition

The condition of a parameter was chosen to be either low, normal, or high. However, since there is no clear limit from which to say a parameter was no longer too low but normal, or no longer normal but too high, three rules were written for each parameter that give the certainty of its value being in each of the three conditions, low, normal and high.

3.3 Fault Models

The Fault Models were described using a Baye's rule for each fault. The Baye's rule algorithm basically corresponds to a pattern recognition technique where the certainty of recognizing a fault is based on the certainty of recognizing the specified parameter's conditions which are obtained using the previously described rules. Additionally the relative importance of each parameter condition is encoded in two constant factors, LS and LN, called "Logical sufficiency" and "Logical necessity". They represent how sufficient (important) and how necessary (indispensible) each particular condition is for the fault being considered.

Before discussing an example it is important to say that specifying the correct pattern, and the sufficiency and necessity of each element in it, is the most important part of any fault diagnosis system. It is also the most difficult to set exactly. The fault models included in this work were established with help from many engineers with whom they were discussed and also from data given in the manufacturers manuals and other references. This is the part of the system where expert knowledge is essential.

4. SOME EXAMPLES OF FAULT MODELS

The Fault Models together with the condition curves, are the most important part of a diagnosis system. In each Fault Model a "syndrome" or "pattern" was established. Each element

in the pattern has a sufficiency to prove it and a necessity
to disprove it.

The selection and exact nature of the Fault Models depends
upon which measurements and other kinds of observations are
available. Naturally, the more different types of information
that are available, the more precise the models can be, and
the larger is the variety of faults that can be identified.
If we consider the amount of information a human "fault fin-
der" can get in a very short time, it is not surprising he has
a higher rate of success than any known computer program. For
example, at a quick glance he can establish how well a machine
was maintaned, how well the repairs were made, the quality of
the equipment and installation etc. All this information is
important and should be employed. This is one of the reasons
to use Expert Systems. For example, in a gas turbine control
and surveillance system there may be over 100 separate items
of information from one machine. It is therefore an almost
hopeless task for an engineer to rapidly diagnose a problem.

A typical fault model is seen in Figure 2, together with the
rules that combine the necessary inputs logically together
using Baye's rule. This forms part of the program source
code. The model is written in an "expert system language"
which is very like normal English.

At run time, before the user has answered any question, all
the Models start with a certainty of zero, but the program
calculates the maximum and minimum certainty they can achieve.
In Figure 3a the current certainty of the Failure Models can
be seen marked with an asterisk, and the possible movements of
their certainty as dashed lines. At any time during the consultation, the user can ask for this diagram which is called
the Histogram.

The maximum and minimum certainty as well as the current certainty of a Model vary as the user answers more questions, in
this way he can establish the influence that a particular
answer has on the certainty of a Model. In other words, the

Histogram represents the "best" answer the expert system can give based on the available data.

Figure 3b shows the diagnosis after answering only one question. Note the dashed lines only reach + 4 in the top line of Figure 3b for CYLINDER-UNIT-CONDITION. This model represents good condition, whereas the the other models are faults.

The conclusion at this early stage of the diagnosis is therefore that the cylinder unit is <u>NOT</u> normal, although more evidence is required to confirm this hypothesis.

Figure 3c shows an example of the final diagnosis when all available information has been evaluated.

5. CONCLUSION

To sum up we can say that the expert system approach provided of powerful tool for diagnosing and ranking of faults. However, the value and reliability of the results depends on the quality of the expert knowledge that was fed into the computer. A considerable amount of testing is therefore required using data from machines with known faults before the full potential of any expert diagnosis system can be properly evaluated. We are, however, convinced that expert system based automated fault diagnosis will become commercially available in the not too distant future.

```
MODEL WORN_PISTON_RINGS
DESC " THE PISTON RINGS ARE WORN OUT OR DAMAGED , BLOWBY SYMPTOMS "
BAYES LOW_MIP                         LS 1.5   LN 0.67
      LOW_P_COMP                      LS 7.5   LN 0.09
      LOW_P_MAX                       LS 4.5   LN 0.13
      NORM_A_P_MAX                    LS 1     LN 0.09
      LOW_P_EXP                       LS 3     LN 0.22
      HIGH_T_EX                       LS 4.5   LN 0.22
      SINGLE                          LS 1     LN 0.09
      HIGH_PR_ST                      LS 1.5   LN 1


MODEL LOW_P_COMP
DESC " THE COMPRESSION PRESSURE RATIO IS TOO LOW FOR GIVEN AMBIENT
        AIR PRESSURE AND SCAVENGING AIR PRESSURE. "
DOWNSLOPE  P_COMP  HLV_P_COMP  IDL_P_COMP

SPACEFUNC IDL_P_COMP
DESC " THIS IS THE EXPECTED  P_COMP  VALUE "
FUNC  P_SCAV_ABS * 32.02 - P_AMB

SPACEFUNC LHV_P_COMP
DESC " THIS IS THE LOWER  P_COMP  VALUE THAT IS CONSIDERED TOO HIGH "
FUNC  IDL_P_COMP + BSV_P_COMP

SPACEFUNC HLV_P_COMP
DESC " THIS IS THE HIGHER  P_COMP  VALUE THAT IS CONSIDERED TOO LOW "
FUNC  IDL_P_COMP - BSV_P_COMP

SPACEFUNC BSV_P_COMP
DESC " THIS THE  P_COMP  DEVIATION "
FUNC  0.02 * IDL_P_COMP

SPACEFUNC P_SCAV_ABS
DESC " THE ABSOLUTE SCAVENGING AIR PRESSURE IN bar UNITS "
FUNC  P_SCAV * 0.980665 + P_AMB

SPACEFUNC P_SCAV_ABS
DESC " THE ABSOLUTE SCAVENGING AIR PRESSURE IN bar UNITS "
FUNC  P_SCAV * 0.980665 + P_AMB

EVIDENCE P_COMP
DESC "WHAT IS THE COMPRESSION PRESSURE (bar) ? "
HELP "
-PLEASE ANSWER GIVING THE MAXIMUM PRESSURE IN THE CYLINDER BEFORE THE
FUEL IGNITES (PRESSURE DUE ONLY TO THE COMPRESSION WORK OF THE PISTON)
IN bar UNITS.
-THE MAXIMUM PRESSURE BEFORE THE FUEL IGNITES IS SEEN IN THE CYLINDER
PRESSURE DIAGRAM AS A SHORT LEVELING OF THE CYLINDER PRESSURE BEFORE
THE STEEP RISE DUE TO THE COMBUSTION PROCESS.
      "
VALUE  45  130
```

Fig. 2 EXAMPLE OF FAULT MODELS AND RELATED RULES

```
-----------------------------------------------------------------

   Models              Certainty:  -5 -4 -3 -2 -1  0  1  2  3  4  5

CYLINDER_UNIT_CONDIT             ----------------<*>----------------
EARLY_INJECTION                  ----------------<*>----------------
LATE_INJECTION                   ----------------<*>----------------
INJECTOR_LOOSE                   ----------------<*>----------------
INJECTOR_STICKS                  ----------------<*>----------------
LEAKY_EX_VALVE                   ----------------<*>----------------
WORN_PISTON_RINGS                ----------------<*>----------------
WORN_PLUNGER                     ----------------<*>----------------
LOW_FUEL_VISCOSITY               ----------------<*>----------------
HIGH_FUEL_VISCOSITY              ----------------<*>----------------
BAD_FUEL_QUALITIES               ----------------<*>----------------


-----------------------------------------------------------------

A)   Initial model status


-----------------------------------------------------------------

   Models              Certainty:  -5 -4 -3 -2 -1  0  1  2  3  4  5

CYLINDER_UNIT_CONDIT             <*>--------------------------------
EARLY_INJECTION                  ----------------<*>----------------
LATE_INJECTION                   ----------------<*>----------------
INJECTOR_LOOSE                   ----------------<*>----------------
INJECTOR_STICKS                  ----------------<*>----------------
LEAKY_EX_VALVE                   --------------<*>----------------
WORN_PISTON_RINGS                ----------------<*>----------------
WORN_PLUNGER                     ----------------<*>----------------
LOW_FUEL_VISCOSITY               ---<*>-----------------------------
HIGH_FUEL_VISCOSITY              ---<*>-----------------------------
BAD_FUEL_QUALITIES               ---<*>-----------------------------


-----------------------------------------------------------------

B)   Status after one question answered


-----------------------------------------------------------------

   Models              Certainty:  -5 -4 -3 -2 -1  0  1  2  3  4  5

CYLINDER_UNIT_CONDIT             <*>
EARLY_INJECTION                  <*>
LATE_INJECTION                   <*>
INJECTOR_LOOSE                                                <*>
INJECTOR_STICKS                       <*>
LEAKY_EX_VALVE                   <*>
WORN_PISTON_RINGS                   <*>
WORN_PLUNGER                     ---<*>----------
LOW_FUEL_VISCOSITY               ---<*>----------
HIGH_FUEL_VISCOSITY              <*>
BAD_FUEL_QUALITIES                                      <*>


-----------------------------------------------------------------

C)   Final diagnosis based on available evidence

Fig. 3  EXAMPLES OF FAULT DIAGNOSIS HISTOGRAMS
```

KNOWLEDGE – BASED CLASSIFICATION WITH INTERACTIVE GRAPHICS

K. – F. Kraiss
Forschungsinstitut fur Anthropotechnik
Werthoven, West Germany

Interpretation and classification of sensor data in many applications can still not be automated and must be performed by
human operators. Representative examples are sonardata as well
as infrared or X-rayed pictures. The task of the operator in
such applications is on one hand the extraction of features
from the picture and on the other hand the interpretation of
these features using an internal knowledge base.

This paper describes a concept for a two-screen workstation
that supports operators in both tasks. One screen offers tools
for interactive feature extraction like, e.g., filtering and
grey-scale modification. The second screen provides aiding in
feature interpretation using an expert system. Interaction with
the system is in both cases performed by direct manipulation.
In addition, a graphical surface to the expert system has been
deviced that improves user insight into the inference process.

INTRODUCTION

Even after dramatic advancement in automatic pattern recognition and scene analysis in recent years, human visual capabilities remain unmatched in many applications. The ability of the
human to recognize forms and structures very easily, even in
threshold conditions and if heavily distorted, often can not
yet be duplicated by technical devices. On the other hand human
limitations in remembering, evaluating and combining large
amounts of information are notorious.

NATO ASI Series, Vol. F21
Intelligent Decision Support in Process Environments
Edited by E. Hollnagel et al.
© Springer-Verlag Berlin Heidelberg 1986

In view of these facts it is hypothesized that interactive pattern recognition should be superior to the purely human or automatic approach. This should be especially true if human information processing deficiencies are compensated for by the provision of decision making and operating aids. An experimental setup to test this approach to interactive pattern classification has been developed. The selected paradigm addresses the detection, recognition and identification of ships from passive sonar sensor data (LOFAR-grams).

DESIGN PRINCIPLES AND CHARACTERISTICS OF DECISION AIDS

The deficiencies of human information processing as described, e.g., by Moray (1985) may be compensated for by several approaches. Among these are advanced ways of information presentation and data manipulation, automated knowledge based reasoning, decision theoretical approaches, pattern recognition, learning systems, simulation and prediction as well as searching in graphs (Kraiss 1985).

In applying the above principles the basic purpose of a decision aid should be kept in mind. According to Barnes (1980) a decision aid is "a mechanism that extends the intellect of the decision maker and puts structure to what would otherwise be a confusing situation rather than a device which makes decisions according to preset rules which are programmed without any form of intervention by the decision maker".

In order to make sure that a decision aid can serve the purpose outlined above it should
- be interactive
- be transparent and self explaining
- bring structure to a decision situation
- unload the operator from routine work
- leave the final decision to the operator
- be tuned to the cognitive abilities and the training status of the user.

Decision aids not designed according to these guidelines may be rejected by the user or, even more dangerous, may stimulate a feeling of confidence that is not justified.

THE PASSIVE SONAR CLASSIFICATION PARADIGM

For the pattern classification task dealt with in this paper so called LOFAR - grams (low frequency analysing and recording) are used (see figure 1). The data is collected by underwater hydrophones. Subsequently, a Fourier analysis of the noise is performed in the interval 0 - 200 Hz and the spectral components are presented as a grey-scale time frequency plot. To generate a graph the acoustic sensor data is continuously written as a time-frequency plot.

Useful information to be detected by the human observer in the LOFAR-gram are spectral lines emitted from rotating parts aboard a noisy ship as, e.g., from propeller blades, shafts gears etc. Also contained in the picture are inference patterns

Figure 1: Example LOFAR-gram

due to reflections, particularly in shallow waters. From the spatial arrangement and the intensity of the spectral lines the experienced operator can deduce the type, direction and speed of a possible target. This, however requires special visual capabilities for feature extraction and long training and expert knowledge for feature interpretation. Even after long experience however target classification remains a very difficult and risky task.

Recent approaches to automate sonar classification are based on production rule systems (Drazovich et al. (1979), Nii, and Feigenbaum (1978)). As it appears now, the performance of these systems is still worse than that of human operators.

DECISION AIDING IN SONAR CLASSIFICATION

Since sonar classification can not be automated - at least for the moment - efforts are underway to facilitate the task by building expert systems that can be used by operators as intelligent assistants (Funk et al. 1984).

Interaction with such expert systems often poses a problem for the human operator. The transparency of the system inference mechanism usually is very obscure to the user even if features for self explanation are included. The main reason seems to be that explanations mostly take the form of printouts of the productions used in an inference process. Those listings usually are hard to check on line. This difficulty to control a computer proposal may stimulate a tendency for complacency on the part of the operators. In order to avoid tedious work that mostly turns out to be in vain, he may accept aiding without checking it for correctness.

In order to overcome those deficiencies a novel interface concept for decision aids has been worked out that makes extensive use of visualization and direct manipulation techniques. This design hopefully will improve the transparency of the aiding procedure to an extend that online checking becomes reasonably simple.

The proposed workstation for sonar operators consists of two color CRT-screens. One screen shows the LOFAR-gram as depicted in fig.1. It offers interactice graphics like windows and pull-down menus to facilitate feature extraction. The tools available for feature extraction include picture processing like, e.g., filtering and gray-scale modifications.

Since the interpretation of features found in the LOFAR-gram often is difficult as mentioned earlier, the user gets decision support from an expert system that is accessed via a second screen (The productions contained in the knowledge base will not be discussed in this paper). For this expert system a graphical user surface has been developed that enables inter-action by direct manipulation. As may be seen in figure 2 the screen contains a fact window (FAKTEN), a hypotheses window (Hypothesen), a mode window (BETRIEBSARTEN) and a communication and prompting area.

To store a new fact "FAKTEN SPEICHERN" is selected with a pointer device from the mode window and subsequently the appro-priate fact is marked in the fact window. Facts stored this way are painted green. As a next step the deduce operating mode ("SCHLUSSFOLGERN") performs forward reasoning using the facts known so far. A result of a successful deduction is indicated in the hypothesis window by green (accept) or red (reject) painting of a priori hypotheses.

Diagnose (DIAGNOSTIZIEREN) means backward reasoning. In this mode facts presented in the prompting area have to be accepted or rejected by the operator. In order to do so it may turn out to be necessary to go back to the left screen and look again for a particular feature in the LOFAR-gram. If accepted the respective fact gets a green color in the facts window, other-wise the color becomes red.

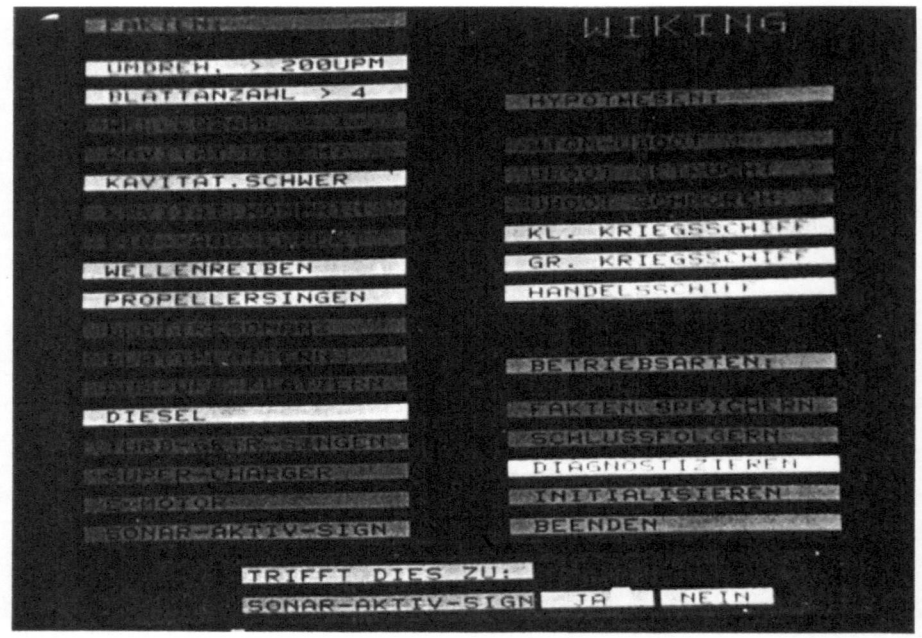

Figure 2: Graphically interactive interface for a knowledge-
based passive sonar classification system.
Interaction is by direct manipulation using a mouse
as pointer device.

During the described dialog with this aiding system the user
is guided through subsequent steps of goal directed feature
extraction and knowledge assisted feature interpretation. It is
very easy to observe on-line what information is used and how
one hypotheses after the other is rejected as new information
becomes available. This continues until finally one hypotheses
can be accepted, i.e, classification is successful. It may also
happen that all a priori hypotheses must be rejected since not
enough data or knowledge is available.

SUMMARY

A graphically interactive interface for a decision aiding
system in the pattern classification domain has been presented.
It appears that this concept satisfies most of the guidelines
for decision aids as stated in this paper.

The combination of interactive feature extraction together with
knowledge-based feature interpretation offers largely improved
transparency of the inference process. The context between the
facts used and the conclusions reached is immediately and continuously visible.

At this moment of time a pilot version of this concept has been
implemented. First experiences with the system appear to be
very promising.

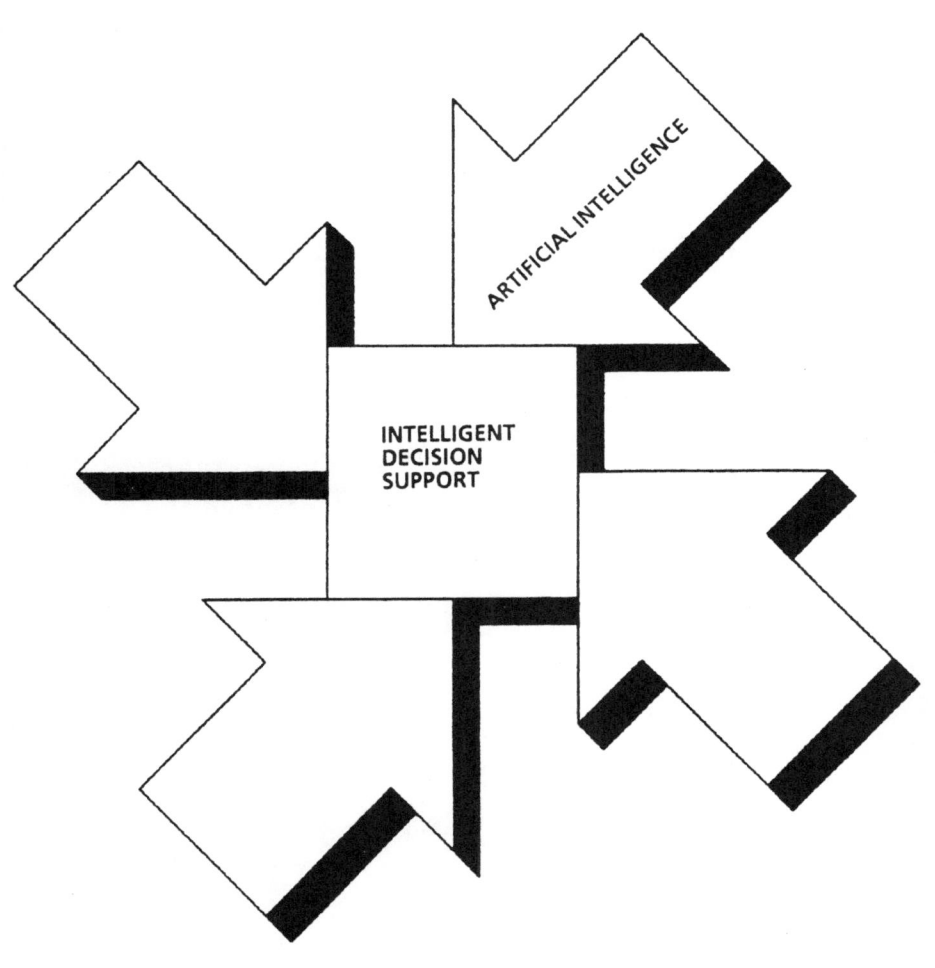

ARTIFICIAL INTELLIGENCE

INTELLIGENT
DECISION
SUPPORT

INTELLIGENT DECISION AIDS FOR PROCESS ENVIRONMENTS: AN EXPERT SYSTEM APPROACH

Massimo Gallanti
CISE s.p.a., Milano, Italy

Giovanni Guida
Progetto di Intelligenza Artificiale
Dipartemento di Elettronica, Politecnico di Milano
Milano, Italy

The paper is devoted to illustrate the impact of expert system technology on the design of intelligent decision aids for process environments. It first introduces the basic concept of a process environment and discusses its main features. The fundamental points of expert system technology are later briefly illustrated, focusing on design and application issues. The impact of this newly emerging technique on the design of intelligent decision support systems for process environment applications is then analyzed, and a state-of-the-art of current applications is presented. A case study concerning the design and implementation of the PROP system devoted to support the operator of a thermal power plant in monitoring the pollution of cycle water, is reported and discussed in detail. The paper concludes with an assessment of the role of intelligent decision aids in process environment applications.

1. INTRODUCTION

The impact of expert system technology in application environments has been large and profound in recent years. The success of several projects in such areas as, for example, medical diagnosis, chemical analysis, geological prospection, and computer system configuration (Hayes-Roth, Waterman, and Lenat, 1983; Waterman, 1986), has pushed the industrial world to experiment application of this newly emerging technology in more challenging and demanding fields. Applications to real-time tasks and to process environments are among the most promising areas presently under consideration.

The extension of the focus of attention to these new fields has raised, in turn, several new research problems for the artificial intelligence specialists: the current technology is, in fact, not powerful enough to allow construction of robust, real-size applications in such challenging fields as process control, plant diagnosis, on-line supervision of discrete processes, and real-time control of complex machineries (Sauers and Walsh, 1983; Waltz, 1983).

This paper is devoted to illustrate some of the most remarkable tools offered by expert system technology to the designers of

NATO ASI Series, Vol. F21
Intelligent Decision Support in Process Environments
Edited by E. Hollnagel et al.
© Springer-Verlag Berlin Heidelberg 1986

applications in process environments, and to discuss the major challenges that this new area poses both to artificial intelligence researchers and to system designers. More precisely, section two introduces the notion of process environment and discusses some of its basic features. Section three illustrates the major points of expert system technology, including a taxonomy of application areas and a discussion on expert system design methodologies and tools. Section four analyzes some of the most promising opportunities for application of the expert system approach in process environments and discusses the expected benefits. Section five presents a brief survey of industrial and research projects. Section six is devoted to the discussion of a case study concerning the PROP system for continuous, on-line pollution monitoring of the cycle water in a thermal power plant. Finally, section seven presents some conclusive remarks and gives a brief insight on promising research issues.

Other classes of applications of expert system technology that are of great interest for industry but do not directly relate to process environments (such as computer aided design, computer aided engineering, off-line diagnosis and computer aided testing, production planning and control, computer aided manufacturing, flexible manufacturing systems, etc.) will not be covered in this paper (Kempf, 1985).

2. PROCESS ENVIRONMENT ANALYZED

In order to provide a definition of the scope of our investigation and to establish a clear reference framework for the following discussion, we introduce in this section the concept of process environment from the general point of view of information processing.

A process environment can be considered as a composite system comprising the basic parts depicted in Figure 1, namely: a process, a sensory and measurement system, a process control system, an interface, and a group of experts (operators, maintenance staff, and specialists).

The process is characterized by several features that include:

- complexity of structure, i.e. it is made up of several interacting subsystems;

- complexity of description of its parts and their internal operation (often requiring sophisticated mathematical models);

- complexity and variety of involved physical and engineering models: electrical, thermodynamical, nuclear, mechanical, etc.;

- time-dependency, i.e. its dynamic behavior can hardly be analyzed without explicitly taking into account the representation of time relations;

- speed of evolution, that often imposes real-time constraints to the interacting systems (process control system and experts) in order to ensure timely intervention.

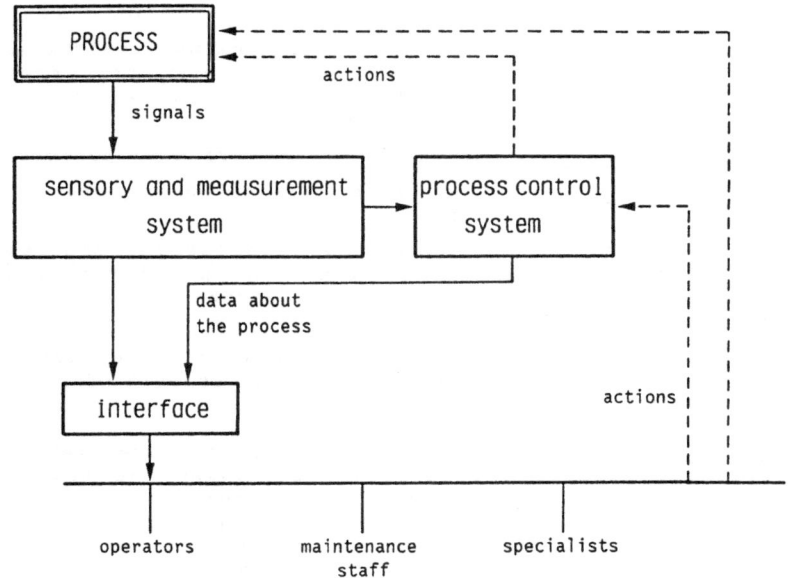

Fig. 1. A schema of process environment.

The <u>sensory and measurement system</u> is devoted to collect data directly from the plant (often through a complex distributed architecture) and to supply them to the process control system and to the experts. It features three basic characteristics, namely:

- the large number of data collected from the process (due to the large number of measurement points and of measurements done in a time unit);

- the uncertainity of measurements (due to spurious readings, sensor degradation, etc.);

- the possible overlapping and conflict between different observations.

Therefore, one of the major problems for the process control system and the experts is validation, integration, and interpretation of the various data collected form the plant. In fact, the perception activity, which is aimed at correctly observing process operation at the needed level of abstraction, is one of the main problems to be faced in a process control environment: it requires transformation of numeric data collected at the process into symbolic conceptual representations.

The <u>process control system</u> comprises the traditional systems (usually computer-based) used for on-line process control, also including possible off-line decision support tools such as numerical simulation programs.

The <u>interface</u> is devoted to channel data coming from the sensory and measurement system and from the process control system to the experts (mainly the operators) in such a way as to meet ergonomic and human engineering requirements.

The group of <u>experts</u> comprises three categories of persons:

- the <u>operators</u>, who have direct responsibility for process surveillance and operation;

- the <u>maintenance staff</u>, that intervenes at the process site upon detection of abnormal situations in order to diagnose the malfunction or fault and operate adequate repair interventions;

- the <u>specialists</u> of the process (often including the system designers), who can supply specific support for solving complex technical problems.

The basic dynamics of the process environment above sketched is, in principle, simple. The operators supervise the behavior of the process in order to control that it occurs according to a desired pattern. They have access to the process only through the interface, and they must be capable of correctly interpreting a large amount of signals, data, and alarms coming from the plant. During normal operation, the activity of the operators is well defined: parameter setting, operation procedures, controls, corrections to operative conditions according to external events, and surveillance actions belong to the usual practice of their job. Whenever an abnormal or critical situation occurs, the operators try to assess the problem, to formulate a diagnosis, and to intervene with appropriate actions on the process in order to re-establish a normal (optimal) operative situation. They resort to the assistance of the maintenance staff and of the specialists whenever necessary.

The experts (operators, maintenance staff, and specialists) have to face in their job several complex tasks that require different kinds of capabilities and expertise. A partial list of such tasks and of the kind of reasoning paradigms involved is reported in Figure 2 (Hayes-Roth, Waterman, and Lenat, 1983; Clancey, 1985).

Moreover, several types of knowledge are needed for accomplishing these tasks:

- models of the process, including the process control system, at different levels of abstraction:

  - shallow models, based on empirical associations between data and partial conclusions of interest (e.g., situation-action, operation-effect, symptom-diagnosis);

  - deep causal models, mostly qualitative in nature, describing process structure and behavior and relying on general physical laws and design principles (Hart, 1982; Bobrow, 1984; Chandrasekaran and Milne, 1985);

TASKS	REASONING PARADIGMS
data interpretation	pattern matching simple classification hypothetical reasoning
surveillance and supervision	pattern matching heuristic classification
tuning and optimization	prediction forward reasoning planning
malfunction and fault detection	pattern matching heuristic classification
diagnosis	heuristic classification forward and backward reasoning hypothetical reasoning truth maintenance opportunistic search
critical event management	simple classification prediction forward and backward reasoning hypothetical reasoning opportunistic search configuration planning
maintenance and repair	prediction forward reasoning configuration planning

Fig. 2. Experts' tasks in a process environment.

- quantitative mathematical models, often supported by numerical simulation programs;
- data coming from the process (through continuous measurements, specific tests, etc.);
- expert knowledge, mainly heuristic, fragmentary, partial and uncertain, concerning the professional and technical skills of the operators, maintenance staff, and specialists.

The role of the experts in a process environment is generally of primary importance and, in several situations, critical. Therefore, experts often constitute a real bottleneck for the correct and effective operation of a process. In fact, skilled experts are rare, not easy to hire, and, in any case, they cost a lot. Novices need long training in order to become experts, and existing experts are often unwilling or unavailable to transfer their expertise to young colleagues. Moreover, experts are generally not available 24 hours a day, and often they can hardly be reached when critical situations occur. Finally, the human cognitive process in an industrial process environment may be often affected by several detriments that can result in an unproper, unefficient, or unreliable behavior. Among these we mention (Meijer, 1983b):

- psychological: level of stress and after-effects, social tensions, phobia, mind-sets, complexes;

- physiological: physical conditions, drugs and after-effects, alcoholic consumption, age, human limitations;

- lack of confidence: unreliable hardware (e.g., excessive failure or alarm histories), distrust of colleagues;

- ineffective/inappropriate aids: superfluous and ambiguous information, ambiguous directions, unexpected responses, poor human factors engineering;

- environmental conditions: too hot, too cold, too humid, too dark, too much noise, distracting sounds, conversations.

Therefore, it would be highly desirable to support (or even partially substitute) the experts and their cognitive processes with intelligent and powerful computer-based assistants. Most of the problems involved in the above mentioned tasks exhibit, however, a very high complexity and are therefore hardly tractable with traditional approaches. Moreover, they are often inherently ill-defined and under-constrained. There is seldom any known or standard solution technique for them: there may be rules of thumb, or standards of good practice in approaching a problem, but seldom is there any guaranteed and viable solution process (and there is hardly ever an optimal or unique solution) (Kempf, 1985). Therefore, it seems appropriate to investigate whether and how the new tools offered by expert system technology can be profitably used for designing and implementing intelligent decision aids for process environments .

3. THE EXPERT SYSTEM APPROACH

This section is devoted to introduce a brief survey on expert system technology, and to discuss its relevance to the design of intelligent tools for supporting the experts who operate in a process environment.

An expert system is a computer system that can provide expert-level problem solving capabilities in specialized subject domains. Over the last fifteen years artificial intelligence researchers have achieved considerable success in the development of basic methodologies and techniques in this newly emerging area and in applying them to several challenging problems of real interest that could not be conveniently faced with more traditional approaches because of their high complexity. For example, several applications in the fields of medical diagnosis and therapy, mineral prospecting, computer system configuration, chemical structure elucidation, symbolic mathematics, electronic circuit analysis, and education are now well known and testify to the increasing success of the expert system approach (Hayes-Roth, Waterman, and Lenat, 1983).

In fact, expert system technology (Davis and King, 1976; Buchanan and Duda, 1983) offers a basic advantage over traditional programming: it enables the programmer to tackle problem solving tasks at a higher level of abstraction and in a more flexible, effective, and natural way. The usual activities of problem analysis, invention of a solution algorithm, and programming are replaced, in the expert system approach, by the representation of knowledge about the application domain (including available resources, constraints, and problem solving skills), of the specific problem to be solved, and of the goal to be achieved. Responsibility about <u>how</u> to use knowledge in order to solve a given problem is left to the expert system: the programmer has only to represent <u>what</u> is known about the problem and is likely to be relevant to <u>its</u> solution. This turns out to allow much more complex and demanding problems to be successfully faced with reasonable effort. The concept of knowledge-based problem solving appears, therefore, as a new way of programming a computer, suitable to large classes of challenging application domains, where usual programming methods fail.

The general architecture of an expert system is shown in Figure 3.

The core part of the system is constituted by a knowledge-based problem solver which includes:

- a <u>knowledge base</u> that encompasses knowledge about the problem domain;

- a <u>data base</u> that contains a definition of the specific problem needed to be solved, and also serves as a working memory for system operation;

- an <u>inference engine</u> which is a specialized problem solving program capable of appropriately using the content of the knowledge base for constructing a solution to the problem considered.

Several different knowledge representation methods may be used for encoding knowledge about the problem domain, including production rules, frames, semantic nets and logic, and the knowledge base may be internally structured in various ways so as to facilitate knowledge retrieval and use. Also the inference engine may include several sophisticated reasoning paradigms, such as forward and backward chaining, property inheritance, truth maintenance, hypothetical reasoning and analogy, and complex control strategies may be used to direct system operation, including agenda and blackboard mechanisms, conflict resolution, heuristic state space search, planning, and procedural control. Special purpose knowledge (called meta-knowledge) may be used for encoding criteria about how to effectively use domain knowledge and may have an important role in determining the control-level operation of the inference engine.

Two fundamental modules, external to the knowledge-based problem

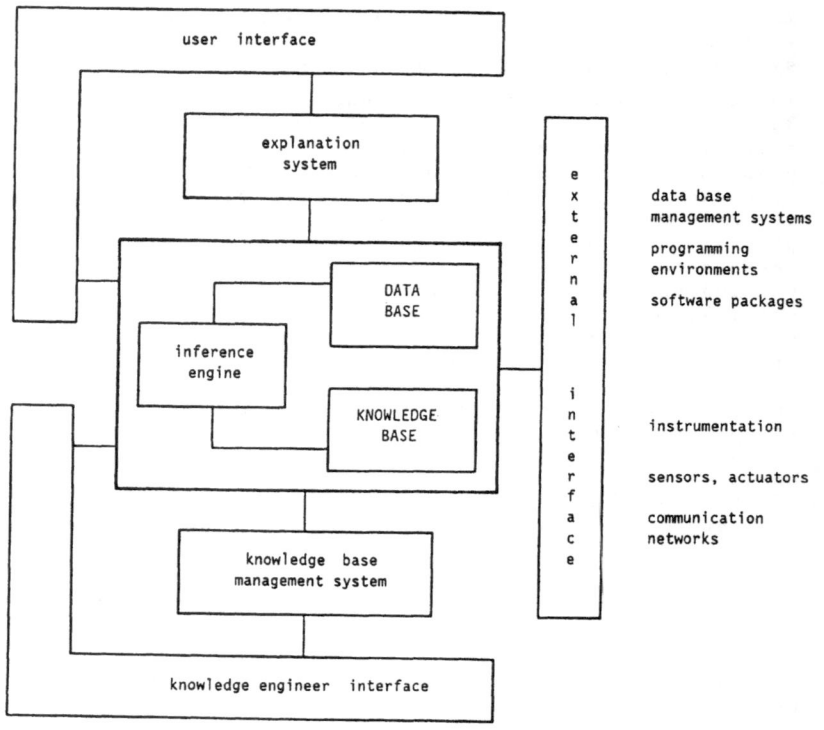

Fig. 3. General architecture of an expert system.

solver, are aimed at supporting system construction and use. Namely,

- an <u>explanation system</u> is devoted to explain and justify the system behavior in such a way as to make the reasoning mechanisms of the expert system transparent to the user;

- a <u>knowledge base management system</u> is designed to allow effective knowledge elicitation from the experts and easy creation, debugging, updating, extending, and inquiring of the knowledge base.

Three interfaces complete the architecture of an expert system:

- a <u>user interface</u>, which enables effective and graceful interaction between the user and the system;

- a <u>knowledge engineer interface</u>, devoted to support the communication of the knowledge engineer, and possibly also of the experts, with the knowledge base management system;

- an <u>external interface</u>, devoted to channel the expert system to the outside world: data base management systems, traditional programming environments, software packages, instrumentation, sensors, actuators, communication networks, etc.

A deep technical illustration of the structure and mode of operation of an expert system, including details on architectures, knowledge representation methods, and reasoning paradigms, is beyond the aim of this paper. For a general introduction to expert system technology the reader can refer to (Hayes-Roth, Waterman, and Lenat, 1983; Weiss and Kulikowski, 1983; Harmon and King, 1985; Waterman, 1986). In the sequel of this section we will discuss the main features of the expert system technology, and we will and briefly illustrate the life-cycle of an expert system, focusing on some of its most important points.

Several technical features make the expert system approach specially useful for several complex and demanding applications; among these we mention:

- it is appropriate from the cognitive point of view for modeling several features of human mental processes and for simulating human performance on specific tasks through a program which embodies a theory of the observed behavior;

- it provides a mechanization of the reasoning process, with knowledge about the problem domain clearly separated from the inference program;

- it allows knowledge to be represented in a highly declarative way, without bothering of its use;

- it allows several, different knowledge representation paradigms to be adopted and used together, ranging from highly declarative to purely procedurals ones;

- it enables fragmentary, ill-structured, approximate, incomplete, uncertain, heuristic, judgmental knowledge to be easily described and used, in addition to formal knowledge of established theories;

- it supports implementation of non-deterministic search and control strategies and it covers several different reasoning paradigms;

- it is robust, i.e. it can operate with uncertain or incomplete knowledge, and it can support graceful degradation at the borders of the specific competence domain;

- it is transparent, i.e. it can provide natural explanations and justifications of its specific line of reasoning and answer queries about its knowledge;

- it is flexible, i.e. there is the possibility of incrementally creating, debugging, and updating very large knowledge bases;

- it is effective, i.e. it can operate with reasonable efficiency in complex domains;

- it supports an effective design methodology, thus allowing the programmer to develop quite large and complex applications with a limited effort.

The design and construction of an expert system includes six main phases that are reported below (Harmon and King, 1985).

1. Selection of an appropriate domain:
   - identifying an appropriate problem domain and a specific task to tackle, taking into account the features of involved knowledge and reasoning paradigms
   - assuring the availability of qualified and cooperative domain experts
   - identifying a tentative approach to the problem and analyzing it with the domain experts
   - preparing a preliminary development plan, including needed resources and time schedule
   - analyzing expected risk, costs, and payoff of the project

2. Selection of an appropriate expert system development environment:
   - identifying the requirements and constraints of the demonstrator, prototype, and target system
   - analyzing the available software tools suitable for developing the expert system
   - selecting the appropriate hardware and software environment

3. Development of a demonstrator system, i.e. a small-size expert system devoted to check the suitability of the chosen approach, to assess possible technical problems, and to show through sample cases the expected performance and benefits of the expert systems under development:
   - analyzing the problem domain and the specific task
   - specifying system requirements and performance evaluation criteria
   - eliciting knowledge from the experts
   - developing an initial implementation
   - testing with selected cases and completing the demonstrator
   - evaluating the demonstrator and developing a detailed design for the complete system

4. Development of a prototype system, i.e. a complete expert system developed up to a point where it can be put at work on real-size problems and its capabilities and performance can be actually evaluated:
   - implementing the core structure of the complete prototype system (inference engine, explanation system, and knowledge base management system)
   - acquiring knowledge from domain experts and constructing the knowledge base

- designing and tailoring appropriate user, knowledge engineer, and external interfaces

- testing and extending the knowledge base until the desired performance is achieved

5. Development of the target system, i.e. the final system engineered and optimized in such a way as to effectively run in the real operation environment:

    - engineering and optimization of the prototype system

    - transfer to the real application environment

6. Maintenance, updating, and extension:

    - recovery of bugs in the knowledge base (conflicts, incompleteness, etc.) and refining of its structure and content

    - updating of the knowledge base according to changes in the problem domain

    - extension and restructuring of the knowledge base in order to cope with new requirements.

It is apparent from the above illustrated life-cycle, that the development of an expert system involves two basic technical points; namely, the design and construction of the programs implementing the inference engine, the explanation system, the knowledge base management system and the interfaces, and the construction of the knowledge base. Of these two activities, the latter, involving knowledge elicitation and acquisition, is generally recognized as the major bottleneck of expert system development. It includes five main steps(Hayes-Roth, Waterman, and Lenat, 1983):

- identification: analyzing problem characteristics and identifying the basic concepts that are needed for knowledge representation;

- conceptualization: eliciting, organizing, and representing knowledge at a conceptual level;

- formalization: designing formal structures to organize knowledge inside the expert system;

- implementation: formulating knowledge using the designed mechanisms;

- testing: validating (debugging, extending, updating, etc.) knowledge represented in the system.

These stages of knowledge acquisition are not clear-cut, well-defined, or even independent. They basically aim at eliciting knowledge from domain experts and encoding it in a form suitable for the expert system. This transfer and transformation of domain knowledge and problem-solving expertise from a knowledge source to a computer program is the heart of the expert system development process. This task is the primary job of the knowledge engineer. Through an extended series of interactions with the domain experts, he defines the problem to be attacked, discovers the basic concepts involved, and expresses the relationships existing among them. Although work

is rapidly progressing on automating this critical phase of the
expert system development process, at present knowledge
engineers must rely on their own skill and insight to guide the
knowledge acquisition activity.

The construction of the software packages necessary to implement
the building blocks shown in Figure 3 is the other major
technical point in the development of an expert system. In
addition to traditional programming languages (such as Pascal,
Lisp, C, Prolog, etc.) and to high-level environments for
artificial intelligence programming (such as Interlisp D,
Poplog, MRS, Smalltalk, LOOPS, etc.), several high-level expert
system building tools (or expert system shells) are now
available on the market, that are designed to facilitate the
rapid development of expert systems in such a way as to reduce
the global design and development effort. Such tools incorporate
several components of an expert system, including: methods for
knowledge representation, reasoning paradigms, control
strategies, knowledge acquisition and debugging systems,
explanation and justification systems, etc. They offer two basic
advantages to expert systems designers (Harmon and King, 1985):

- they provide for rapid system development by offering a
  substantial amount of computer code that would otherwise
  need to be designed, written, tested, debugged, and
  maintained;

- they provide specific techniques for handling knowledge
  representation, inference, and control that help knowledge
  engineers to model the salient characteristics of a
  particular class of problems.

However, each tool can be applied only to specific, although
sufficiently large, classes of applications. The problem of
matching a given task with the appropriate tool is not always
simple, and existing tools do not cover all possible application
requirements. It is responsibility of the expert system designer
to verify whether he can rely on an available tool that fits his
current application, or he has to build up a new environment
using traditional, general purpose artificial intelligence
programming systems.

Existing knowledge engineering tools fall into three major
categories:

- small expert system building tools, that can be run on
  personal computers, and are designed to facilitate the
  development of systems of small size (less than 400 rules),
  including fast prototypes, demonstrators, and simple
  applications for easy tasks [e.g., ES/P ADVISOR, M.1,
  EXPERT-EASE, SAVOIR, NEXPERT, EX-TRAN];

- large, narrow, specific expert system building tools, that
  run on large computers or dedicated artificial intelligence
  machines, and are designed to build large, powerful
  applications [e.g., S.1, ENVISAGE, EXPERT, TIMM, RULE
  MASTER, KES];

- large, broad, hybrid expert system building tools, that run
  on large computers or dedicated artificial intelligence
  machines, and are designed to include the features of

several different design paradigms, thus allowing a greater flexibility and leaving the system designer free of taking several significant technical decisions [e.g., KEE, ART, KNOWLEDGE CRAFT].

An annotated list of available tools is reported in (Harmon and King, 1985).

To sum up, expert system technology, though young and therefore still developing and far from having reached the level of an established discipline, offers today a set of sound and sufficiently reliable methods and tools that can be effectively used in several concrete application domains. These tools are in a sense a supplement to traditional information processing techniques, and can be profitably used in several complex cases where the latter fail. Clearly, expert system technology does not enable a computer to accomplish any task that can not be accomplished through traditional programming as well. The advantages it offers mainly consist in the relative ease with which certain performances can be achieved and in the extent to which they can be achieved in a natural, transparent, and flexible way (Alty and Guida, 1985).

4. EXPERT SYSTEMS FOR PROCESS APPLICATIONS

Taking into account the analysis of process environment and expert system technology presented in the previous sections, we can now try to identify a class of promising applications and discuss their expected benefits.

Expert system technology can offer powerful tools for developing intelligent assistants (or decision support systems) for the experts who operate in a process environment, that can support (and possibly partially substitute) them in their complex job. With reference to the schema of a process environment reported in Figure 1, expert system capabilities can be profitably embedded in the interface system, in such a way as to turn it into an expert interface (Alty and Guida, 1985) that can provide competent support and assistance. This approach leaves the rest of the system unaffected, and can be applied in existing process environments, without the need for restructuring the process control system or even the process itself.

The expected performance of such an expert interface includes a wide spectrum of capabilities and of man-machine interaction paradigms, which range from purely informative to substantially normative attitudes. Four main classes of tasks for an expert interface in a process environment can be identified (Meijer, 1983b):

- receiving data directly from the sensory and measurement system and the process control system, and interpreting them;

- continuously monitoring system operation, identifying possible abnormal situations, diagnosing malfunctions and faults, proposing appropriate interventions, and supervising their execution;

- managing a dialogue with the experts (specially the operators) in order to:

- acquire information about the process not available from on-line systems;

- inform the operator about process state, incipient abnormal situations, alarms, critical events, etc.;

- discuss doubtful cases and possible alternative solutions to technically complex problems generally involving the maintenance staff and the specialists;

- support the operator in decision making in uncertain and ill-defined situations, that are inherently or subjectively risky;

- aid the operator in performing critical tasks under often unfavorable stress conditions (such as alarm analysis, management of critical events, plant start-up and shut-down, economic optimization, etc.);

- propose appropriate interventions for optimization, repair, and maintenance;

- answer possible queries and provide justifications of its own reasoning process;

- tutoring and training novices or untrained technical staff.

The major benefits that are expected from the use of an expert interface with the capabilities listed above include:

- storing of well organized, documented, portable, and usable specific technical know-how;

- better work organization, safer working conditions, and more rational structuring of complex operative procedures;

- better use of technical personnel and improvement of quality and professional content of work;

- more automation and less dependency of the production process on the availability of highly specialized and experienced technical personnel;

- reduction in costs and improvement of quality and reliability of surveillance, diagnosis, repair, and maintenance activities;

- better efficiency and reduction of unavailability times for plants and machineries;

- better care of the plant and machinery, less maintenance, and less storage costs for spare parts;

- increase of quality and quantity of product.

5. A SURVEY OF THE AREA

A brief survey of the major projects on expert system applications in process environments developed in recent years is reported below.

REACTOR (Nelson, 1982) - developed at EG&G Idaho Inc., P.O. Box 1625, Idaho Falls, ID 83415, USA - is an expert system devoted to assist operators in the diagnosis and treatment of nuclear power plant (pressurized water reactor) accidents. It uses for

accident diagnosis both event-oriented and function-oriented
strategies, which are incorporated in production rules and
response trees respectively. The integration of these two
mechanisms of knowledge representation provides a powerful
combination for handling emergency situations. In fact, in many
plant diagnosis problems, it is difficult to determine the exact
cause of the fault. When this occurs, it is helpful to be able
to use function-oriented techniques to deal with the situation.

The system developed by Underwood (1982) - at the Georgia
Institute of Technology, School of Information and Computer
Science, Atlanta, GA, USA - is an experimental computer-based
nuclear power plant consultant for pressurized water reactors.
The system can interpret observations of a particular plant
situation in terms of a commonsense algorithm network model that
characterizes the normal and abnormal events of the reactor
plant. It operates in a basically forward chaining mode and
incorporates data on frequency of component failures and
abnormal events to aid determining certainty of competing
hypotheses. The system can diagnose the type of plant accident
and recommend appropriate interventions.

The work developed by Lusk and Stratton (1983) - at Argonne
National Laboratory, USA - concerns the potential use of expert
system technology in allowing nuclear power plant operators to
have high-level expertise readily available, specifically to
permit cooling to be directed to portions of the plant under
various conditions of equipment failure. A major contribution of
this work is the use of deep knowledge concerning the physical
nature and the basic engineering features of the process to
develop expert rules automatically.

Fortin, Rooney, and Bristol (1983) - at Foxboro, USA - consider
an alarm advisor application. They use both deep causal
knowledge as well as shallow heuristic knowledge to develop
domain specific reasoning and to control the overall deduction
process.

CFES (Meijer, 1983a) - developed at Combustion Engineering Inc.,
Instrumentation & Controls Engineering, Nuclear Power Systems,
Windsor, CT, USA - is an expert system aimed at assisting the
operator of a nuclear power plant to monitor and accomplish
critical functions (for safety, efficiency, security, etc.). It
identifies to the operator the state of the plant and the
critical functions in general, the paths that are available to
accomplish the critical functions, and the best path to select
for reaching a particular goal. CFES embeds an almost complete
machine-implementation of the cognitive process of an operator,
thus serving as an intelligent companion to aid in the decision
making process.

Sakuguchi and Matsumoto (1983) - at Mitsubishi, Japan - consider
the problem of restoring the normal operating state of an
electric power system, following the occurrence of a fault. Both
problems of diagnosis and intervention planning are addressed by
means of a rule-based approach.

HAPS is a novel expert system architecture developed by Sauers
and Walsh (1983) - at Martin Marietta Corporation, Artificial
Intelligence Unit, P.O. Box 179, Denver, CO 80201, USA - that
incorporates several features designed to facilitate the

implementation of large expert systems operating in real-time environments. HAPS includes modifiable sets of control strategies and conflict resolution strategies which make the system responsive to changes in its environment. It can also handle specific mechanisms, such as a production rule compiler, to increase the overall level of system efficiency.

Yamada and Motoda (1983) [see also: Motoda, Yamada, and Yoshida (1984)] - at Hitachi Ltd., Energy Research Laboratory, 1168 Moriyamacho, Hitachi, Ibaraki 316, Japan - consider the application of a diagnostic expert system to a nuclear power plant (boiling water reactor). The system integrates both event-oriented and function-oriented knowledge within a complex representation schema that includes production rules and frames. Inference mechanisms are based on forward chaining (with best first search) and resolution. A prototype version of the expert system is presently running in a simulated environment.

PDS (Fox, Lowenfeld, and Kleinosky, 1983; Fox, 1984) - developed at Carnegie-Mellon University, Intelligent Systems Laboratory, The Robotics Institute, Pittsburgh, PA 15213, USA, and at Westinghouse Electric Corporation, Pittsburgh, PA, USA - addresses the problem of on-line, real-time diagnosis of malfunctions in machine processes. The operation of the system is based on information directly acquired from distributed sensors attached to the machine and, therefore, includes a complex data interpretation activity in addition to diagnosis capabilities. The basic issues of spurious readings and sensor degradation are faced through new techniques called retrospective analysis and meta-diagnosis. The inference procedure utilized by PDS is backward chaining.

WASTE (Fox, 1984) - developed at Carnegie-Mellon University, Intelligent Systems Laboratory, The Robotics Institute, Pittsburgh, PA 15213, USA - is an expert system aimed at the domain of chemical wastewater treatment, specially at diagnosing faults such as clogged pipes and valves and broken pumps.

PICON (Moore, Hawkinson, Knickerbocker, and Churchman, 1984) - developed at LISP MACHINE Inc., 6033 West Century Blvd., Suite 900, Los Angeles, CA 90045, USA - is an expert system which performs forward and backward chaining inference in a real-time environment, with dynamic data obtained from measurements taken from a distributed process control system. Inferential and procedural heuristics give real-time alarm and control advice to the process operator. The first application of PICON is in a refinery. The expert system package can is easily extendable to a broad range of real-time applications for process control. It is one of the most recent and successful attempts in designing a general purpose tool for process control applications.

EX (Ali, Scharnhorst, and Chi, 1984) - developed at the University of Tennessee Space Institute, Knowledge Engineering Laboratory, Tullahoma, TE, USA - is a rule-based expert system for management of a coal-burning magnetohydrodynamic power plant. It is devoted to assist operators in monitoring and decision making activities, and, in addition, to serve as a consultant for process diagnosis. EX uses a modified form of backward chaining which includes the capability of dynamically generating new top-level hypotheses. It is presently running in a prototype version (concerning the first steps of the ignition sequence) in a simulated environment. It will be implemented at the real plant in the next future to supplement and possibly replace an existing traditional software package, that collects information on-line from various sensors, displays it through a graphic terminal in the control room, and assists operators to set control parameters.

SIDDA (Decreton, 1984) – developed at CEN/SCK Nuclear Research Center, Department of Technology and Energy, 2400 Mol, Belgium – is an expert system for managing a complex network, which centralizes the measurements and control operations of several nuclear experiments and processes. It is devoted to perform two main tasks: automatic development of specific measurements and control procedures, and automatic diagnosis of failures or malfunctions. SIDDA is a rule-based system which utilizes both forward and backward chaining strategies. Its use proved to bring an important saving of work and time as well as a great increase in system user-friendliness and overall reliability.

The work by Hery (1985) – at Electricity of France, Clamart Cedex, France – concerns a prototype expert system for conduct of a nuclear power plant (pressurized water reactor). The system operates in forward chaining and handles rules in first-order logic. The expert system is coupled to a plant simulator and its main operational use is computer-aided training.

LES (Scarl, Jamieson, and Delaune, 1985) – developed at NASA, Kennedy Space Center, FL 32889, USA – is an expert system devoted to detect and locate faults during loading of fluid oxygen in the external tank of the Space Shuttle. The controlled apparatus is quite complex and comprises both discrete and analog devices with a mixture of electronic, pneumatic, hydraulic, and mechanical control. LES processes data acquired from sensors, assesses the presence of faults, and decides whether it is necessary to stop fuel loading.

SAAP-2 (Klebau, Baldeweg, Fiedler, and Lindner, 1985) – developed at the Akademie der Wissenschaften der DDR, Zentralinstitut fuer Kernforschung, P.O. Box 19, Rossendorf, 8051 Dresden, GDR – is an expert system for disturbance analysis developed for application in a nuclear power plant. The system, which has been designed to meet real time constraints, collects data directly from plant sensors, and extracts from them all relevant findings for malfunction detection and diagnosis. SAAP-2 uses a data-driven inference mechanism which operates on knowledge compiled into a causal network in order to improve run-time efficiency. The operator interface has been designed according to specific functional and ergonomic man-machine communication criteria, in order to ensure end-user acceptability.

PROP (Gallanti, Guida, Spampinato, and Stefanini, 1985) – developed at CISE s.p.a., P.O. Box 12081, 20134 Milano, Italy – is an expert system for on-line monitoring of the pollution of cycle water in a thermal power plant. Its main features will be discussed in detail in the following section as a case study.

6. A CASE STUDY: THE PROP SYSTEM

In this section we analyze in some detail an application to
on-line pollution control of the cycle water in a thermal power
plant, developed at CISE s.p.a. (Milano, Italy) under a research
contract of ENEL (Ente Nazionale per l' Energia Elettrica –
Direzione Studi e Ricerche) (Gallanti, Guida, Spampinato, and
Stefanini, 1985).

A simplified schema of a thermal power plant is reported in
Figure 4.

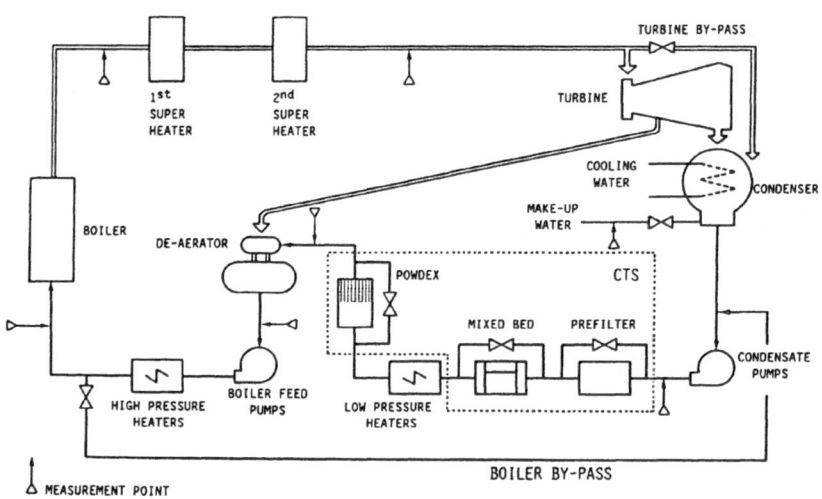

Fig. 4. Simplified schema of a thermal power plant.

In order to ensure the correct operation of the plant and to
preserve the materials constituting some critical subparts, the
cycle water must be kept as pure as possible. In fact, possible
pollutants dissolved in the cycle water may react with the plant
in different ways, mainly at high temperature, giving both
oxidation products and acids that can damage the tubes, the
walls of the boiler, and the rotating parts of the turbine.
Therefore, a chemical treatment subsystem (CTS) is included in
the plant and is used by an operator to clean the cycle water
whenever it is affected by some pollutant. The CTS includes two
different kinds of filters:

- the mixed bed, containing ion exchanging resins, that
  operates a chemical filtering;

- the prefilter and the powdex, containing cellulose and
  pulverized resins respectively, that remove oxides through a
  physical filtering.

The insertion level of the filters is regulated by the operator
depending on the nature and quantity of the pollutants present
in the cycle water.

Pollutants of different kinds can enter into the cycle:

- marine or river water carrying salts and sand (through the condenser);
- air (in particular, oxygen and carbonic dioxide);
- cellulose and resins released by the CTS itself.

Pollution phenomena are controlled by continuous measuring of several chemical parameters (total conductivity, acid conductivity, oxygen concentration, etc.) at several points of the cycle (see Figure 4). The control equipment currently installed in the plant can automatically open the valves that by-pass the critical subsystems (e.g., the boiler or turbine) when one of the above mentioned parameters at a single measurement point increases beyond a fixed threshold level in order to avoid immediate damages.

The task of the operator is to prevent the pollution reaching this limit situation by properly activating the CTS as soon as some sensors show abnormal values. In fact, when a limit situation occurs, some components of the plant may have already been damaged, and, in any case, the by-pass of a critical subsystem causes a significant loss of power of the plant.
During working time a chemical staff can advise the operator on the most appropriate actions to be taken to control an incipient pollution phenomenon; during the night, week-end, and holidays the operator must follow the instructions in a handbook.

The operative procedure above described has often proved to be unsatisfactory for several reasons:

- The operator usually guesses that a pollution phenomenon is occurring only when the sensors show highly abnormal values, so that the corrective action is not as timely as it should be.

- Sometimes the presence of a pollutant can be detected only by correlating the temporal trends of different parameters. These correlations require expertise not available to the operator.

- The precise diagnosis of a cause of pollution is often difficult and may require a lot of reasoning on data collected from the plant and ad-hoc measurements, tests, and observations. Thus the operator often prefers a timely but poorly specific, and to some extent uncertain, intervention rather than more goal-oriented actions that can only be identified through a time consuming and complex investigation.

- The operator is inclined to reason in terms of short-term effects. Therefore, he often underestimates slight pollutions, that, however, can generate long-term damages.

- When the operator has to manage a situation without the advice of the chemical staff, he is often in trouble as the procedures specified in the handbook are too simple and schematic and turn out to be useless in complex cases.

In order to face these problems the possibility of using an intelligent computer-based system that can directly acquire chemical parameters from the sensors on the plant and give advice to the operator has been investigated.

This led to the design and implementation of PROP, a real-time expert system that can assist the CTS operator. The aim of the system is to avoid damages to critical subsystems and to limit plant unavailability caused by chemical agents, by means of an early fault detection and diagnosis and an appropriate intervention.

PROP directly acquires data from the plant through sensors at a fixed time rate. These measurements concern general information on the plant (e.g., flow rate, generated power, etc.), chemical parameters (e.g., acid conductivity, oxygen concentration, etc.), and states of some subsystems (e.g., mixed bed insertion, de-aerator valve position, etc.). Moreover, through the keyboard, the operator can introduce further information that cannot be directly acquired by the sensors (e.g., results of off-line chemical analyzes, etc.).

As soon as an anomaly is detected, PROP displays on the operator console a set of hypotheses that can explain the malfunctioning. Afterwards, the system keeps the evolving situation under control, tries to focus on the most plausible hypothesis, and suggests proper interventions. After the situation has been brought to normality, PROP closes the diagnosis/intervention session, informs the operator as to the success of the undertaken interventions, and resumes normal monitoring activity.

At the time the PROP project has been started (early 1984), the possibility of relying for the development of a fast prototype on existing high-level expert system building tools has been evaluated. From this investigation it resulted that none of the products available on the market at that time could adequately cope with the requirements of the specific application. In particular:

- The process to be monitored is dynamic and continuous, and, therefore, the diagnostic activity is largely based upon on-line observations of the temporal evolution of the system. The available tools, mostly devoted to off-line consultation tasks, did not offer adequate mechanisms to deal with time representation and time-dependent reasoning.

- The types of knowledge necessary to represent the process and the diagnostic activity include, along with declarative and fragmentary knowledge, also structured, procedural, and formal knowledge. None of the considered tools did provide suitable formalisms for representing and using both these types of knowledge in a natural and effective way.

Therefore, it has been decided to implement PROP from scratch, using a general purpose programming language (Franz LISP) and basic artificial intelligence techniques.

The general architecture of PROP is reported in Figure 5.

Three levels of knowledge representation are used to express different aspects of knowledge about the process and the expertise of the operators and chemical staff:

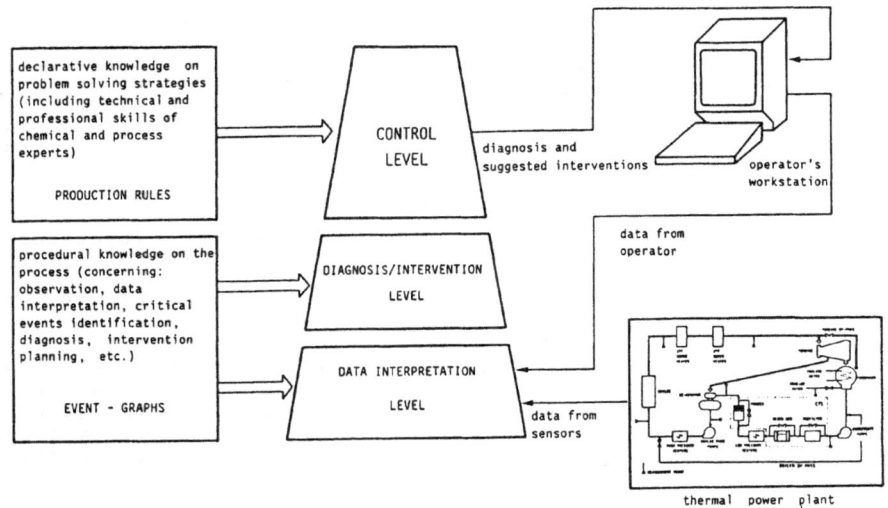

Fig. 5. General architecture of PROP.

1. data interpretation level, that comprises knowledge about
   how to observe the plant in order to translate parameter
   values and trends available from the sensory system into
   symbolic conceptual representation;

2. diagnosis/intervention level, that embodies knowledge about
   hypotheses, expectations, and recovery actions relating to
   each specific pollution situation;

3. control level, that manages relationships and conflicts
   among different competing diagnosis/intervention hypotheses,
   which are generated and carried out in parallel when an
   abnormal situation is detected.

Data interpretation and diagnosis/intervention knowledge is
naturally formulated in a procedural language based on
event-graphs. Control knowledge is more declarative,
fragmentary, and judgmental in nature, and is represented by
means of production rules.

A more detailed illustration of the technical aspects of PROP is
reported in (Gallanti, Guida, Spampinato, and Stefanini, 1985).

The PROP system is implemented in Franz LISP on a SUN-2
workstation, and it has been tested on recorded event patterns
comprising a wide variety of operational situations. A pilot
on-site installation of PROP is presently being developed.

7. CONCLUSIONS

The application of expert system technology to process

environments shows the typical features of a newly emerging field. The number of initiatives is limited and concrete results so far obtained are still scarce. Nevertheless, its impact seems to be large and profound:

- it promotes a better understanding of process environments, of their features, limitations, deficiencies, and requirements;

- it proposes a new sharing of tasks between man and machine;

- it indicates new roles for the experts (operators, maintenance staff, and specialists);

- it discloses a new knowledge-based perspective for process system analysis and design.

The major difficulties that presently limit the development of expert system applications in process environments are basically due to the state-of-the-art of expert system technology (Sauers and Walsh, 1983; Waltz, 1983). Specific research areas that are expected to have a major role in overcoming current difficulties and limitations include, among others:

- modeling of complex physical systems, including mechanisms for representing deep causal knowledge derived from physical laws and design principles (Bobrow, 1984; Chandrasekaran and Milne, 1985);

- heterogeneous (declarative and procedural, fragmentary and structured, etc.) knowledge representation and use (Georgeff and Bonollo, 1983; Gallanti, Guida, Spampinato, and Stefanini, 1985);

- temporal reasoning, including the capability of representing time relations and inferencing on them (McDermott, 1982; Allen 1983, 1984; Allen and Hayes, 1985);

- real-time inferencing, i.e. the capability for the expert system of running fast enough to ensure the strict response time requested for a correct interaction with the process (Sauers and Walsh, 1983);

- integration of expert system technology with traditional simulation techniques, control theory, and decision making models.

Several research efforts are at present ongoing on these challenging topics, and some of them are rapidly progressing. A detailed analysis of these issues is, however, outside the scope of this paper.

ACKNOWLEDGMENTS

The construction of the knowledge base of PROP (mentioned in the case study reported in section six) has been possible thanks to the experts of several ENEL Departments (Construction, Operation, and Research & Development Departments) who have cooperated with CISE in the frame of on internal ENEL working group. We are indebted to the above Departments for the support and contribution offered. In particular, we would like to thank Giovanni Quadri of ENEL-DCO, Flavio Bacci of ENEL-DPT, and Raffaele Pascali of ENEL-DSR.

A MODEL OF AIR COMBAT DECISIONS

Roger W. Schvaneveldt
New Mexico State University, Las Cruces, New Mexico (USA)

Timothy E. Goldsmith
University of New Mexico, Alberquerque, New Mexico, USA

1. Abstract

This paper reviews the current status of a project designed
to develop a simulation model of decision making by expert fighter
pilots in air-to-air combat. The model builds on our previous
work on measuring and representing conceptual structures by using
those structures to model the underlying processes involved in
flying high-speed tactical aircraft. The Air Combat Expert
Simulation (ACES) system incorporates selection rules written in
PROLOG to determine which of 17 basic fighter maneuvers to execute
given a description of an airspace with two competing aircraft
(T-38's). A representation of the aircraft and their flight
dynamics is written in Pascal. The aircraft are displayed in
three-dimensional graphics. The simulation allows maneuvers to be
selected for either aircraft by either a user or ACES. The
maneuvers are mapped onto inputs to dynamic flight equations that
update the airspace as maneuvers are executed. The selection of
maneuvers by ACES compares favorably with the selections made by
expert fighter pilots. Future directions for the model and
applications in fighter lead-in training are discussed.

2. Introduction

This report summarizes the first year of a project designed
to simulate decision making by highly skilled experts in a dynamic
environment. The model, Air Combat Expert Simulation (ACES), is
an expert system (Buchanan & Feigenbaum, 1978; Duda, Gaschnig, &
Hart, 1979; Nilsson, 1980; Shortliffe, 1976; Winston, 1979) that
simulates maneuver selection by fighter pilots in air combat.

With ACES, we intend to investigate several aspects of expert
system design and the problems of knowledge acquisition, knowledge
representation, knowledge use, and system evaluation. A primary
objective of the project is to better understand the nature of the
cognitive skills underlying pilot performance. Planning and
decision-making skills are of particular interest. We also hope
to understand how these skills are acquired and what distinguishes
an expert fighter pilot from a mediocre performer.

Although ACES is an expert system, there are some important
differences between ACES and most other artificial intelligence
(AI) expert systems. First, expert systems have dealt primarily
with static tasks in which the particular problem at hand does not

NATO ASI Series. Vol. F21
Intelligent Decision Support in Process Environments
Edited by E. Hollnagel et al.
© Springer-Verlag Berlin Heidelberg 1986

change much over time. In contrast, ACES is concerned with a dynamic, rapidly changing environment. It is likely that existing formalisms for representing and using knowledge in expert systems will prove inadequate in dynamic environments. Second, an important goal of ACES is to consider the psychological validity of certain aspects of its computational formalism (cf. Hollnagel, 1985). In contrast, most expert systems are motivated primarily by performance criteria and are not intended to be psychological models. Our initial efforts, however, have been largely guided by the goal of achieving performance levels that approximate those of expert pilots. The next phase of our work will be more concerned with issues of psychological validity.

We have employed techniques of knowledge engineering in AI (cf. Hayes-Roth, Waterman, and Lenat, 1978) and expertise research in cognitive psychology (cf. Gammack & Young, 1985) to understand the cognitive skills underlying pilot performance. Scaling procedures have been used to identify and represent the conceptual structures of pilots with various levels of expertise (Schvan-eveldt, et al. 1982, in press). From discussions with expert pilots and the results of psychological scaling with network structures (Schvaneveldt, Durso, & Dearholt, 1985) and multidimensional spatial representations (Kruskal, 1964; Shepard, 1962a,b), we have identified information critical for selecting maneuvers, strategies for maneuvering against opponents, and means for evaluating relative positional advantages.

Another goal of ACES is to teach student pilots the cognitive skills required in selecting appropriate maneuvers. We envision ACES serving as a desktop training system to supplement students' academic training. It should be emphasized that ACES is not intended to teach the perceptual-motor skills required to actually maneuver aircraft. Instead, its teaching potential lies in its ability to convey the mapping of particular conditions onto appropriate actions. The extent to which training with ACES is successful will provide another means of validating the model.

3. The ACES Model

ACES is a computer simulation model of the cognitive skills pilots employ in air-combat maneuvering (ACM). ACM occurs when two or more opposing pilots attempt to obtain an advantageous position over one another. The problem domain of ACM is cognitively rich and offers much of the complexity found in other problem solving tasks such as chess. Some of the components of ACM include predicting what an opponent will do next, planning a course of action, executing actions in a timely manner, and continuously evaluating and updating a plan of action.

A set of fundamental flying maneuvers forms the basis of ACM. This set consists of eight offensive and six defensive maneuvers. The major objectives of offensive maneuvers are to: gain energy to effect closure on an opponent, control overtake to prevent over-shooting an opponent and thereby losing an offensive position, and decrease the angle between one's nose position and an opponent's tail. Defensive maneuvers are performed to: decrease closure, cause an opponent to overshoot, and reduce an opponent's nose position advantage.

Two airplanes in combat are either in a neutral position or one has a tactical advantage over the other. There are two major factors which determine a tactical advantage: the airplanes' relative positions and their relative energy states. A neutral state might exist, for example, if both airplanes are meeting head on or if each airplane is behind the other and heading in opposite directions. A non-neutral state exists if one airplane is offensive and the other is defensive. An example of an offensive position is that of an airplane flying behind its opponent with both airplanes heading in the same direction. The airplane in front is defensive. An airplane's energy state is determined by its airspeed and its altitude. In general, a high energy state relative to one's opponent is an advantage.

The decision-making process of selecting air-combat maneuvers occurs in a dynamically changing environment. Fighter pilots continually evaluate a situation, choose a course of action, and execute and modify these actions. However, in addition to the real-time decision processing that occurs in the airplane, fighter pilots also reason about air combat maneuvering at a more abstract level. Pilots routinely review actual air combat engagements in debriefing sessions. Typically these sessions involve describing the air combat arena at a single point in time and discussing what actions would be appropriate. Moreover, during their training, fighter pilots are required to select actions for airplanes involved in a hypothetical air combat engagement. Hence, fighter pilots are accustomed to evaluating air-to-air combat engagements on the basis of abstract descriptions.

Represented within ACES is an airspace state consisting of two opposing aircraft of equal capabilities (T-38s). Each airspace state is a snapshot in time describing the positions, orientations, and airspeeds of the aircraft. ACES begins by assigning values to each aircraft creating an initial state. Next, each aircraft cycles through a series of stages that include selection, action, and updating. During selection, an action is chosen that is deemed appropriate for the aircraft to perform at that time. Currently, the major focus of ACES is on selecting appropriate actions. Action selection is guided by a model of the planning and decision making skills of an expert pilot. After an action has been selected, the aircraft performs the chosen action. ACES then updates the airspace state to reflect the new aircraft values. ACES has each aircraft take turns cycling through these three stages.

Also represented within ACES are flight equations to describe the aerodynamic characteristics of high-performance aircraft. Each aircraft action is eventually specified as a set of inputs to these equations. After an action is performed, the equations give the aircraft's resulting flight characteristics. This provides a realistic task environment in which ACES operates.

ACES is based on a production system architecture (Anderson, 1976, 1982; Newell & Simon, 1972, Young, 1979). The database represents an airspace state at a particular point in time. It contains information useful for selecting maneuvers. Some examples of items in the database include an aircraft's position, orientation and airspeed. ACES' production rules are if-then

statements that describe conditions for performing particular
maneuvers. Our current efforts are directed at introducing goal-
structured actions into the production rules. We eventually
intend to include the formation and evaluation of plans within
ACES' production system.

3.1. System Details

Representation of the airspace. Each airspace state is
defined by a record of information that describes the conditions
of the aircraft at each point in time. ACES stores information
about each aircraft, including location (a point in three-
dimensional space), orientation (pitch, bank, and heading), and
airspeed. Relative information such as closure rate, angle-off,
and range is also stored. Additional information such as aspect
angle, g s, and clock position is derived from the stored
information. ACES maintains a second-by-second record of the
airspace states as it operates.

Database. The data base for ACES' production system consists
of the airspace information that is considered by the production
rules. The database currently consists of three lists of data
items. One list contains information pertaining to CHEVY (one
airplane), one list contains information pertaining to FORD
(another airplane), and one list contains information relative to
both airplanes. Table 1 shows a sample database.

Table 1

A Sample Database

Aircraft Values		Meaning
CHEVY	FORD	
7000	7000	Altitude (feet)
0	165	Aspect Angle
340	360	Airspeed (knots)
Behind	Front-of	3-9 Position
0	0	Pitch (angle)
12	6	Clock Position

Relative Values	
0	Altitude Difference (feet)
9000	Range (feet)
19	Angle Off
-18	Closure Rate (feet per second)

The CHEVY and FORD lists contain the same data elements.
Most of these are self-explanatory, but some of the elements are
not generally familiar. Aspect angle is the angle between the
line-of-sight from an aircraft and the longitudinal axis of the
opponent. Aspect angles vary from 0 to 180 degrees. Low aspect
indicates that an airplane is behind the opponent, a desirable
position in air-to-air combat. An airplane's 3-9 plane is an
imaginary geometric plane that extends perpendicular to its
longitudinal axis at its wing line. An airplane is always in one
of three positions relative its opponent's 3-9 plane: behind,
front-of, or on. The relative list contains four data elements

that are relative to CHEVY and FORD. Angle off is the angle
between two airplanes longitudinal axes, or more simply their
heading difference. Closure is the rate at which the two
airplanes are moving toward (positive values) or away (negative
values) from one another.

Production rules. ACES has been designed to select an
appropriate action for either aircraft given a particular airspace
state. The set of actions considered by ACES include twelve basic
flying maneuvers plus five additional actions including two for
deploying weapons. Each production rule in ACES corresponds to
one of the seventeen actions known to ACES. Each action has
several production rules. Production rules are represented in
micro-PROLOG as single clauses. The head of each clause specifies
a particular maneuver and the body of the clause specifies the
conditions (see Table 1) under which that maneuver may be
selected.

Interpreter. The ACES interpreter functions in much the same
way as in other production systems. Given a description of the
state of the world, the interpreter checks each production rule
against the database. If more than one maneuver has conditions
satisfied, a conflict resolution strategy determines which
production rule is chosen. The particular conflict resolution
strategy was chosen for its potential psychological relevance.
Conflict resolution occurs by ordering the list of production
rules along the dimension of selectivity. Rules with conditions
that have high generality (low selectivity) are given high
priority. The generality of a rule is determined by the number of
data items its conditions reference; fewer data items indicate
more general rules.

The psychological appeal of this strategy is based on the
need for pilots to select maneuvers very quickly in actual air
combat. Perhaps for the pilot, the maneuver chosen, in the case
where multiple maneuvers may be appropriate, is the one which can
be verified most quickly. In ACES, the amount of processing
required to check the conditions of a production rule against the
data base is determined by the number of data items referenced by
the rule. Production rules that are high in generality can be
checked more quickly than rules low in generality. The conflict
resolution strategy, then, is to be viewed as representative of
part of the actual processes employed in the selection of
maneuvers by pilots.

User interface. The current airspace state is described to
an ACES user in both tabular and graphic forms as shown in Figure
1. The table provides information specific to each aircraft, such
as heading, airspeed, and altitude and also information relative
to both aircraft, such as range and closure rate. The graphics
display depicts a three-dimensional image of the two aircraft.
The reference point for viewing the airspace can be changed by the
user to allow for various viewing orientations and distances. The
information table and graphics are displayed simultaneously. The
user can select maneuvers for either of the aircraft or allow ACES
to select for one or both aircraft. The user may also rerun the
execution of a maneuver, changing viewpoints if desired. In
addition to the standard maneuvers, the user may also select a
standard level turn for either aircraft.

	chevy	ford
ias	393	412
g	3.9	1.4
hed	349	50
fpa	-1	-12
att	4	-10
bnk	-78	-17
asp	167	60
clk	8	12
alt	12966	13107
max	9	13
pot	120	120
man	bt	ly

range	7703
ang off	62
closure	201
time	5

t)p r)t l)t c)ok
i)nit p)se n)ops
s)el d)isp x)it

N↑

Figure 1

3.2. Performance of ACES

The overall performance of ACES was tested by having both the
program and expert fighter pilots select maneuvers for a set of
airspace environments. The selections were then compared. Pilots
made selections using an earlier version of ACES that did not
include the graphics display. The tabular display is shown in
Table 2.

Table 2

Description of CHEVY and FORD in the Current Airspace

	CHEVY	FORD
ALTITUDE	7000	7000
KIAS	340	360
ASPECT ANGLE	0	165
3-9 PLANE	BEHIND FORD'S	FRONT-OF CHEVY'S
ATTITUDE	0	0
CLOCK POSITION	FORD AT 12	CHEVY AT 6

Relative Values in the Airspace

RANGE	9000
ANGLE OFF	19
CLOSURE RATE	-18

Actions

A ACCELERATION	E LAG-ROLL	I BREAK-TURN	M HI-AOA-ROLL
B BARREL-ROLL	F QUARTER-PLANE	J NOSE-LOW-TURN	N SCISSORS
C LOW-YOYO	G SEPARATION	K OPT-TURN-VERT	O REVERSAL
D HIGH-YOYO	H DEFENSIVE-TURN	L LEAD-TURN	P FIRE-MISSILE
			Q FIRE-GUNS

Eight air combat scenarios were developed in consultation
with an expert from Holloman Air Force Base. Each scenario
consisted of from 3 to 10 individual states, or airspaces. The
eight scenarios were divided into two groups. Scenarios A through
F contained 22 airspaces and scenarios G through H contained 18
airspaces. The scenarios were constructed so that the separate
airspaces would appear to be realistic continuations of an air
combat engagement. The scenarios represent a wide range of
tactical positions. For example, in some of the scenarios CHEVY
had a clear advantage, in others FORD had a clear tactical
advantage, and in still others the aircraft had neutral positions.

The airspace environments were evaluated by seven Air
National Guard Pilots (GPs) volunteers at Kirtland Air Force Base.
Scenarios A through E were given to 5 GPs and scenarios F through
H were given to 2 GPs. All of the GPs were familiar with the
maneuvers used in the study. An IBM personal computer displayed
the airspace environments to the pilots and collected responses.

Each pilot read a set of task instructions on the CRT. The
experimenter clarified the task if necessary. Pilots were shown a

description of an airspace and then asked to select maneuvers for
CHEVY. An example of a display shown to the pilots, including the
set of maneuvers from which to select, is shown in Table 2.
Pilots selected a maneuver by pressing the letter on the computer
keyboard that was adjacent to that maneuver. For each display,
pilots were asked to select the most appropriate maneuver for
CHEVY. After responding, the same display remained and pilots
were asked to select the second best maneuver for CHEVY, if
another maneuver was appropriate. If a pilot selected a second
maneuver, he was then asked to select the third best maneuver for
CHEVY, again only if another maneuver was appropriate. At any
point after selecting the first maneuver the pilot could indicate
that no other maneuver was appropriate.

Each of the states examined by the pilots was also evaluated
by ACES. Maneuvers that had their conditions satisfied were
placed in a conflict set. The conflict resolution strategy rank
ordered the set. The three highest ranked maneuvers were then
compared to the pilots' selections.

The analysis of ACES' performance centers around a comparison
of the maneuvers selected by ACES with the maneuvers selected by
the GPs for a given airspace. Similarly, the performance of each
GP is analyzed by comparing his selections with the selections of
other GPs. Analyses of this nature allow ACES to be evaluated in
an absolute sense and also with respect to individual GPs.

To compare an individual's (the term individual will refer to
either a GP or ACES) selections with the selections of a group of
individuals, a composite list of maneuvers is generated from the
group's selections. Each composite consists of maneuvers selected
for a particular airspace. To develop a composite, each maneuver
selected by a GP for a given airspace is weighted by the order of
selection. Composites were computed from all of the GPs who
evaluated a particular state. The performance of ACES was then
compared to these composites. Composites were also computed from
a group of GPs with a single GP removed. The selections made by
the removed GP were compared to this composite.

The first analysis examined the number of maneuvers selected
in common between an individual and the composite. The number of
maneuvers shared between these two maneuver lists was computed for
each airspace and for each individual. The results of this
analysis are shown in Table 3.

Based on an overall analysis of matching maneuvers, ACES'
performance was quite good and generally approximated the
performance of the GPs. The probability of obtaining by chance
the number of matches observed for ACES was less than .001. A
similar claim can be made for all of the GPs except GP3. GP3's
poorer performance resulted from his choosing only one maneuver
for each airspace. Thus, with this selection strategy the best he
could have attained was 33%.

On examining the number of maneuvers selected by ACES, it was
discovered that ACES selected fewer maneuvers overall than the
GPs. On closer inspection, a common set of maneuvers was
routinely found in the group composite with which ACES was
compared but did not occur on ACES' conflict resolution list.

Apparently, the conditions of some of the production rules were
too narrowly defined. One explanation for this may be that the
production rules of ACES were developed according to classic
textbook definitions of air combat maneuvers. However, the GPs at
Kirtland used certain of these maneuvers more generally.

Table 3

The Extent to Which an Individual's Selections
Matched the Selections of His Comparison Group's Composite

	INDIVIDUAL	PER CENT	PROBABILITY*
	GP1	56.1	<.001
	GP2	45.5	<.001
	GP3	24.2	.09
SCENARIOS	GP4	36.4	<.001
A through E	GP5	57.6	<.001
	ACES	45.5	<.001
	ACES2	72.3	<.001
	GP6	74.1	<.001
SCENARIOS	GP7	75.9	<.001
F through H	ACES	63.0	<.001
	ACES2	90.7	<.001

*Probability of obtaining at least the number of matches by chance

---

As an example, ACCELERATION was initially designated to be
selected by an airplane in an offensive position and with his
opponent clearly defensive. The conditions for ACCELERATION
stipulated that the selector be behind his opponent's 3-9 plane
and that his opponent be in front of his own 3-9 plane. The GPs
selected ACCELERATION in a variety of other combat situations.

To evaluate the importance of relaxing the conditions on some
of the selection rules, a new version of ACES (designated ACES2)
was developed. ACES2 attempted to capture a broader meaning of
certain of the maneuvers. This was accomplished by revising the
conditions of the production rules for those maneuvers that
appeared to have been too restrictively defined. Based on
conversations with GPs at Kirtland, and on the actual data
obtained during the testing phase of ACES, the conditions of the
production rules of seven maneuvers were broadened for ACES2. The
other ten maneuvers remained unchanged along with all other
aspects of ACES.

ACES2 was presented with both sets of combat scenarios. Not
surprisingly, ACES2 generally selected more maneuvers for a given
airspace than ACES. In 83 percent of the airspaces, ACES2
selected three maneuvers. The GP's composite lists contained
three maneuvers in 75 percent of the airspaces. In comparison,
ACES had selected three maneuvers in only 35 percent of the
airspaces. The number of maneuvers reaching the conflict set is
clearly more reasonable with ACES2. ACES2's performance was
evaluated with the same composites used to evaluate ACES. The
results are shown in Table 3. ACES2 had 27 percent more matches
than ACES.

A second analysis determined whether the first maneuver on a group's composite list occurred at all in an individual's list of maneuvers. The results of this analysis are found in Table 4.

Table 4

The Extent to Which an Individual's Selections
Contained the First Selection of His Comparison Group's Composite

	INDIVIDUAL	PER CENT	PROBABILITY*
	GP1	63.6	<.001
	GP2	72.7	<.001
	GP3	45.5	<.001
SCENARIOS	GP4	50.0	<.001
A through E	GP5	77.3	<.001
	ACES	63.6	<.001
	ACES2	77.3	<.001
	GP6	61.1	<.001
SCENARIOS	GP7	72.2	<.001
F through H	ACES	61.1	<.001
	ACES2	83.3	<.001

*Probability of obtaining at least the number of matches by chance

---

The optimal maneuver for a given airspace, as specified by the composite list, was selected by ACES, and also by ACES2, in a majority of the airspaces. The performance of ACES compares favorably with that of the GPs in this respect. Taken together, the above analyses indicate that ACES' selections matched the GPs' selections about as well as the GPs' matched one another. The even better showing of ACES2 suggests that the scope of certain of ACES' maneuvers is too narrow. However, the broader interpretation of the maneuvers designed into ACES2 may be specific to the GPs at Kirtland. Apparently there are regional differences in the interpretation of air combat maneuvers by U.S. fighter pilots. To broaden ACES' rules simply on the basis of the GPs' selections at Kirtland Air Force Base may be premature.

In conclusion, the pattern-matching architecture employed in ACES appears to be appropriate for modeling the selection of air-combat maneuvers in a simulated task domain. The actual maneuvers selected by ACES corresponded closely to those selected by the GPs. In addition, some support for the psychological validity of the conflict resolution strategy employed in ACES was obtained. Additional work in this area will investigate the correspondence of ACES and decision making by pilots in more detail.

4. Future Directions

We are currently taking ACES in several new directions. One effort is to incorporate planning and goal structuring into ACES' selection process. We shall also evaluate ACES as a teaching tool during fighter lead-in training by comparing the flying perform-

ance of students exposed to ACES as a supplement to their academic
work with a control group. A third line of development will
investigate the value of representing maneuvers in terms of
sequences of more basic actions (e.g., pull-up, unload, roll out).
With this new representation, selection rules can be defined at
either the maneuver level or the basic action level. Our work in
planning will benefit from this new level of representation. Some
planning is concerned with goals which require low-level control
such as gaining energy.

5. Acknowledgement

This work is supported by Contract No. F33615-84-C-0072 from
the U.S. Air Force Human Resources Laboratory. We wish to thank
Richard Tucker of Holloman Air Force Base, New Mexico for his air
combat expertise and Lt. John Brunderman of Williams Air Force
Base, Arizona for his assistance with the flight equations.

ARTIFICIAL INTELLIGENCE AND COGNITIVE TECHNOLOGY: FOUNDATIONS AND PERSPECTIVES

Michael J. Coombs
Department of Computer Science
University of Strathclyde
Glasgow, United Kingdom

Introduction

A major concern of presentations has been the analysis of process control tasks with respect to operator performance and it has been broadly concluded that increasing the "intelligence" of systems will enhance human abilities. The vision of a process run co-operatively between a human and a computer as interacting intelligent agents has left the realm of science fiction and is entertained as both possible and desirable by engineering professionals.

The irony is that this interest in sharing control has occurred just at the time when control theory has reached sufficient maturity to fully automate much of our world. We have, however, lost our taste for full automation, and wish to find means of engaging our machines as partners. We want to be able to monitor their reasoning, converse with them over motives, plans, inferences and conclusions, test our judgements against them, and in the event of a disagreement, be in a position to impose our own, human will.

The foundation for belief in the viability of automated intelligent decision support is, of course, the widely publicized success of artificial intelligence research in the area of expert systems. The notion of implementing in detail the actual problem-solving and advisory behaviour of a human expert within a program is very attractive in an area such as process control. The expert will always be on call, will not be affected by fatigue, information overload or adverse environments, and will not leave the company at an inconvenient time. In contrast to numeric control systems, processing decisions will, in theory, be readily understandable to operators without higher mathematics, and conversely may benefit from the direct inclusion of knowledge derived from the operator's practical experience. However, before we get carried away by possibilities, it is important to make some assessment of the ability of artificial intelligence to support the required cognitive technology, and to lay plans for research where it falls short. This is the objective of the paper.

NATO ASI Series. Vol. F21
Intelligent Decision Support in Process Environments
Edited by E. Hollnagel et al.
© Springer-Verlag Berlin Heidelberg 1986

The paper will first characterize the artificial intelligence approach to modelling cognition, particularly with respect to expert problem-solving. Secondly, there will be a discussion of the limitations of current expert system design for the implementation of systems where user understanding is at least as important as problem solution. Finally, methods will be proposed for confronting these limitations using techniques drawn from cognitive science and computational linguistics.

Artificial Intelligence and Artificial Reasoning

Recently, I participated in a university/industry study on the use of artificial intelligence (AI) in command-and-control which required a definition of AI to "focus the selection of important applications". Definitions are, however, difficult because AI illustrates a statement by Newell, one of the founders of the subject, that:

> "Scientific fields emerge as the concerns of scientists congeal around some phenomena. Sciences are not defined, they are recognized" (Newell, 1973, p1).

The phenomena around which AI has developed are mental, essentially human and commonly taken to require intelligence: natural language understanding, complex problem-solving, visual interpretation. The characteristic means of expressing theories is the computer program, and characteristic ideas are the computational notions of symbolic representation and interpretation. However, the notion of what is a good computational model of an intelligent process is far from clear. Pragmatic theorists want effective functionality, but do not necessarily want to simulate human methods; psychological idealists search for the "essential" form of human processes; psychological realists seek to model cognition in detail, including human mental limitations.

My claim is that the psychological idealists have provided the "foundation strategy" for AI research behind expert systems, with some support from the pragmatic theorists. Moreover, this strategy has inherent weaknesses for the design of decision support technology. To begin the discussion, I therefore wish to describe the program which became the basic pattern for psychological idealist problem-solvers - Newell, Shaw and Simon's Logic Theorist (Newell & Simon, 1972).

Logic Theorist (LT) was presented to the Dartmouth Conference in 1956 at which the name "artificial intelligence" was first used. The program aims to explore the way a mathematician is able to prove a theorem in predicate logic "even though he does not know when he starts how, or if, he is going to succeed". The objective is thus to design a theorem prover using the knowledge and procedures employed by human mathematicians; this contrasts with a pragmatic theorist's theorem prover which may use knowledge structures and methods not easily "brain computable" but proven to be valid according to the rules of logic - the application of the resolution principle to Horn clauses by a Prolog system, for example.

Axioms and theorems used in LT are taken from Russell and Whitehead's "Principia Mathematica", an expert work in the domain. There are five axioms, - e.g. (A v A) → A, A → (B v A), (A -> B) -> ((C v A) -> (C v B)) - while a simple theorem might be (p → ~p) → ~ p.

Proofs are conducted using the common human strategy of reasoning backwards from the proposed theorem. At each stage, rules-of-thumb - heuristics - are applied in an attempt to convert the problem to one of the axioms. Following human reasoning, heuristics are not guaranteed to produce a solution, although good ones will do so on most occasions. Heuristics include:

Detachment operator: IF problems B & axiom A -> B, THEN prove A.

Forward chaining operator: IF problem A -> C & axiom A -> B, THEN prove A -> C

Backward chaining operator: IF problem A -> C & axiom B -> C, THEN prove A -> B

Substitution test: IF theorem X contains variable A & an expression p, THEN substitute p for all occurrences of A

Replacement test: IF A and B are expressions connected by X (e.g. A -> B), THEN replace X by its definition (i.e. ~A v B)

The above heuristics are applied blindly and exhaustively with the aim of reducing the problem expression to a single axiom. The initial problem is first paired with each of the axioms, the Substitution and Replacement tests being applied in an attempt to generate a match. If these tests fail, the expression is declared as an "open problem". Being the only open problem, it is taken and submitted to each of the operators in turn, in search of a new expression which will pass the tests. Expressions generated by the operators which do not pass the tests are placed on a list of candidate open problems. Once all operators have been tried, open problems are removed from the list in first-on/first-off order and the cycle repeated with a new application of operators. A successful match between a problem expression and an axiom signals a solution, and the sequence of open problem/operator pairs leading to the solution constitute a proof.

Both conceptually and structurally, LT set in 1956 the framework for the cognitive technology of expert systems. LT emulates expert reasoning in the constrained domain of logical proof, using expression reduction and testing knowledge. Application of this knowledge generates a set of expression states, implementing a top-down state-space search for a path from the original expression to an axiom, this path representing a problem-solution. Expert system architectural principles are also present. LT execution is founded on a basic inference engine (conducting a top-down, breadth-first search) which applies heuristic knowledge (implicit productions) to a data-base representing the current state of solution (the open problem list). Moreover, problems of implementing an heuristic reasoning system within this architecture established the research issues central to AI.

The main concern of heuristic reasoning is to control the application of atomic units of knowledge to ensure a solution will be found (given available data), to minimize solution paths, and to minimize false trails. Heuristics may be incomplete (fail to identify theoretically derivable solutions), may not guarantee an optimum solution, and may generate many failure paths before eventually reaching a solution. Research on the control of heuristic knowledge took two directions.

Work in the 60s was dominated by the pragmatic theorist approach, involving a syntactic analysis of domains and the formulation of a small number of powerful heuristics to exploit the syntax. This usually involved standardizing and simplifying knowledge structures, and abstracting heuristics. Prolog systems illustrate this approach to theorem proving, which use the much simplified syntax of Horn clauses (universally quantified implications with a single conclusion and multiple ANDed conditions) and a single inference rule (the resolution principle).

The 70s saw the growing realization that abstraction was in itself unlikely to tame realistic knowledge domains, within which human experts excelled. This encouraged the injection of psychological realist thinking into AI, and more attention to the knowledge and strategies used by humans. Systems were allowed to contain large, relatively unconstrained, bodies of knowledge.

However, it was now no longer possible to guarantee either the efficiency or even success of execution. The expert system researchers approach to these difficulties was to open systems to user's semantic intuitions. If open problems are regarded as processing goals, and heuristics are regarded as methods of both satisfying goals and setting new ones, a trace of execution may be interpreted by the user as an explanation for a result. The user may evaluate the semantic sequence derived from the trace with reference to his understanding of the world, identify missing or erroneous information, convert this to system syntax and so modify the system heuristics.

The main attraction of expert systems as decision aids has been in exploiting this semantic window to a knowledge-based reasoning machine. However, problem-solving power and explanatory power appear to impose incompatible requirements on system design.

The Methods and Limitations of Expert Systems

Explanation and knowledge acquisition facilities are presented as defining features of expert systems, and are responsible for much of the interest. In addition to problem-solving, the technology promises to enhance user understanding of a domain by supporting personal knowledge-based experiments. The knowledge-base may be modified and results, traced in human terms, evaluated against human expectations. One example of the failure of this ambition - the tutorial use of MYCIN - is well documented (Clancey, 1981), blame being attributed to missing knowledge concerning domain structure and reasoning strategy which are important in teaching. However, there addition proved far from simple and required the radical modification of the original problem solver.

Clancey's findings question whether a problem-solving architecture will support other cognitive functions. A partial answer may be obtained from study of expert activities within industry, where it was found (Coombs & Alty, 1984) that individuals played three different roles: problem-solvers, tutors and consultants. Moreover, while close parallels existed between the behaviour of specialists in the first two roles and their automatic counterparts, we could find none for the latter. However, it was the latter role which was most valued and which industrialists most wished to automate.

As problem-solvers and tutors, experts pursued some clearly defined goal in their own domain, employing pre-defined classes of evidence and well exercised, conscious strategies. Much of the skill lay in matching the strategy of evidence collection to domain, the expert using his experience to optimize the activity, knowing which tasks to follow, which alternatives to consider and which to abandon under different conditions of evidence. The control of exchanges was principally with the specialist, as a problem-solver collecting evidence through questioning and as a tutor through evaluating performance. In both cases, the user is in some sense the patient of the expert's activity.

The third role, that of consultant, was the most frequently employed and the most valued. Here an expert acted as a knowledge broker, helping to integrate his knowledge with another expert over some domain problem. There was often no clear solution path, nor was the problem goal clearly located within the knowledge of a single individual. Indeed, the problem could be seen as covered by some new knowledge domain arising as the intersection of subsets of knowledge held by both participants and not having a separate identity prior to the problem.

An excellent description of the consultancy role is given by Hawkins (1983) in the area of petroleum geology. Although Hawkins distinguishes between the expert consultants and subject specialists as two types of scientific professional (not simply different roles played by one professional group), his account of interactions parallels ours.

Believing a stratum of a particular sandstone to have been reached in drilling an oil well, the user may wish to know its depth. Information may be obtained from a geologist, who will examine the features of rock fragments, and a log analyst, who will apply geophysical methods to assess the density of formations along the bore hole. Each method of data collection is subject to error (geological samples may be distorted by the cutting process and geophysical measures may be affected by conditions in the bore hole). The expert's role is to translate the implications of findings between specialists in an attempt to negotiate a model of the drilling event of interest which satisfies them both and which yields an answer to the user's query. The experts we observed played both a specialist and an expert role in interaction with a second specialist (to use Hawkins' terms).

The essential skill is one of model building by information integration, not of applying a pre-defined model. The expert uses his knowledge of the drill site and the drilling process to build a frame within which geological and log findings may be refined and integrated. During integration, the implications of data from one specialist must be translated so as to come under the scrutiny of the other, and so be rejected or prepared for inclusion. The model itself thus becomes the full answer to the user's query, from which specific aspects may be developed to address specific user utterances. The query "What is fine-grained sandstone doing in these cuttings?" may require attention to the presence of "fine-grained sandstone in the cuttings", "sandstone in the cuttings" or "fine-grains in the sandstone". Running the model provides an explanation for the answer.

My claim is not that current expert system architectures would be unable to support consultancy functions. It is rather that they are unsuited to combining the required set of tasks: learning during problem-solution, propagation of effects between interacting specialists, conflict resolution and interpretation of user queries. Consultancy programs using conventional architectures would thus be unnecessarily complex. To support this claim we must characterize the requirements and form of expert systems.

Expert system research follows in the AI tradition of heuristic problem-solving. Having found that many humanly computable problem-solving tasks could not be automated by methods of exhaustive enumeration, research concentrated on methods of reducing search by crafting heuristic knowledge to exploit the structure of the problem domain. Attention was thus directed at uncovering and representing the syntax of solution spaces, and developing general "control" procedures which maximized the likelihood of a solution within that syntax. This formalization process thus converts the semantic features of the problem domain necessary for the solution of a given problem (or class of problems) into a language of entities, structures and procedures.

This has a number of implications. Any traces of system execution will essentially be accounts of control decisions (e.g. "I will not explore option X because I have too many processes running at this point"), without any record of their semantic justification or implication. Moreover, while with a constrained problem some semantics may be recoverable from entity names, structure names and the inclusion of semantic meta-knowledge (e.g. Davis, 1982 - TEIRESIAS), this is not easy and will not necessarily be understandable to the user; some distortion of the domain may have been necessary in fitting it to a convenient syntax. In addition, the syntax chosen for convenient computer execution may not be human computable. Finally, unconstrained knowledge acquisition is difficult to implement given the risk that the new information may not fit the existing representational and control structures.

Expert system control structures may be classified in terms of the complexity of the knowledge structures within which they operate (Stefik, 1982). The simplest is one which factors into a number of independent sections. Here, a generate-and-test strategy may apply, with the requirement that some heuristic be found to recognize and stop the development of unsuccessful problem states. However, there are many problems where no such evaluator can be found. In this case, strategies involving abstraction of the problem space have been successful. These vary in complexity, depending upon the interaction between subproblems. Where there is no interaciton, a fixed sequence of abstract steps will be in order, while heavy interaction may require the conservative development of a number of subproblems at one time, and if this fails, intelligent guessing. In even more complex cases, many different problem-solving strategies (and/or representations) must be kept in play.

Within conventional expert system design, the control regime provides a syntax within which decision procedures are expressed. The method of implementing a particular control regime will depend on the programming paradigm selected. Rule-based approaches provide a simple illustration where it is often necessary to decide which of several competing rules should be used to generate the next problem.

A common strategy is to select the rule which tests more specific features of the current problem description. However, it proves very difficult to produce a reliable definition of specificity (Sauers, 1986). In "subsumption", for example, RULE1 is said to be subsumed by RULE2 if: 1) each of the antecedents of RULE1 maps to an antecedent in RULE2; 2) each antecedent in RULE1 is more specific than its related antecendent in RULE2; 3) each variable instantiation in RULE1 is more specific than its related instantiation in RULE2. Suppose we have two rules:

RULE1: If is-a X penguin   RULE2: If is-a X bird
        Then waddle X   Then fly X

and know that:

1.   is-a harry bird
2.   is-a harry penguin

RULE1 is more specific than RULE2 because "penguin" is more specific than "bird". However, it cannot be recognized only on the syntactic features of the two rules, and so would not be able to determine whether Harry waddles or flies. To do this we would have to change RULE2 to:

RULE2: If is-a X bird
          not(is-a X penguin)
          not(is-a X ostrich)
        Then fly X

which is a cumbersome expression of a simple fact that "penguins are exceptional birds in that they do not fly".

Psychological Realism in Consultancy System Design

Our concern is with the adequacy of AI technology for the implementation of consultancy systems. Such systems solve problems by helping users integrate knowledge from different subject areas over the problem domain. Through active involvement, they come to "understand" the solution in two senses: 1) they learn to "mentally model" the physical or logical relations which justify it; 2) they learn to derive the model from underlying primitives. However, this requires the computational system to employ processes closely related to human conscious processing. The system must develop and refine expert domain models in order to derive a solution, not just interpret the results of expert decision making in the mode of conventional expert systems.

Problem-solving techniques based on modelling are not new to AI. Diagnostic aids in medicine and electronics have used model-based techniques for some time (e.g. CADUCEUS - Pople, 1982; Davis, 1984), and the late 1970s saw many explorations of qualitative reasoning in the physical sciences based closely on human protocols (e.g. deKleer, 1975). However, there are significant differences between this work and consultancy system requirements. First, diagnostic systems usually use a pre-defined set of models to generate conclusions from user evidence. The emphasis is on the conclusions of reasoning, in the manner of conventional expert systems, rather than equally on the method. Models are "used", not "built", and thus systems cannot easily learn from experience and are not optimized for human extension. Secondly, work on qualitative reasoning either explores idealized representations which are too abstract to reflect human thought, or reflect too truthfully the warts of human thinking. A consultancy system, however, should be capable of constructing concrete models which closely reflect local physical reality, but be cognizant of abstract semantic relationships which may be used to control model construction and evaluation.

The proposed architecture answers user queries by the construction and evaluation of models. These models are built from primitive domain facts, which are interpreted using a procedural semantics. This "constructive" approach to cognition is well supported in psychology as fundamental to memory, formal reasoning and natural language understanding (Bransford, Barclay and Franks, 1972; Johnson-Laird, 1983). The common idea is that conceptual objects are represented as named procedures which are used to construct models out of the meanings (syntactically related terms) of expressions. Cognitive functions may then be described as operations upon these models. Some of these ideas can be seen in early AI research into natural language understanding (Winograd, 1972).

At Strathclyde we have been exploring model manipulative approaches to problem-solving and tutoring. Both areas employ similar procedures, which may be illustrated by a basic session with a prototype consultancy system advising on questions of quantificational reasoning over spatial relations. The system (a "spatial" expert) interacts with a user, who may be regarded as a different type of specialist in spatial reasoning. The question posed by the user, but standing before both agents, is:

In SITUATION1, where (p1) some C are right-of B1,
            and   (p2)   no A are right-of B1

        must   (c1)   all C be right-of some A?

SITUATION1 concerns an arrangement of 5 entities:

A1 left-of B1
C1 above   B1
A2 left-of A1
C2 left-of B1

Following procedure observed both by ourselves (Coombs & Alty, 1984) and by Hawkins (1983), the expert takes the initiative and tests whether the question is meaningful by attempting to model the described situation (all statements contained in the question). The models are built by activating the procedural semantic representations of all terms to compose arrangements of entities. These must comply with both the spatial constraints of SITUATION1 and the quantificational constraints of the question premises. The first model built satisfies SITUATION1 but not p1 and c1; the second satisfies all constraints. Thus the question is meaningful.

Q1:   C1   Q2:   C1
       C2 A2 A1 B1   A2 C2 A1 B1

Having validated the question, the system seeks an answer within its own resources. The answering procedure will depend on the question type, although in general this will involve a search for models which are counter-examples of question goals. In other words, the question will be answered by an attempt to falsify it. Search for counter-examples appears to underly not only much human reasoning, but also the strategies required for answering many different types of question (e.g. questions concerning causal antecedents and causal consequences - Lehnert, 1978). In the present case, the question involves logical entailment and thus the conclusion must be true of all possible models of SITUATION1. Models are generated by recursive revision, using the procedural semantics for relations. This gives three possible models.

M1:   C1   M2:   C1   M3:   C1
     A2 C2 A1 B1   C2 A2 A1 B1   A2 A1 C2 B1

The question asked "must all C be right of some A?". This condition is clearly true of M1 and M3, but not of M2, and thus the answer must be "no".

So far, model manipulation has provided an expensive means of doing theorem proving and would be better handled by resolution in Prolog, or even by an LT-like system. The real advantages only appear when a conclusion is challenged. Let us assume that the second spatial expert has a strong intuition that his conclusion is correct (he was simply seeking confirmation of a decision already taken). The system is thus asked to justify its result. With consultancy, the justification process is not passive, but is used as an opportunity either to learn from the user or to tutor him.

An expert's justification procedure will be determined by the original question, plus any qualifications present in the justification request and any knowledge of the user's concepts and goals. In the present instance, a

common strategy is to trace the construction of the counter example model.
At any point the user may challenge a move, and assert what he would have
expected. Discrepancies between actual and expected moves are used to
identify differences between the two experts semantics for the relation
driving the move. Once uncovered, these can be used either to tutor in the
system's view or may be remembered for future questions asked by the same
user (representing beliefs about the user - Wilks & Bien, 1983).

Transforming M1 into M2 requires swapping the positions of C2 and A2.
However, the present user rejects the swap, arguing that C2 and A2 should
be made to occupy the same position.

```
    USER:           C1
          A2/C2 A1 B1
```

This identifies the semantics of the "left-of" relation for investigation.
The system accesses the procedural definition, and seeks the minimal
modifications which would permit the user's placement. In the prototype,
which places entities according to X/Y co-ordinates, and seeks the next
available slot on the appropriate co-ordinate to fulfil the relation, the
minimal modification is to allow more than one entity to fill a slot. The
system then evaluates this semantic by seeking to falsify the original
question using it. This generates four models:

```
J1:         C1  J2:         C1  J3:         C1  J4:         C1
   A2 A1/C2 B1       A2 C2 A1 B1      A2/C2 A1 B1      C2 A2 A1 B1
```

the final model J4 again falsifying the conclusion.

This indicates that there must be some other difference in the semantics of
"left-of" between the two experts, and so the procedure is repeated with
the new models. Transformations between J3 and J4 are elaborated. Here the
user notices that the model J3 is in fact incorrect, C2 being remote from
B1. To him, "left-of" means "immediately left-of", hence all models but J1
are unacceptable. The expert system accordingly modifies its semantics,
verifies that J1 is the only possible interpretation of the question given
these semantics and accepts that "all C are right-of some A".

Conclusions

The example given in the last section illustrates a typical interaction with
a consultant. A number of differences should be noted with conventional

expert systems. Firstly, processing strategies are related to question type and are designed to parallel familiar human strategies. The question type is determined from a parse of the question, the words "all", "some" and "must" being significant in our example. However, if C1 and C2 had been real world objects (e.g. moving vehicles) separated by a barrier B1, the question might have been "can C1 and C2 touch?". This would have been parsed as a causal consequent question, the significant term being "touch".

A second difference is that model building forces relations implicit or ambiguous in a set of assertions to be made explicit, and so provide a context for the human evaluation of models. In addition to the relations specified in the assertions of SITUATION1, a user may observe of M3, for example, that "C2 right-of (A1 & A2)" and that "C2 not left-of A1". Moreover, it may be concluded of all models that "B1 right-of (A1 & A2 & C2)" and that "C1 above (A1 & A2 & C2)". With deeper procedural processing, it might be concluded that "B1 does not move". It should also be noted that model manipulation provides a method for undertaking non-monotonic reasoning, "C2 right-of A1" being true of M1 and M2 but not of M3.

Additional differences include providing the user with direct access to the semantics of entities and relations used by the system. The advantages of this are illustrated above during the justification stage of a consultation, when the system is attempting to learn about the user's point of view. Less obviously, the paradigm invites experimentation with constraints to keep models to a brain computable size and to enable users to select the entities, relations used within models. Finally, the semantic nature of the approach invites the use of psychologically valid control structures, employing notions based on memory (least-effort) and natural language processing (metaphorical relationship).

One difficulty with model manipulation approaches is the very large number of models which often need to be constructed to address question concerning entailment. While parallel computation may provide one solution, we have found that experts often make judicious use of domain knowledge to constrain possibilities. In the example given above, a "next-to" constraint on the "left-of" relation reduces possible models from 4 to 1. Indeed, much of domain expertise seems to involve sophisticated knowledge of such constraints; the expert knows when to constrain a theoretical scientists abstractions. At this point, research into model manipulative approaches coincides with work on conventional paradigms in the formulation of expert meta-knowledge and its use in control. Moreover, the implementation of meta-control knowledge requires the use of the very AI programming techniques used by Stefik (1982) - above - to classify expert systems.

HUMAN AND MACHINE KNOWLEDGE IN INTELLIGENT SYSTEMS

Edward Walker and Albert Stevens
BBN Laboratories, Inc.
Cambridge, Massachusetts, USA

1. Introduction

The discussion below is driven by examples of large and small intelligent systems implemented by a variety of researchers at BBN Laboratories over the past few years. Two motivations underlie this case history style of presentation. The first is *ad hominem*: we find the case method felicitous for expositing our work. The second, and more debatable motivation is our belief that AI is primarily an experimental science in which ideas and concrete contexts go together.

1.1. Abstract

In the course of the material presented here, we hope to support three general precepts for designing and developing intelligent human machine systems. The first is that for the types of systems we discuss, knowledge must be explicitly divided between machine and human in order to make best use of the capabilities of each. Such a division acknowledges that the state of the art in AI is such that only some aspects of intelligence are mechanizable. The second precept is that the many dimensions along which knowledge is represented must be made explicit in order to provide conceptual congruence between the machine and human intelligence of a given system. And the third is that the division of knowledge between human and machine and the complex character and detail of multi-dimensional representations of knowledge place particular importance on the design and implementation of the interface through which human and machine communicate.

The systems built recently at BBN have come to adhere to a standard organization which reflects these precepts. They are designed as interactive human-

NATO ASI Series, Vol. F21
Intelligent Decision Support in Process Environments
Edited by E. Hollnagel et al.
© Springer-Verlag Berlin Heidelberg 1986

machine systems; the intelligence allocated to the machine is organized as a multi-dimensional central knowledge representation with specialized computational and inferencing modules; and the interfaces for these systems permit humans to recognize, establish, or inspect the distinctions and organizing principles of the knowledge representation, as well as select and manage the application of the computational and inferencing components of the system for accomplishing a particular task.

2. Dividing Knowledge between Human and Machine

It is not a particularly subtle principle of system design to allocate tasks to the subsystems most suited to their execution. Given this principle, it is not surprising that tasks for which machines are better suited; such as, search, simulation, correlation, consistency checking, and accessing or maintaining large knowledge representations, are assigned to machine intelligence, while those which humans perform well; such as, spatial or temporal analysis, recognizing distinctions and similarities, or selecting among inference techniques are, for the time being, left to humans.

In an actual system, the line of demarcation between human and machine intelligence generally reflects the state of the art in AI understanding of the kinds of intelligence in the system. What is understood is on the machine side of the line and is, incidentally, no longer considered AI. What is not even partly understood is on the human side of the line and is not yet AI. Those aspects of intelligence which lie near the dividing line are in the process of being understood, either in the field at large or in the developmental course of the system at hand. These are AI; at least for now.

2.1. The Division of Knowledge: An Example

Consider the task of designing a telecommunications network.

In simple terms, the problem is to interconnect many sites with a network that provides an acceptable level of performance and is reliable and cost effective. Generally the first step is to connect sites to concentrators so that the group presents enough traffic to use expensive long—distance connections efficiently. The concentrators are then interconnected to form a trunking, or backbone network.

In practice, there is a complex trade—off of cost, reliability, performance, and various idiosyncratic "other" factors, and the job of designing a network requires a rather experienced expert, who uses a variety of computational tools to calculate cost, traffic flows, performance, and so forth.

The problem of connecting individual sites to concentrators——the access design problem——is relatively well understood, and currently available clustering algorithms provide access design that require minimal adjustment by human experts. However, algorithms have not yet been developed for designing the backbone network, although network designers have developed a variety of computer techniques for evaluating candidate designs, and they have accumulated considerable expertise for recognizing types of networks, as well as for criticizing designs and improving them.

This division is reflected in a network design system under development at BBN. A considerable amount of machine intelligence has been developed for arriving at an access design automatically, but backbone design has been allocated to the human designer, at least for the time being. The allocation of access design to the machine has resulted from a gradual shift of intelligence from human to machine. In the early stages of system development, human designers retained a considerable amount of expertise about how to use the clustering algorithms and adjust their results. System commands began to appear which reconnected sites that were closer to a nearby concentrator than to the one to which they connected by the algorithm. Heuristics for selecting the size of clusters became parameters for algorithms. The access design phase has become largely automatic.

Getting to the starting point of backbone design, a minimally connected network, can be difficult for humans, particularly for large networks. The task requires that at least one connection be made to each node in the network, while observing a variety of constraints. Creating an initial topology and adjusting it to meet design criteria is one of the central tasks of network design, and it remains largely a human task. Our system is beginning to approach acceptable performance for providing an initial topology, and getting started may be one of the first aspects of the backbone design task to be shifted from human to machine intelligence.

For adjustment task, the system provides a comprehensive design environment in which knowledge about the state of the design is conveyed via a sophisticated interface described below. Thus, the knowledge required for network design remains largely in human form, although we intend that the design environment of our system will create a testbed for acquiring and evaluating machine components of the backbone design task.

3. Representing Knowledge and Functionality in Multi-Level Systems

The design algorithms mentioned above constitute only one component of the machine knowledge represented in our network design system. The system also contains a large central representation of knowledge about network objects and their characteristics. This representation includes descriptions of the cost, capacity, location, etc. of network devices and of the characteristics of other objects such as the communication protocols the network employs, the tariffs which govern the cost of connecting sites, the nature and amount of the traffic in the network, etc. The explicit representation of such knowledge establishes conceptual congruence between the human and machine. The machine and the human know the same things about network objects.

3.1. System Organization

The organization of the Steamer system, which has been described in the literature, provides a palpable set of examples. Figure 1 shows the system organization in very schematic form. A central frame representation contains a large body of essentially static information which includes a formal description of the objects, principles and procedures that go to make up a steam in a particular state. Along with this store of knowledge, there are a variety of specialized program modules which support the use of the knowledge in a variety of applications.

Figure 1 about here.

3.2. Central Representation

The central store of knowledge contains several important kinds of representation. The first is a representation of the important concepts used in the domain. While Steamer is about a conspicuously physical world, the are many important abstract concepts as well. These include the notions of a "prime mover", a "line", a "positive displacement pump", a "valve", etc., all of which are used to model specific objects in the domain and construct new abstract objects. Besides concepts for abstract and concrete devices or systems, there are analogous abstractions of plant operating procedures. These include such notions as "align a line", "start a motor-driven pump", "light off a positive displacement pump", and so forth.

The second kind of representation in the system is that of the actual system being modeled and the specific procedure being executed. This representation includes such concepts as "Fuel Oil Service Pump 1A" in a particular state such as "suction valve open, discharge valve closed, and pump on". All the steps in the procedure being executed and constraints on the order in which they should apply, based on abstract procedure concepts, would apply. Figure 2 illustrates the interaction of abstract and concrete representations of structural and behavioral concepts.

Figure 2 about here.

3.3. Special Computation and Inference Procedures

Around the central representation of concepts is organized a set of computational and inferential modules which embody the system's ability to act in various application environments. These include modules for presenting graphic and textual views or descriptions, procedure monitoring and configuration checking components, and simulations of the operation of the steam plant.

Steamer graphics include animated visual models of systems and subsystems which are directly coupled to the simulation of the plant and can be manipulated by the user of Steamer to investigate the operation of the plant. Textual presentation of information contained in the central knowledge base is structured and permits students to browse in a so-called hypertext fashion to the depth of description they desire. (Steamer's graphic and textual presentation facilities form part of the discussion of intelligent interfaces below.)

The procedure monitoring and configuration checking components are intelligent. A very simple inference process makes heavy use of the information in the central knowledge representation to monitor students executing procedures or trying to construct a plant system or subsystem. The monitoring process notes each step in the procedure being executed by a student and compares that step to the corresponding step in all abstract procedures which apply. Students are notified of any violations of abstract procedures. If the student is constructing a control system, for example, the central representation applies its knowledge that a control system must measure some parameter external to it. If the student's system does not contain some provision for measuring an external parameter, Steamer will advise him of that lack.

Steamer includes both a quantitative simulation converted from a FORTRAN simulator used by the Navy to drive a plastic mock-up of a steam plant, and a qualitative simulation based on information stored with each object type. Using this stored information about functional behavior expressed in qualitative terms, the system can determine how a collection of objects will behave in response to some perturbation.

3.4. Dimensions of Representation

It is worth noting that the system we have described represents knowledge at multiple levels in a variety of dimensions. In particular, we have noted that concepts about steam plant objects and about procedures for controlling subsystems range from those which are quite abstract-- "prime mover"-- to those which are quite concrete-- "fuel oil service pump 1A". The information stored about the behavior of objects also varies from the quite concrete characteristics involved in the quantitative simulation to that involved in the less well understood qualitative simulation. A third continuum ranging from utterly static to dynamic properties of objects is also represented.

4. Intelligent Interfaces

Given the view that an effective human–machine system, at least for the kinds of tasks being discussed here, is one in which there is a cooperative division of knowledge and labor between human and machine, and given the complexity and degree of detail of the knowledge representations which underlie such systems' performance; it is clear that the communications interface between human and machine is a critical component of the system. The interface must make it possible for the human to access any information needed for the task at hand, as well as convey to the machine any information provided by the human.

Besides mere access to information, the interface must provide appropriate modalities and techniques for presentation and explanation. For most applications this implies not only a capability to arrange the presentation from a variety of user perspectives, but also the capacity to make use of a rich variety of modalities and presentation techniques, while observing sound principles of graphic design, text composition, and so forth. Such an interface is hard to build, but its importance to the performance of the system can hardly be overestimated.

4.1. Perspectives on Knowledge

Representations of knowledge such as that described above for the Steamer system allow one to adopt a variety of perspectives for viewing the facts, structural relationships and behavioral interactions which go to make up the knowledge represented in an intelligent system. In the case of an intelligent tutoring system like Steamer, the most obvious viewpoints are those of the student, the instructor and the subject matter expert, as indicated in Figure 3. As the Figure indicates, some aspects of a particular user's viewpoint probably require specific interface features; however, these specialized interfaces build on a general facility for interacting with multiple representations of knowledge. For large systems, the number of perspectives will be large, and efficient, multipurpose use of knowledge will be a requirement of the central representation.

Figure 3 about here.

The overall requirements of a set of expert systems being planned and developed under the auspices of the DARPA Strategic Computing Program provide an illustration of this point. A number of research organizations, including BBN Labs, are participating in research and development projects directed toward the kind of intelligent systems required for the management of the US Navy's Pacific Fleet, systems whose function ranges from the logistical aspects of managing the real—time deployment of several hundred ships, to evaluation of their current tactical or strategic capabilities, to real time simulation of fleet maneuvers.

The basic knowledge required for these systems obviously includes information about the capabilities, physical characteristics, and equipment of units-- a more detailed version of the data in Jane's Fighting Ships--and their position and state of readiness. However, although different types of users required similar data, they view those data from quite different perspectives. For example, a logistics officer's interest in a ship's schedule of port calls may be quite different from that of an officer attempting to reschedule the ship to serve a new mission. Both the interface and the underlying knowledge must be organized to provide such perspectives.

4.2. Spatial, Textual, and Auditory Presentation Modalities

The network design system discussed earlier provides an especially useful example of the value of multiple modalities of presentation. A great deal of information concerning the location, configuration, utilization, name, and other characteristics of components of the network must be accessible at some stage of the design process. To make all information available simultaneously or to restrict usage to one modality or perspective at a time would be either confusing or clumsy. However, it is apparent that each modality is particularly suited to particular presentation problems.

The solution we have adopted in the network design system is to provide both geographical and tabular forms of the data available and to provide a variety of mechanisms for controlling the operation of the interface to gain a closer look at the data or convey detailed specific information within a general context. Thus, use of the color dimension is used to show relative utilization within a geographic display of network topology; pop-up menus allow user to ask for additional detail and choose modality from either graphical or tabular displays of the same object; a preliminary version of a general mechanism for selecting sets of objects to be presented has been provided; and a structured text description of network components like that in Steamer has been implemented. While we have not dealt specifically with interface design principles, we acknowledge that observing such principles is a critical requirement in the implementation of the interface.

The basic characteristics of this integrated interface are shown in Figure 4.
Figure 4 about here.

We have observed that certain interface features are broadly characteristic the particular application of the system. The tabular form of data in the network design interface conveys exact information about utilization, location, etc. and, because the lines in the table reflect a rank ordering by utilization, the relative utilization of each device. The graphic presentation, on the other hand, makes connectivity and proximity apparent. We have recently added a graph output capability to the system which can be used to make comparisons among several design criteria in order to assess the trade-off between such parameters as cost and utilization.

5. Conclusions

The discussion above illustrated the need to divide knowledge between the human and machine components of an intelligent system, and the need for multi-dimensional representations of knowledge. With representations and interfaces like those described above, it is possible to construct a presentation of information which either resembles the physical arrangement of objects in order to convey physical relationships or ignores the physical arrangement of an actual system in order to illustrate functional relationships with greater conceptual clarity. These capabilities are particularly useful in developing and maintaining intelligent systems for design, monitoring and control. The views of the system can be made active and can reflect both the actual and conceptual organization of the system, and they can provide a conceptual view of the operation of the system, the flow material within the system, or the effect of control measures.

Such human-machine systems require powerful and well managed interfaces capable of conveying the great complexity and detailed specific content of the knowledge contained in the system. In fact, we believe that our experience argues for a completely transparent interface-- one that provides effective access to all information represented in the system, as well as an effective means for acting on that information. Controlling the richness such interfaces to avoid clutter and overload of the user is a challenging problem. However, an active, intelligent interface, linked strongly to a detailed central representation of knowledge, allows system designers to choose exactly the presentation required by the communicative intent of the system and the special abilities of the user.

As Mark Twain said about the choice of language, "The difference between the right word and the almost right word is the difference between lightning and a lightning bug."

Steamer Organization

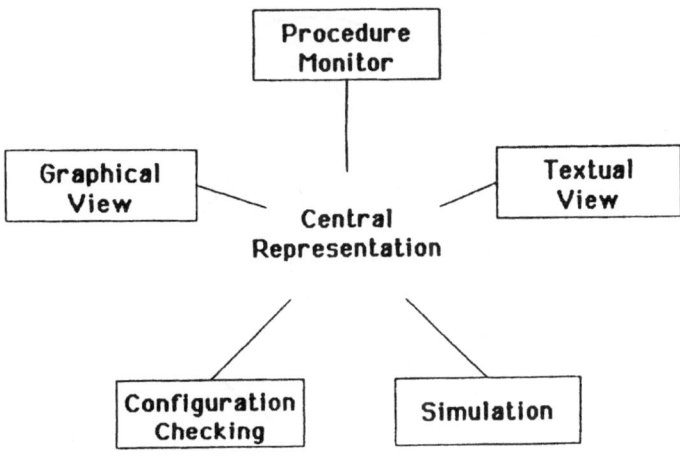

Figure 1.

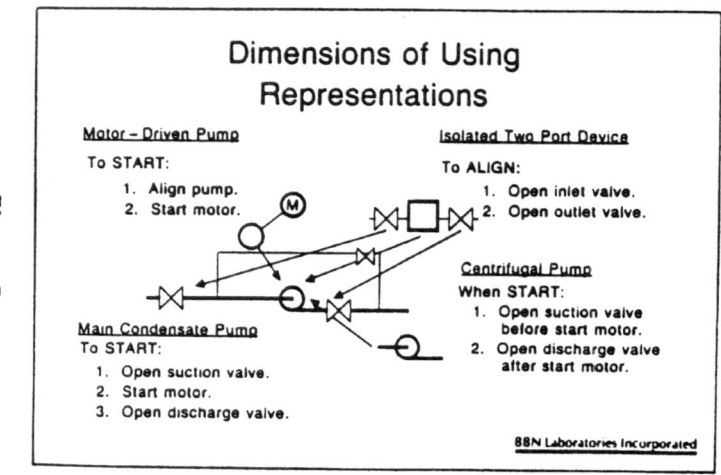

Figure 2.

Architecture of an Expert Instructional System

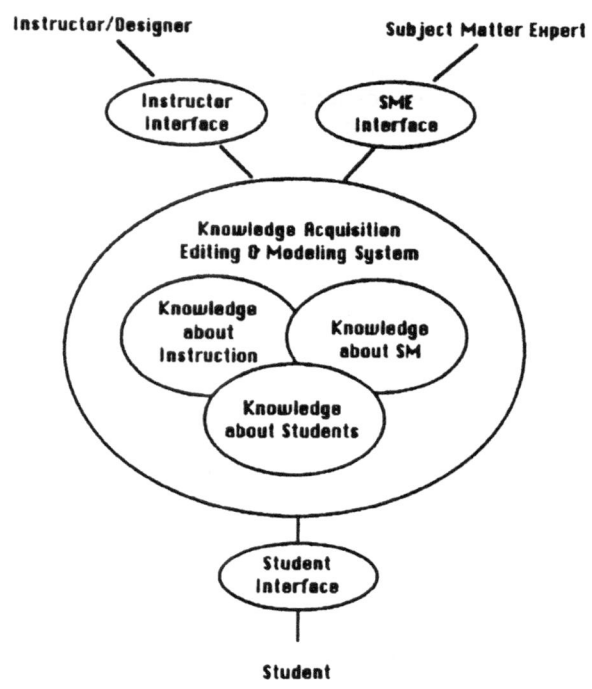

Figure 3.

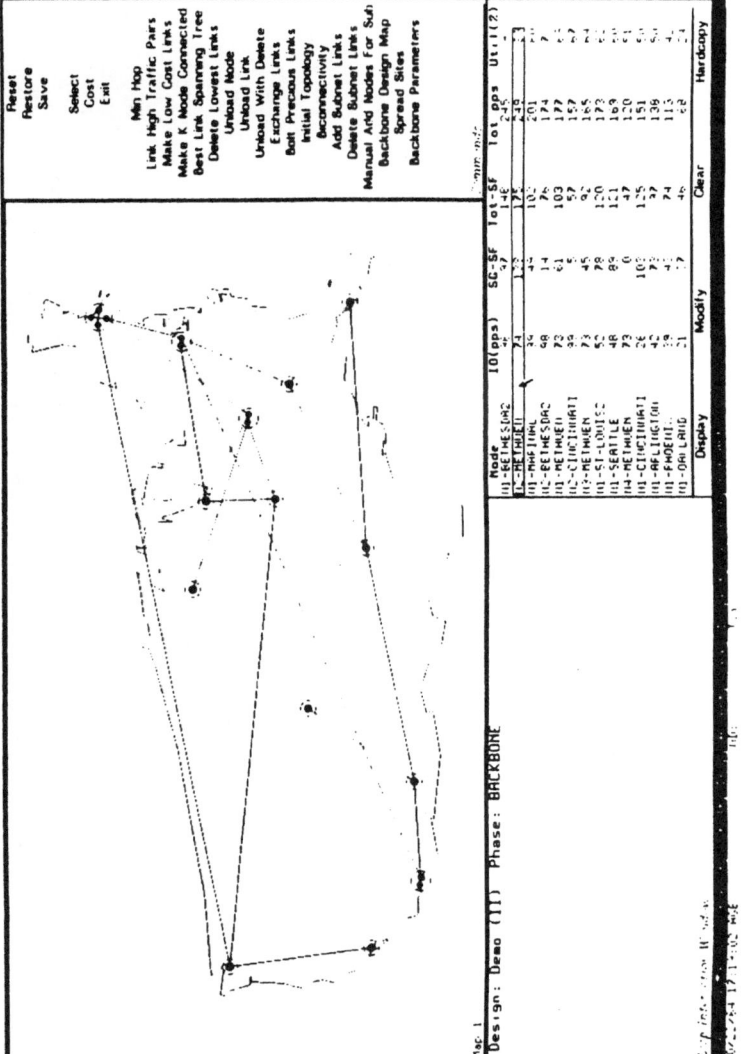

Figure 4.

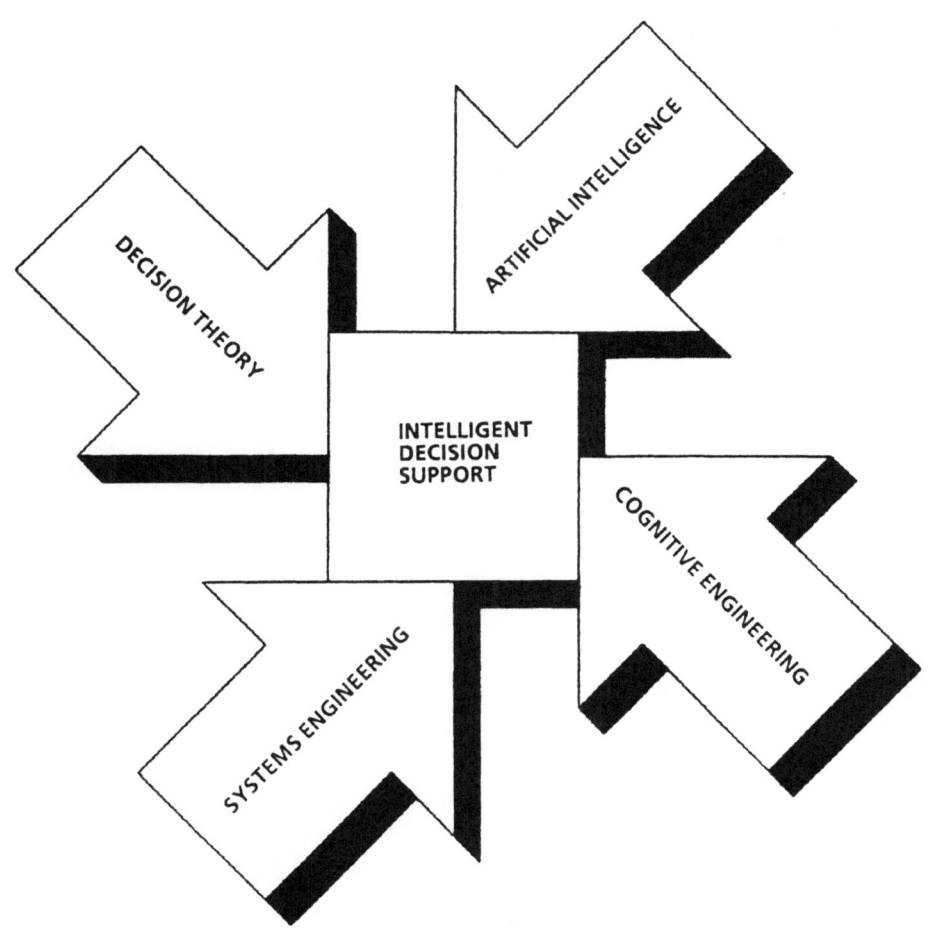

DECISION THEORY

ARTIFICIAL INTELLIGENCE

INTELLIGENT
DECISION
SUPPORT

COGNITIVE ENGINEERING

SYSTEMS ENGINEERING

CONCLUDING COMMENTS

PANEL DISCUSSION ON DECISION THEORY

Edited by George Apostolakis

Mechanical, Aerospace and Nuclear Engineering Department
University of California, Los Angeles, CA, USA

This essay is a summary of the views that were expressed at the
NATO Advanced Study Institute on Intelligent Decision Aids in
Process Environments during a panel discussion. The panelists
were D. Dubois, Université Paul Sabatier, Toulouse; B. Fischhoff,
Decision Research, Eugene; P. Garbolino, Scuola Normale Superiore,
Pisa; G. Volta, CEC-Joint Research Center, Ispra; and W. Wagenaar,
University of Leiden, Leiden. The panel was chaired by the
author, who also assumes full responsibility for the contents
of this summary.

A normative decision theory is recognized as one that is based
on a set of axioms that guarantee self consistent (coherent)
conclusions and inferences. Such a theory has a clearly defined
goal, e.g., the maximization of expected utility, and establishes
a set of rules that guide the decision maker to that goal. Compliance with the rules of a normative theory is often identified
with rationality.

This definition of normative theories and, especially, the
identification of the axiomatic formulation with rationality are
considered by some investigators as unsatisfactory. There are
many decision problems that are very different, they argue, and
to attempt to solve all of them by the same theory of axiomatic
"rational" behavior is too restrictive. For example, the decision problem of an individual as to whether to carry an umbrella
in anticipation of rain is very different from the decision a
whole society may have to make regarding the acceptability of a
technology like nuclear power. The problem of what constitutes
rational behavior of groups of decision makers has, of course,
long been recognized as an important one.

NATO ASI Series, Vol. F21
Intelligent Decision Support in Process Environments
Edited by E. Hollnagel et al.
© Springer-Verlag Berlin Heidelberg 1986

In view of these concerns, the concept of a prescriptive decision theory is advanced. In lieu of placing the emphasis on an axiomatic formulation, such a theory would establish procedures through which a decision would be reached. An example is the "due process" of the judicial system.

Prescriptive and normative theories, by formalizing the decision making process, are restricted and, in many respects, simplified versions of the actual decision problems that people deal with in real life. The goals of   the decision makers are not always easy to identify and do not necessarily coincide with those of the formal theories. For example, some decision makers may consider it of the utmost importance not to have to regret a decision in the future and to avoid embarrassment (or, possibly ridicule), rather than to maximize the expected utility.

A further problem of formal theories in real settings has to do with the input requirements which may either get out of hand or may not be available in the required form. A well-known example of the latter case is from the history of the development of expert systems like PROSPECTOR and MYCIN, where people found that the experts (e.g., medical doctors) were very reluctant (or even refused) to assign probabilities to complementary events that summed to unity. While many objections can be raised against the ad hoc methods for handling uncertainty that have been developed as a result, the fact remains that the input data that the formal theory requires are unavailable.

Descriptive (or behavioral) decision theories investigate the decision making processes of people in real situations. It is not clear, however, whether these theories simply describe people's behavior or try to explain it. Furthermore, the lack of predictive qualities makes some researchers wonder whether they even deserve to be called theories.

In light of these problems that both prescriptive and descriptive theories have a natural question is whether they could be com-bined to provide useful guidance in real life problems. In this context, it appears that a prescriptive theory could serve as a point of departure. Its restrictions, requirements and goals should constantly be questioned and modified, as necessary, to include people's attitudes and concerns. For example, Bayesian decision theory requires that people express their knowledge in terms of probabilities. Descriptive theories tell us that such a process is very unreliable and subject to distortions. These findings can be used to assure that the required inputs are free of unintended biases. Thus, the descriptive and prescriptive approaches complement each other in this case.

There are cases, however, where the two approaches are in con-flict and there is no guidance as to what the decision maker ought to do. For example, Prospect Theory finds that people use preference functions on probabilities, a fact that is not allowed in Bayesian Decision Theory, where preferences are defined only over the space of outcomes.

In the context of developing intelligent decision aids to operators of process plants, normative decision theories are of little, if any, use. Descriptive theories, on the other hand, could provide useful input, although they have been largely ignored by the developers of such aids. These issues are, of course, among the subjects that are explored by the participants of this Advanced Study Institute.

PANEL DISCUSSION ON SYSTEMS ENGINEERING

Edited By Sergio Garriba

CESNEF – Politecnico di Milano
Milano, Italy

1. <u>The main issues</u>. It is important to have a realistic appreciation of what intelligent decision aids can and cannot be expected to accomplish in process industries. Accordingly, the panel discussion concentrates mainly on decision making by plant operators in the various circumstances that may occur. In this respect, four inter-related issues can be distinguished and dealt with in a sequence. First, is the need of producing some kind of knowledge representation for decision making in realistic environments. Second, is the meaning of descriptive and prescriptive models for decision making, in order to assign a proper role to intelligent aids. Third, is the assessment of values that different intelligent decision aids may have in improving the operation of engineered plants and systems. Fourth, is the identification of cases and situations where intelligent decision-support systems (IDSS) would find their best use.

2. <u>Process industries as a dynamic environment</u>. Everybody seems to recognize that a distinguishing character of decision making in process industries is the evolutionary and dynamic nature of their environment. Decision tasks would be a component of a larger decision problem that does not simply terminate with the choice of an alternative and the observation of a final outcome. Instead, the decision maker will obtain the outcome and continue decision making. Johannsen remarks that a chemical plant operator when facing an accident, is not likely to run a single monitoring or inspection test and then apply a single action. Depending upon the response to the initial action or remedy, a long sequence of subsequent action and remedy decisions may occur. In fact not only are there opportunities for more data collection, but the outcome feedback from earlier remedy decisions may itself provide valuable guidance for improving the strategy.
Operation and control of complex systems basically depend on the means for dealing with the complexity. In its turn, the observed complexity has relations with the degree of resolution applied for information search and with the technology of system-to-operator interface. Engineered systems of process industries are complex in terms of large numbers of information sources and devices for performing fundamental control actions. Therefore, it can be agreed with Jens Rasmussen when he argues that a way to deal with these systems is to structure hierarchically the situation and "thereby transfer the problem to a level with less resolution".
In practice, decision aids must rely upon various simplifying assumptions (like conditional independence), or utilize

NATO ASI Series. Vol. F21
Intelligent Decision Support in Process Environments
Edited by E. Hollnagel et al.
© Springer-Verlag Berlin Heidelberg 1986

potentially biased inputs (like subjective probabilities and measures of uncertainty). In complex systems it may be difficult to analytically evaluate the impact of these defects on the quality of the final decision. Mancini underlines how simulation experiments would help in identifying situations where the consequences are serious, versus those where decisions are relatively unaffected. The role of feedback in dynamic tasks is particularly important since its presence or absence can dramatically affect the accuracy required of human judgment. Consider, for instance, the analogy between judgment and aiming at a target. It is more difficult to recover from an emergency situation from a distance than to continuously monitor and control the physical process.

On the other hand, simulation experiments would help in identifying situations where decision makers could seek out the use of decision aids. Decision makers already make intuitive evaluations about the cost versus benefit trade-offs associated with different decision strategies. In this frame, they can gain advantage from a full screening performed by fast-running computer codes in which many sequences of system behavior and operator responses can be generated. This screening would permit the identification of peculiarities in the type of intelligent decision aid which can be designed and eventually tested.

3. How to model decision making. Opinions of panelists converge on the need of improving the comprehension and representation of decision mechanisms in process environments. Specifically, De Keyser emphasizes the problem of describing how humans make their decisions in the highly dynamic environments that are encountered in process industries. Normative decision aids are usually built around a prescriptive but rigid problem structure, referred to as a decision analysis model. This model may not be compatible with the evolutionary approach to system development, which is characteristic of artificial intelligence. Two aspects deserve mention. First, in expert systems, once general decisions have been made regarding the basic control procedures and the organization of the rule base, the knowledge base can be incrementally ameliorated by adding, modifying, or deleting individual production rules. Second, expert systems encode knowledge in the form of production rules. This fact makes it relatively easy to construct a user interface that adopts only terms and references familiar to the users.

Øwre also points out that "models" offered today are of a purely descriptive nature, and very little is known of what exactly happens in the process of decision making of the human operator. Therefore human operators and experts must be asked to explain what a kind of theories, heuristics, facts, rules and procedures they base their decisions upon. The knowledge engineer must try to formulate tasks that can make it possible to analyze, criticize and further develop his/her opinions. The use of task determinants of performance is essential since they serve as a tool for creating more knowledge.

Johannsen notes that there is a clear synergy between the

rule-based program architectures adopted in artificial intelligence (and expert systems), and the more conventional hierarchically organized programs which reflect the prescriptive problem-structuring techniques. This conceptual synergy can then be used with regard to strategies followed in the introduction of intelligent decision aids, to assess their sensitivity to variations in the structure of the environment and task determinants. Strategies to be investigated should be those of interest to behavioral researchers.

Knowing the task determinants of performance can be useful to improve decision making by changing the context, rather than the process of decision making. The perspective of jointly optimizing the task structure and the decision strategy has a value in the design of computer-based decision support systems. These systems indeed, change both the structure of the decision environment and the strategies adopted for decision making. Typically, information is presented in new different formats and sophisticated capabilities are provided.

In this context, it must be remarked that taking away the man from the control loop, when not necessary, could result in a worst overall performance. There are certainly areas where we today see IDSS developed that could be remedied in a more manual or traditional way. For instance, we should weight detection aids against better design of control rooms, diagnosis aids against better training of operators, guidance aids against revision and improvement of organization and content of the written procedures.

4. Performance measures to evaluate intelligent decision aids. A problem surfaces which refers to the criteria and techniques for evaluating the performance of intelligent decision-support systems versus "non-intelligent" system configurations, or other intelligent supports. Since there is no agreed upon definition of intelligent behavior, it may be worthwhile to quote the view expressed by De Keyser, saying that the degree of success attained when introducing intelligent aids is highly dependent upon the mesaures used to evaluate the overall performance.

According to Kraiss the goal in applying intelligent decision-support systems generally must be to enable the operator to make the transition from knowledge-based to rule-based behavior. In other words, a performance measure of the intelligent aids must reflect this transition. A possible approach to this end may be the concept of user perceived complexity as put forward recently by Kieras and Polson. Basing on a task/knowledge analysis, they attempt to quantify the cognitive load for particular tasks and designs.

Øwre remarks that the goal is to match human behavior in terms of problem-solving capability, rather than to match only the final results that are looked for. Thus, emphasis should be placed on the sequence of actions taken. Because of the obvious limits in the available resources, a feasible solution would be to carry out performance measures in as close to real time environments as possible, like in full- scope simulator

control rooms. Comparison tests seem to be a natural choice.
Types of performance measures that can be adopted would be :
(i) number of errors made in carrying out certain tasks,
number of omissions and commissions; (ii) time used in accomplishing specific tasks, such as detection, diagnosis, location, implementation; (iii) if one could measure the "threat
to safety", this would also be a good performance measure. It
has to be considered, however, that this approach though
practical, may not be always satisfactory due to the reduced
spectrum of situations involved (particularly, few scenarios
of accident, pre-thought and pre-taught sequences).

5. Priority areas to apply intelligent decision aids. In
panelists' opinion there are definite areas in which decision
aids are easier and more direct to implement and have been
implemented already. Examples are predictor displays in vehicle guidance, knowledge-based pattern classification and military tactical decision making.
    As for control in process industries, applications other
than nuclear power seem more prone to an early introduction of
intelligent decision-support systems because of less compli
cacy and fewer inter-relations. The difference among control
rooms encountered in typical process environments is mostly
reflected in the safety aspect. Therefore, it is very likely
that licensing authorities will set stronger requirements on
installation of intelligent decision aids in nuclear control
rooms than one would find in coal-fired power stations, and
chemical or steel industry.
    Kraiss notes that a major characteristic influencing the
design of an IDSS for a particular process is whether this
process has "safe" fallback modes in case of failures, which
allow for "graceful degradation". Power plants, for instance,
can be shut down in several steps in case of emergency, while
a flying airplane must either compensate a failure or land
right away. It is obvious, that the requirements for an
intelligent decision aid must be very different in the two
cases. It seems meaningful to recall that nuclear power plants
in the Federal Republic of Germany are designed in such a way,
that the operating personnel has at leat 30 minutes to analyze
a failure in the process. During this time interval the
nuclear reactor is running automatically.
    Another peculiar character of nuclear power plants refers
to the design of safety functions. Operation of nuclear plants
shows centralization and hierarchical, or cascade procedures.
Mancini stresses the fact that many safety barriers are
contemplated with possibility of accidents crossing several of
these. Furthermore, accidents are subject to several uncertainties in their accident consequence evaluation. Therefore,
it seems that IDSS in the nuclear sector are much more needed,
but difficult to implement.
    Another difference between nuclear and non-nuclear plants
is that in the nuclear case the risk for the environment is
much more felt than in other industries. In these industries
safety is still an "internal" dimension of the product and of
the production process. As a result, the structure of an
intelligent decision-support system in the nuclear case would

be multilevel and much more articulated, involving decisions at different stages. Needless to recall that the condition of real-time operation could raise additional obstacles. All such considerations given, the trade-off between higher degrees of automation and human operator tasks ought to be the subject of future research.

A last point concerns the endowment of human resources required for the successful introduction, and use of IDSS. The coming into service of IDSS will certainly change process plant operation. If a correct equilibrium between automation and operator aid is reached, one could foresee that safety of the plant (under almost all events) can be treated as an attribute of production and not as a separate issue. This fact could entail modifications in the structure of the crew. Particularly, some roles of plant operators would be eased by alleviating burdens coming from certain tasks.

It is observed by Øwre that one of the purposes of IDSS is to make an already highly skilled person capable of doing his/her job even better. The direct implication is that the qualification of the staff would improve and that new staff background qualification should not be less than earlier staff had. On the other hand, it is very important to understand how operators structure their decisions during emergencies in order to design the proper aid that would avoid wrong decision approaches or possible mistakes. The knowledge to be given to the operators has to be extracted from designers and encoded into the computers.

In the planning stage, design team of IDSS ought to be composed by system designers and end users. Most panelists would also include knowledge engineers and some psychologist. All panelists share the opinion that design procedures must be interactive with designers having a leading role. Designers would interview, get comments and feedbacks as the project goes on. Team composition will change during the life cycle of the system, designers will leave and responsibility will remain on the maintenance teams and end users.

PANEL DISCUSSION ON COGNITIVE ENGINEERING

Edited by Keith Duncan

UWIST, Department of Applied Psychology
Cardiff, Wales

Throughout the papers and discussions expressions like consultant, prosthetic, tool, amplifier, replacement, advisor, assistant have been mentioned. But what can we say about what constitutes decision support, much less 'intelligent' decision support? What knowledge is resident in a decision support system? What is the relationship between the machine and human elements of the 'human IDS ensemble'?

Not unpredictably, there was little enthusiasm to define the intelligent component of decision support. There was widespread agreement that a more productive activity was to focus on what constitutes an effective decision support. Ultimately the goal is not so much to build an 'intelligent' aid as it is to build an aid that helps the human act more intelligently. We must therefore consider what obstacles are to be overcome and what pitfalls are to be avoided on the way to that goal. However, in the course of examining these issues it seems impossible to dodge completely the question of what is intelligence.

The definition of what is an intelligent system and even what is a useful system may lie only with those who interact with the system in an actual work environment. For example, when a new system labelled as intelligent is introduced into an application, the user's expectations may be 'ah, a system that will solve most of our problems', but later the reaction may be disappointment because intelligent in that context refers to a large spectrum of capabilities.

Others emphasised the need for adaptability in effective decision support. This can be adaptability in terms of the domain problems, e.g. a system can provide support in a dynamic environment where the problems to be solved can change in the course of an incident, or in terms of the human part of the ensemble, e.g. adapting to different user styles or to different stages of learning. Another important factor is the ability to anticipate or predict events. Others emphasised building systems that are able to extract more information from less input, that is, to read between the lines or to optimally combine all of the available evidence.

If we think of human or natural intelligence it is clear that human experts are capable of making judgments about their own expertise (although see Wagenaar this volume). This metalevel ability to decide when one's expertise is relevant and the

NATO ASI Series. Vol. F21
Intelligent Decision Support in Process Environments
Edited by E. Hollnagel et al.
© Springer-Verlag Berlin Heidelberg 1986

capability to modify one's response based on this type of knowledge would seem to be a very important feature of intelligence. Present systems are unable to judge their own expertise in the context of a specific problem. It is only when a system is taken outside its original design application that we start to learn about the kinds of problems it solves and the reasons that it can solve those and not other problems. How can or will the machine know when it is outside its range of competence? This is clearly a research question since at this time designers of the most popular support system, expert systems, rarely understand their machine's range or type of competence and the typical design methodology provides little opportunity to discover the machine's boundary conditions. If in the near term the designer and the machine will be unable to judge the machine's capabilities in a specific problem, then how will we support the human's basic domain skill and his opportunity to learn about the machine's competence. Also people often note that AI technologies can produce systems that degrade gracefully at the boundaries of system capability. Just as the human operator must be able to distinguish disturbances in the underlying process from failures in the control systems, the human partner is responsible to detect and intervene when conditions exceed the capabilities of the intelligent machine. To this end there is a tradeoff between graceful degradation (improved machine performance at the task) and clear manifestation that the machine is in trouble (improved human supervision).

While one underlying tone is an enthusiastic survey of possibilities in providing more and more effective decision support, another current is more cautious. Where representatives of the former emphasise new possibilities and capabilities, representatives of the latter understand the obstacles, the lack of knowledge about how to confront the problems of providing effective decision support -- perhaps because they have confronted these difficulties in the course of the history of failures in decision support. There are several delusions which we can fall prey to in the enthusiasm over IDSs. The first is that decision support systems are new; there have already been several generation of support systems. We are not so much on the eve of some dramatic breakthrough as we are part of an evolution and development in the capability and understanding of effective decision support.

It has been said that the one thing you cannot aid is shortage of time. Therefore, there may be limits to powerful decision support systems in dynamic high risk environments, like nuclear power plants; rather greater benefits may accrue from focusing efforts off-line, so to speak, before one gets to the time critical situations, in maintenance activities, test and calibration activities, and design activities. The objective would be to reduce the number of problems that occur in these areas, problems that cause challenges and that complicates time critical responses, for example, in control rooms.

While our power to create more intelligent support systems is on the rise, really effective support may not need to be so intelligent, for example, relatively simple memory aids may reduce one very frequent category of human error -- omissions. Similarly, there is a case to be made for starting small -- to build very small scale but useful systems. To accomplish these modest goals, information from several experts will have to be brought together perhaps for the first time into a single system. So you have a very small system, used by only a few people in the company, but the experience of trying to bring together various types of expertise is very useful, first in the sense that new things are discovered or communicated more broadly, and second that the results makes a small but clear improvement in the process under consideration.

Ultimately the modifier 'intelligent' may simply be a name that we attach to one stage of capability in the production of powerfuldecision support systems. It is completely arbitrary to define systems circa 1985 as intelligent; we can only say that it is the most advanced stage of systems we have and that by convention we call such systems intelligent, much as one past stage of architecture is called modern.

Early attempts to provide operator support in process control were quickly rejected as disabling and it is unclear whether more recent attempts such as the safety parameter display systems in the nuclear power plants are positive or ineffectual. Similar attempts to provide support that sometimes proved to create new problems are evident in other industries, e.g. aircraft.

Perhaps a final delusion is that IDSs will fail less than the human in the loop does today (or not fail in new and potentially worse ways than people do). Support systems can be enabling or disabling, and there is a long history of so-called aids that turned out under actual working conditions to fail to support or even to hinder people from carrying out their job. A related point is that the capability to provide powerful decision support may tempt us to provide layers of band-aid solutions to systems that are intrinsically hostile to human rather than to engage in more fundamental redesign of the system. Early enthusiasm has often failed to match practical successes and todays expert systems seem to be more of a success in laboratories than in actual work environments.

Whatever our understanding of what is effective decision support, it is clear that support systems will be built. The question or problem for the person in some work environment will be how to make use of this new device that someone has so kindly dropped into his lap. The human part of the ensemble can choose to use a support system as a crutch, work around the hindrances it creates, or use it as an instrument to enhance his capabilities in the context of his work, or a mixture of these. For example, acceptability may not be an issue per se, but only a sign of the usefulness of the device in the actual context of work. If there is a problem to be solved and an instrument is available that helps solve it, then people will more than accept that device, they will generally embrace it. It was pointed out that the problem is not getting people to accept useful devices, but that many systems that are offered to them are not useful.

The problem of providing effective decision support is fundamentally the question how does the designer decide what will be useful in a particular application. Can researchers provide that designer with concepts and techniques to determine what will be useful or are we condemned to simply build what can practically be built and wait for the judgment of experience? If principle-driven design of IDSs is possible, then we will need to move in the short run towards a taxonomy of decision support systems and the kinds of problems that they can address. For example, Patrick Humphries and Dina Berkeley have started to develop such a taxonomy. Their level 1 support systems are basically table lookup devices. In level 2 systems there is a fixed model where you can vary the inputs to that model and see the effects (e.g. a spreadsheet). Level 3 support systems allow you to change the structure of the model but within the same general class of models, while level 4 systems allow you to change across classes of models. Now if you take a real situation, say a nuclear power plant, you have a range of decision making tasks which encompass all these levels. During normal operation level 1 support often dominates; the supervisor may have some level 2 decisions and therefore a different support system. In accidents you may not have the same structural entities and therefore need to think about level 3 support and in rare unanticipated circumstances even level 4 support. It is not that this taxonomy is necessarily the best, but that we need to understand forms of decision support and kinds of problems to be addressed and to be able to match the support to the problem type.

But all this is mere talk. The pace of development is quickening. This is a boon to research, but there is a danger of practice outstripping knowledge with the consequences of producing support systems that fail to support. We need to build models and provide empirical results that can help determine what is effective decision support, what should be the relationships between machine and human portions of the ensemble, what obstacles must be overcome and how, what are the pitfalls to be avoided?

PANEL DISCUSSION ON ARTIFICIAL INTELLIGENCE

Edited by Roger Schvaneveldt

Computing Research Laboratory
New Mexico State University, Las Cruces, NM, USA

Different perspectives lead one to see different aspects of the same thing. The papers and discussions at the Advanced Study Institute (ASI) reflect this truism. I intend to draw out some of the varying perspectives on intelligent decision support systems (IDSS) and to examine some of the themes and issues that arise naturally out of these perspectives. I could ask which of the perspectives is the most productive, but that question would likely produce rather unproductive arguments. Instead I shall attempt to profit from taking various perspectives and thereby gaining a more complete view of the problem at hand.

Since our goal is to analyze the present status and future prospects of IDSS, we might look for different perspectives in the components of such systems. These components include computer systems and the people who are to be supported by these systems. Also, an IDSS is intended to accomplish some tasks (or meet some goals). Various perspectives differ in terms of the emphasis placed on each of these three aspects of IDSS (i.e., **computer systems, users,** and **tasks.**

Although these three components help to identify the perspectives presented in the various papers, there is another dimension that distinguishes the papers as well. This dimension reflects the degree to which papers focus on **practical** as opposed

NATO ASI Series, Vol. F21
Intelligent Decision Support in Process Environments
Edited by E. Hollnagel et al.
© Springer-Verlag Berlin Heidelberg 1986

to **theoretical** analyses. Different researchers have different goals. Some are primarily concerned with developing formal, theoretical systems that may have some applications in IDSS. The primary emphasis, however, is on the quality of the theory. Other researchers are primarily concerned with the practical implementation of IDSS. These individuals are more likely to emphasize descriptive analyses of the problem that aim at completeness rather than formal criteria such as coherence, consistency, and computability. The ASI meetings saw numerous discussions about the relative value of formal and practical approaches to IDSS. Many people seem to feel that formal models are not of much value because they oversimplify the problem to the point that the resulting models fail to capture several important aspects of actual IDSS. The counter argument is that formal models help to specify the limits of our systematic understanding of a problem, and formal models have often provided useful tools in the handling of complex problems. This debate is likely to continue as it has for many years. My own view is that we should encourage both formal and practical approaches to the problems associated with IDSS. The practical work helps to define areas that need formal analysis, and the formal work leads to tools that can be implemented in IDSS.

We could combine the theory-practice dimension with the other three components (tasks, users, and computer systems) by allowing the degree of emphasis on each of the three components to define three additional dimensions. Our space of perspectives would then be a four dimensional space. Each of the papers could be represented as a point is this four-dimensional space reflecting the degree to which the paper is concerned with theory vs. practice and the degree of emphasis placed on tasks, users, and computer systems. Now it could be argued that the ideal analysis of IDSS would consist of a complete analysis of task requirements, the needs, abilities, and limitations of users, and would be sufficiently formal to allow effective implementation in today s computer systems (all at an affordable price). Clearly, we are not yet in a position to achieve that! What we gain from recognizing the perspectives of others working on aspects of IDSS is an appreciation of the complexity of a thorough analysis. At this stage of development, we should probably encourage diversity in perspectives so that the various dimensions of IDSS will be adequately investigated. There is also a need for integration of the various perspectives, but we don't appear to be particularly close to that goal at present.

In what follows, I attempt to draw out some of the particular concerns that arise from focusing on each of the three components: tasks, users, and computer systems. If we were to focus on only one of these three components, it probably would be the tasks involved in IDSS. The task dimension appears to be a necessary part of any analysis; at a minimum, a perspective on IDSS must include some analysis of what it is that IDSS are supposed to accomplish. This ASI is concerned with decision making in process environments so abstractly, we shall define the task accordingly. Several of the papers in the ASI (e.g., Hollnagel, 1985; Johannsen, 1985; Rasmussen, 1985) are concerned with detailed analyses of the functional requirements of IDSS. A common theme in the papers concerns the analysis of process control tasks with varying emphases on the quality of decisions, the needs and capabilities of users, and the status and prospects of artificial intelligence (AI) work, particularly expert systems. Different perspectives may be found in these variations in emphasis which are likely reflections of disciplinary concerns. Let us begin with the tasks.

The Task Perspective. Among the common themes in papers analyzing the functional requirements of IDSS are: the complexity of such systems, the range of user expertise, the costliness of errors, the importance of correctly handling infrequently occurring malfunctions, the division of expertise, the allocation of authority and responsibility (Gallanti & Guida, 1985; Rasmussen, 1985).

The tasks involved in process control environments include: acquisition and interpretation of data; monitoring of normal activity; detection of abnormal conditions; reasoning and problem solving involved in diagnosing an abnormal situation; selection of appropriate corrective actions; coordinating the activity of operators; consulting with other operators and supervisory personnel. There are also ancilary tasks such as training and the social aspects of working in groups.

The task perspective encourages such questions as: What are the components of the tasks performed by IDSS? What is the appropriate level and focus of analysis? Given a collection of problem solvers (human and machine), how should the components of tasks be allocated?

An exclusive focus on the tasks involved in IDSS can lead to problems from both the user and computer system perspectives. Since there are alternative ways of analyzing tasks, a useful analysis should reflect the properties of the users of IDSS as well as the technology available for implementing IDSS. Otherwise, the task analysis may yield tasks that are beyond the capabilities of users and/or beyond the technology. Of course, such analyses may be used to push technological development (which is useful), but if the interest is in the development of working IDSS, the analysis cannot go too far beyond present capabilities.

Other perspectives may reflect varying degrees of sophistocation in the analysis of the tasks involved in IDSS. To some degree, the computer-system perspective and the user perspective serve to identify weaknesses in earlier analyses of the purposes of IDSS.

The Computer System Perspective. From the perspective of the computer-system component of IDSS, we find a concern with the construction of computer programs that use knowledge of various kinds to solve problems that are ordinarily the province of human experts. This perspective has concrete (sometimes working) systems to offer (Gallanti & Guida, 1985; Walker & Stevens, 1985). These systems serve as existence proofs for the contention that machines can be programmed to perform tasks that require considerable amounts of knowledge and reasoning. The failure of such systems to handle certain problems may suggest that these problems are particularly difficult. Some examples of such difficult problems include: developing and using good models of the users of the system, knowing when a problem is beyond a system s capabilities, integrating human and machine expertise in a cooperative problem solving environment, failing to appreciate the significance of the context of particular problems, and learning.

The systems can also be criticized for failing to accomodate the real needs, capabilities, limitations, and attitudes of users of the systems (Hollnagel, 1985; Rasmussen, 1985; Woods, 1985). The questions that arise from the computer system perspective tend to emphasize the research and development that will be required to construct new and better computer systems. Some examples of these questions are: What alternative architectures are available for building IDSS (Coombs, 1985; Johannsen, 1985; Walker & Stevens, 1985). What do expert systems have to offer in the domain of decision aiding in process control environments (Gallanti & Guida, 1985; Norman, 1985)? How well do exisiting systems perform? What new capabilities do intelligent systems need to adapt to the process control environment? How might we approach the difficult problems mentioned above?

A recurrent isssue that arises in the context of discussions of expert-system technology concerns the problem of acquiring the knowledge that experts use to solve problems. Knowledge engineering, the label given to the task of coding knowledge in a form appropriate for use by machine, is often said to represent a major bottleneck in developing expert systems. Indeed, determining what experts know seems to be a black art with few systematic procedures. Recognition of this state of affairs has led psychologists to begin to develop more systematic methods for eliciting knowledge from experts (e.g., Gammack & Young, 1985; Schvaneveldt, McDonald, & Cooke, 1985; Schvaneveldt, et al., in press). These methods include procedures that cognitive psychologists have used to determine spatial and network structures underlying judgements of similarity, relatedness, or proximity among sets of concepts or events (cf. Schvaneveldt, Durso, & Dearholt, 1985). As this work develops, we may have a collection of tools that can be used in conjunction with currrent methods such as explanation and consistency checking in expert system development tools.

I suspect that most participants in this ASI appreciate the contribution existing systems have made to the development of IDSS. A major complaint, however, is that insufficient attention has been given to the user and the projected environment in the design of such systems. Another common theme is found in the proposal that IDSS should be modeled on collections of cooperating problem solvers (Fischoff, 1985; Hollnagel, 1985; Woods, 1985). This means that the computer system component should function more in a consultancy role rather than the role of isolated problem solver (Coombs, 1985; Woods, 1985).

The User Perspective The user perspective emphasizes the requirements placed on users of IDSS. The questions asked about users include: What are their limitations? What are their strengths? What causes errors? How do they conceive of the tasks and the systems they are required to control? Do they need automated experts or automated assistants? What are the attitudinal and affective factors? Who (or what) is in charge with IDSS? Who is to blame when things go wrong? What mental models do (should) users develop?

In the context of a computer system, the user perspective leads to such questions as: What is the nature of effective user-system communication? How shall we allow for varying expertise? How can users know about the coverage of a system? How can users know about the method(s) for using a system? What is unique about process environments? In many ways, the concerns of the user perspective on IDSS are the same concerns found in the rapidly growing field of human-computer interaction (HCI). The same issues about effective communication, user models, etc. are found in the HCI literature. Some of the important elements found in many process environments include: emphasis on real time control, extreme complexity of the process environment, and the importance of correctly handling critical events under severe time pressures. Most other user-system problems are similar to those found in the use of any computer system.

The analysis of IDSS, like several areas involving the use of computer systems, leads to a discussion of the importance of modelling human intelligence in constructing intelligent systems. Indeed, the field of AI has long engendered tensions between the modellers of human intelligence and those who simply wish to develop machines that exhibit intelligence whatever the method.

One might argue that with IDSS, we want to develop machines that enhance the human's ability to perform work. Such systems should help to overcome human limitations in memory capacity, attention maintenance, and decision making just as physical tools enhance physical limitations in strength and stamina. It is also true, however, that it is desireable for IDSS to communicate with human users. The systems should be able to explain lines of reasoning, enter into dialogs with users about possible causes of abnormal states and about remedies for the abnormalities. Effective communication depends on shared knowledge.

Is there a contradiction here? One the one hand we want machines that communicate their conclusions and methods. On the other, we want machines that enhance human performance. The machine must understand the user, but how is the user to understand the machine when it uses methods of reasoning not available to the human? Can we expect the system to teach users about its methods? We may be able to develop machines that rationalize their methods for communication purposes while the processes underlying decisions may be quite different from the rationalization. Would this be a desirable property of intelligent machines?

Need for Empirical Evaluation

A topic that I thought was not sufficiently emphasized at the ASI concerns the need for evaluating IDSS. This is a complex problem that may require the development of new methodologies that can deal with the issue of evaluating the effectiveness of complex systems. Evaluation methods from psychology emphasize tightly controlled situations which may not meet the requirements of IDSS evaluation.   Evaluation of AI systems is notoriously unsystematic. Implementation of actual systems will require that serious attention be given to insuring that such systems meet the requirements that lead to their design.

AFTERTHOUGHTS

E. Hollnagel, G. Mancini & D. Woods

As both the written papers and the edited versions of the panel discussions show, a considerable number of issues were raised during this ASI. It is difficult to summarise these on a few pages if one wants to do more than just enumerate them. And to do them all justice could easily produce a small book of its own. The solution chosen here is to discuss three of the main issues in some detail. The choice reflects the opinion of the three ASI directors and should not be seen as the opinion of the ASI participants as a whole. We are, however, convinced that these issues would rank among the most important on anybody's list even though they might not be the top three ones.

ON TOOL AND PROSTHESES

Man is often defined as a tool builder and tool user. Technological development has moved from provision of physical tools (tools that amplify man's capacity for physical work -- strength, speed reach, accuracy), to perceptual tools (extensions to man's perceptual apparatus beyond the limits of sensitivity and capacity of the sensory channels), and now with the arrival of computer technology, intellectual or cognitive tools (tools that increase the human powers for measurement, calculation, inference, and conceptualisation). Although this type has a longer history, AI has increased the interest and capability to provide intellectual tools. The question is then, what is the nature of intellectual tools: Are they something qualitatively new compared to the physical and perceptual tools that we have more experience with? What kinds of intellectual tools are needed? What kinds can be built? How are these tools to be developed? And what can we learn from our experience with physical and perceptual tools to assist in the development of intellectual tools?

If we wish to understand intellectual tools, we first need to define what a tool is. Answering this question requires a clear distinction between a prosthesis and a tool. The former is a replacement for something missing or a remedy for a deficiency, while the latter is a means or an instrument for accomplishing something.

From one point of view, there is the danger of a false dichotomy between tools and prostheses. In one sense all tool use implies a human deficiency with respect to some goal. Thus if we had hard hands, we would not need hammers. But designers may also create situations where people are given tasks they are not very good at. It is then the responsibility of the

NATO ASI Series, Vol. F21
Intelligent Decision Support in Process Environments
Edited by E. Hollnagel et al.
© Springer-Verlag Berlin Heidelberg 1986

designers to provide some kind of support that allows people to fulfill their roles. Human beings are a generalised species and work fairly well in a wide range of environments, almost all man-made today. A tool will amplify existing human capabilities and facilitate further development and adaptation. A prosthesis will make it possible to accomplish a given task, but will do so by replacing a human function with an artificial one.

Our tool building / using skills can be seen as a sort of resource utilisation in the pursuit of a goal. As in the insight problem of classical Gestalt psychology, we need to know what resources are available to help solve the problem. The point is that tool use involves an active role for the tool wielder in some instrumental context. We all know practical examples of the tools that people create to carry out their jobs, even tools to help them work around the hindrances created by the 'tools' they are given. The sense of a prosthesis is entirely too passive to capture this aspect of human tool use. The question for decision support is what this and other characteristics of tool building and tool wielding mean for how we go about providing intellectual tools?

There are several characteristics of tool use which speak to issues of how to build decision support systems. First, the tool user is in control; he wields the tool. Second, there is a repertoire of tools available and often also the material from which to fashion or adapt tools. In order to wield a set of tools the user should know the boundaries and instrumental characteristics of the available material -- what can they do and what are their limits, side effects, preconditions and postconditions. There is more than just tool selection in effective tool use; there is also an element of skill in how the tools are applied to the task. Finally, with high skill levels the tool is no longer a thing outside of the person, an intermediary between the user and the task; instead, the tool becomes an extension of the person.

Wagenaar discusses the overconfidence bias when people at all levels of expertise overestimate how much they know. This bias applies to design problem solving as well as to operational problem solving. Consequently, the designer of an Intelligent Decision Support system is likely to overestimate his ability to capture all relevant aspects of the actual problem solving situation in the behaviour of the Intelligent Decision Support system. This result has often happened with other forms of automation and support systems, and the designer of the system fails to support or even hinders the operator in achieving his goal. For example, operators quickly learn the signatures of automatic controllers -- when they are so unreliable that the task must be performed manually, when they are highly sensitive to disruptions and must be very closely monitored, and when they are adequate to the task and require little supervision -- and adapt their actions to that. This fact emphasises the need to conceive of an Intelligent Decision Support system as an instrument in the hands of the human problem solver.

The characteristic of tool use mentioned above provide a starting point to develop criteria for effective decision support. The work described in the papers by Coombs, Walker, and Woods point to some concepts, techniques for, and examples of building decision instruments. In particular, these papers emphasise a simple fact of psychology that is very important to improving human problem solving -- concrete experience with a task is a first requirement for skill and expertise, even at cognitive tasks. The most important contribution of AI technology to decision support in the long run may be to enhance our ability to obtain inexpensively the concrete experience necessary for effective performance. For example, both the designer of Intelligent Decision Support systems and the operator in some process world faces a complex, ill-defined problem solving task. The best help we can provide to people engaged in these tasks may be to (a) enhance their ability to experiment with possible worlds or possible strategies, (b) enhance their ability to conceptualise by making concrete and visible the abstract or uninspectable (analogous to perceptual tools), and (c) to enhance error tolerance by providing better feedback about the effects / results of actions (not that errors are not made but that error detection and correction is enhanced so that errors do not propagate).

In one sense the difference between a prosthetic and an instrumental perspective on decision support is that with the former we look for deficiencies in the person to be compensated for with 'decision support', while with the latter we look at the task environment to see what must be accomplished to provide instruments that assist in goal achievement. Clearly we need a combination where we compensate for deficiencies, capitalise on how people actually work cognitively, and provide instruments to assist in meeting the cognitive demands of the environment. However, knowing when to do each of these and knowing how to do any of these well is problematic at this stage of the still fledgling discipline of providing practical, effective decision support.

ON MODELS

In order to understand, describe, and eventually design man-machine systems such as Intelligent Decision Support systems, an overall model of the human is needed that takes into account relations between the tasks the human is asked to perform and the limiting conditions (errors) of human behaviour. Academic behavioral scientists have generally addressed this problem from their own premises, which include several limitations such as:

1. Studies have been restricted to laboratory tasks (the book bag and poker chip experiments) or to simplified systems where one can maintain a reasonable degree of experimental

control. The representativeness of the tasks is thereby
diminished.

2. Studies have on the whole neglected to generalise the
existing results to different situations and contexts. The
emphasis has rather been to provide comprehensive
explanations for the particular situation, and account for
and explain all expected (and unexpected) results.

3. When it has taken place, the formulation of theories has
been too much of an episodic and descriptive nature. The
results has been theories that have great explanatory but
little predictive power.

4. There has consequently been a lack of formalism (of
analytical but even of qualitative logical types) in many
models and therefore lack of predictive capabilities. The
theories have thus been of little use in solving the
practical problems found in system design.

Given this state of affairs it is not surprising that
engineers, or more generally practitioners, have satisfied
their need for coherent, consistent, and computable models by
developing their own tools. This approach has, however,
suffered from another set of limitations, as follows:

1. The modelling has been confined to single, independent
human tasks which have been prominent from an analysis of
the practical problems. The selection of tasks has
furthermore often been based on the designers'
understanding of what the practical problems were, rather
than a more unbiased assessment.

2. There has accordingly been a lack of proper context in
model development. In particular, there have been few
attempts to go back to the conceptual foundation and
consider that in the light of the particular instances.

3. There has been an overemphasis on general theories which
could provide the rationale for solutions to specific
design related problems. The concern has been more with
theories that would fit the problems, as the engineers
described them, than with theories that described the
phenomena in depth. Where the behavioral scientitst have
gone into depth and neglected the breadth, the engineers
have done the opposite.

4. There has finally been a lack of awareness of the research
results from the behavioral sciences, or more generally the
established psychological theory which could have
contributed to the engineering tools. Just as behavioral
scientists have been naive or ignorant about the practical
problems, so engineers have been naive about the conceptual
and methodological foundations of empirical research.

Despite these shortcomings, the use of the practitioner's
models have allowed a better definition of the problems and

led to a recognition of the limits of the solutions and the
needs for empirical evaluation.

Having ascertained that, where does one then go? This is a
long and often heated debate. Each camp will emphasise its own
point of view, and only partly recognise the urgency of the
other camp's arguments. This NATO ASI succeeded in providing a
genuine and fruitful mixing of the two camps and of the
several disciplines within them. It remains to be seen whether
the tools provided by AI and advanced computer technology are
the foundations of a common language that can amplify the
interaction between academics and practitioners in the
development of Intelligent Decision Support systems.

ON INTELLIGENCE

When dealing with Intelligent Decision Support systems a
number of questions are almost begging to be asked. The first,
obviously, is the question of what intelligence is. One of the
standard answers in psychology is that intelligence is what is
measured by intelligence tests. Unfortunately, this does not
answers the question any better than Turing's test does (cf.
Hollnagel, this volume). The definition of intelligence is
basically a philosophical matter, but that has not stopped
computer scientists working with AI from proclaiming their own
solutions and from wanting to revolutionise philosophy and
psychology by saving them from their lack of precision and
scientific rigor. In this context we will refrain from
debating the matter further, but simply make the bold
assumption that we in each case can agree on whether a system
is intelligent or not.

Assuming that we know what intelligence is another obvious
question concern the locus of the intelligence. Claiming that
a system is capable of intelligent performance, and knowing
that the system consists of several parts - in this case one
or more persons and one or more computers - inevitably invites
questions about where the intelligence lies. It is, of course,
taken for granted that the human being is intelligent. But as
the use of the term Intelligent Decision Support implies
something more than that, the natural assumption is that some
extra intelligence resides in the computer. This, of course,
invites further questions about what we mean by intelligence
in a computer, whether computers really can be intelligent or
think, and so forth - leading into a discussion that is
decades old and and which shows no sign of coming to an end.

Referring to a Man-Machine System as being intelligent
certainly implies that there is something in it over and above
human intelligence; otherwise every system that included human
beings - as most system do - would deserve the label
'intelligent'. But that an Intelligent Decision Support system
is intelligent does not necessarily mean that something has
been added to the human intelligence. It may also mean that
the human intelligence in the system has been amplified. The

concept of intelligence amplification is far from new but can be found in cybernetics in the 50es (Ashby, 1956). It has, however, been neglected by AI and computer science (as so many other useful cybernetic concepts), and it may therefore be worthwhile to consider it in some detail in relation to Intelligent Decision Support systems.

The idea of intelligence amplification is closely related to the status of the Intelligent Decision Support system as a tool or as a prosthesis. If we consider other aspects of human functioning, the development in man-machine systems has been in both directions. In some cases the goal has been to automate the process and remove the human operator completely from the loop, for instance in most production and assembly lines. In other cases the goal has been to improve the process by amplifying the powers of the human (perceptual and motor) but to retain the operator in the loop. The reason has here been that the operator was necessary, i.e. that he had capabilities which were essential for the task and which could not be replaced by an artifact.

As shown in the discussion above, it is generally detrimental to system functioning to keep the operator for isolated tasks but exclude him from participating in the process as a whole. This is the partwise replacement of operator capabilities by a prosthesis. Rather than amplifying the intelligence in the system such an approach replaces parts of human intelligence with machine 'intelligence' -- which in the long run may attenuate total system intelligence. Instead one should keep the operator continuously attuned to the process and provide the tools that are needed to keep him in control. Just as we can extend our reach, speed, sensitivity, and power in dealing with the physical aspects of the environment, so we can amplify our perceptual and conceptual powers in dealing with the functional environment. Intelligence per se is still a characteristic of human behavior that is poorly understood, despite some spectacular examples in artificial intelligence. However, even if we do not know exactly what intelligence is, we know how it functions and what role it plays in dealing with process environments as well as in daily life. We can therefore from a purely functional understanding improve the conditions for use of human intelligence by amplifying its effect on the environment. Human intelligence should be used to control the truly massive powers of information processing that present computer technology offers, rather than just supplement it where it is still deficient. The combination of concepts and disciplines reported in this book clearly demonstrate that this is not only desirable but also possible.

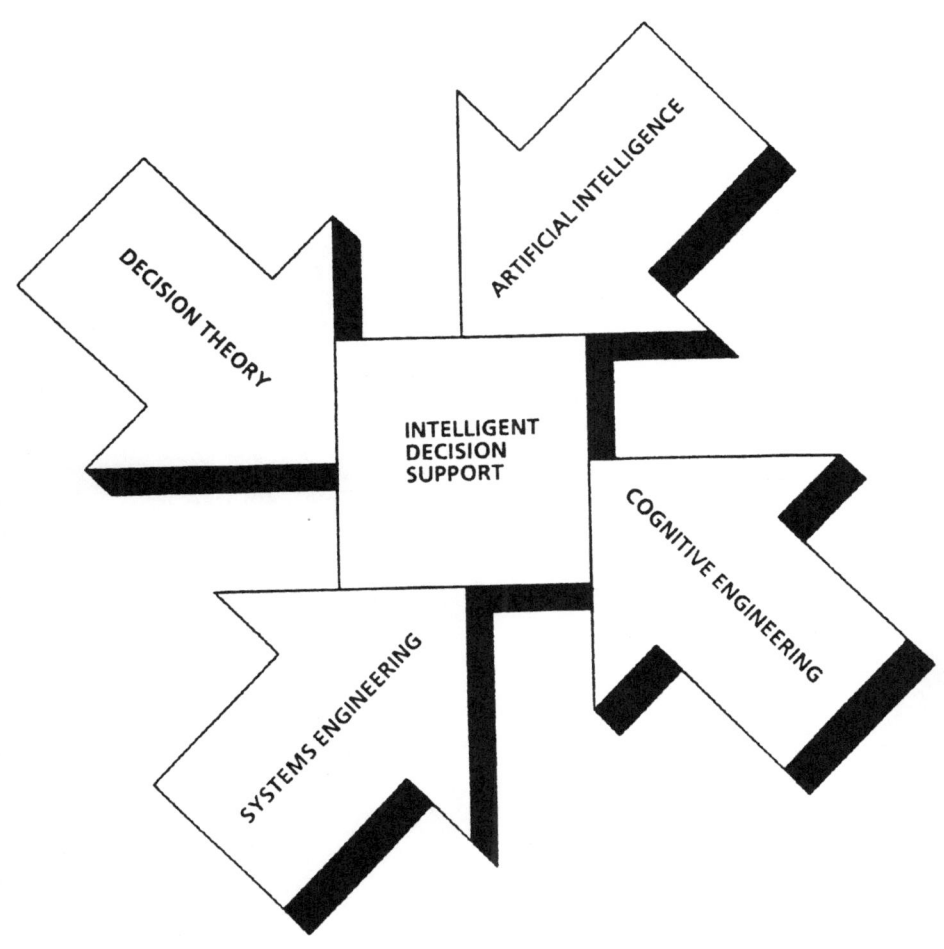

REFERENCES

REFERENCES

Ali, M., Scharnhorst, D. & Chi, S. EX: An expert system for power plant management. Proc. Workshop and Conf. on Applied AI and Knowledge-Based Sys., Stockholm, Sweden, 1984.

Allais, M. Le comportement de l'homme rationnel devant le risque: Critique des postulats et axiomes de l'ecole americaine. Econometrica, 1983, 21, 503-546.

Allen, F. J. Maintaining knowledge about temporal intervals. Com. of the ACM, 1983, 26, 123-154.

Allen, F. J. Towards a general theory of action and time. Artificial Intelligence, 1984, 23, 123-154.

Allen, F. J. & Hayes, P. J. A common-sense theory of time. Proc. 9th IJCAI, Los Angeles, CA, 1985.

Alty, J. & Coombs, M. J. Face-to-face guidance of university computer users - I: A study of advisory services. Int. J. Man-Machine Studies, 1980, 12, 390-406.

Alty, J. L. & Coombs, M. J. Expert systems. Concepts and examples. NCC Publications, 1984.

Alty, J. L., Elzer, P., Holst, O., Johannsen, G. & Savory, S. Literature and user survey of issues related to man-machine interfaces for supervision and control systems (ESPRIT report). CRI A/S, Copenhagen, Denmark, 1985.

Alty, J, L. & Guida, G. The use of rule-based system technology for the design of man-machine systems. Proc. 2nd IFAC Conf. on Analysis, Design, and Evaluation of Man-Machine Sys., Varese, Italy, 1985.

Amendola, A., Mancini, G., Poucet, A. & Reina, G. Dynamic and static models for nuclear reactor operators - needs and application examples. Proc. of IFAC Conf. on Analysis, Design, and Evaluation of Man-Machine Sys., Baden-Baden, BRD, 1982.

Amendola, A., & Reina, G. Event sequences and consequences spectrum: A methodology for probabilistic transient analysis. Nucl. Sci. Eng., 1981, 77, 287-315.

Amendola, A. & Reina, G. DYLAM-1 - A software package for event sequence and consequence spectrum methodology. EUR 9224EN, 1984.

Amendola, A., Reina, G. & Ciceri, F. Dynamic simulation of man-machine interaction in incident control. Proc. 2nd. IFAC Conf. on Analysis, Design, and Evaluation of Man-Machine Sys., Varese, Italy, 1985.

Anderson, J. R. Language, memory and thought. Erlbaum, 1976.

Anderson, J. R. Acquisition of cognitive skill. Psych. Rev., 1982, 89, 369-406.

Annett, J., Duncan, K. D., Stammers, R. B. & Gray, M. J. Task analysis, training information. Paper 6, Department of Employment, London, Her Majesty's Stationary Office, 1971.

Apostolakis, G. E., Salem, S. L. & Wu, J. S. C. A. I. - A computer code for the automated construction of fault-trees. EPRI Report 705, Palo Alto, CA, 1978.

Arbib, M. A. (Ed.) Algebraic theory of machines, languages, and semigroup. Academic Press, 1968.

Argyris, C. & Schon, D. A. Theory in practice: Increasing professional effectiveness. Jessey Bass, 1976.

Aristotle. Topica III, 2, 117a.

Armstrong, J. S. Long-range forecasting. Wiley, in press.

Ashby, W. R. Design for a brain. Chapman & Hall, 1960.

Athans, M. (Ed.), Special issue on large-scale systems and decentralized control. IEEE Trans. Aut. Contr., 1978, AC-23, 105-371.

Avenarius, R. Philosophie als Denken der Welt gemass dem Prinzip des kleinsten Kraftmasses. Prolegomena zu einer Kritik der reinen Erfahrung, III. Leipzig, 1876.

Bailey, R. W. Human performance in engineering. Prentice Hall, 1982.

Bainbridge, L. Le controleur de processus. Bull. de Psych, 1981, 34, 813-832.

Bar-Hillel, M. The base-rate fallacy in probability judgments. Acta Psychologica, 1980, 44, 211-233.

Bar-Hillel, M. Representativeness and fallacies of probability judgment. Acta Psychologica, 1984, 55, 91-107.

Bar-Hillel, M. & Falk, R. Some teasers concerning conditional probabilities. Cognition, 1982, 11, 109-122.

Bar-Hillel, Y. Language and information. Addison-Wesley & Jerusalem Academic Press, 1964.

Bar-Hillel, Y. & Carnap, R. Semantic information. Brit. J. for the Philos. of Science, 1953, 4, 147-157.

Barnes, M. J. Review of five military decision aids. NWCTP6171, Naval Weapons Center, China Lake, USA, 1980.

Baron, S., Muralidharan, R., Lancraft, R. & Zacharias, G. PROCRU: A model for analyzing crew procedures in approach to landing. Tech. Rep. NASA-10035 NASA - Ames, 1980.

Bartlett, F. J. Remembering. Cambridge University Press, 1932.

Bastin, N. Capobianchi, S., Mancini, G. et al. An intelligent interface for accessing a technical data base. Proc. 2nd IFAC Conf. on Analysis, Design, and Evaluation of Man-Machine Sys., Varese, Italy, 1985.

Battig, W. F. & Montague, W. E. Category norms for verbal items in 56 categories: A replication and extension of the Connecticutt category norms. J. of Exp. Psych. Monograph, 1969, 80, 1-46.

Baumol, W. J. Business behavior, value and growth. Macmillan, 1959.

Beach, L. R. & Mitchell, T. W. A contingency model for the selection of decision strategies. Academy of Management Rev., 1978, 3, 439-449.

Beach, L. R., Townes, B. D,. Campbell, F. L. & Keating, G. W. Developing and testing a decision aid for birth planning decisions. Organisational Behavior and Human Performance, 1976, 15, 99-116.

Behn, R. D. & Vaupel, J. W. Quick analysis for busy decision makers. Basic Books, 1982.

Belardo, S., Karwan, K. R. & Wallace, W. A. An investigation of system design - considerations for emergency management decision support. IEEE Trans. on Sys., man, Cyb., 1984, SMC-14, 795-804.

⤵ Belkin, N. J. & Hapeshi, K. Representation and matching of anomalous states of knowledge and document texts. Final report on BLRDD grant SI/G/566. Rutgers University, SCLIS, New Brunswick, NJ, 1985, in preparation.

Belkin, N. J., Hennings, R.-D. & Seeger, T. Simulation of a distributed expert-based information provision mechanism. Inf. Technology: Res. Applications, 1984, 3, 122-141.

Belkin, N. J., Seeger, T. & Wersig, G. Distributed expert problem treatment as a model for information system analysis and design. J. of Inf. Sci., 1983, 5, 153-167.

Berenblut, B. J. & Whitehouse, H. B. A method for monitoring process plants based on a decision table analysis. Chem. Eng., 1977, 318, 175.

Berkeley, D. & Humphreys, P. C. Structuring decision problems and the 'bias heuristic.' Acta Psychologica, 1982, 50, 201-252.

Berry, D. C. & Broadbent, D. E. On the relationship between task performance and associated verbalisable knowledge. The Quart. J. of Exp. Psych., 1984, 36, 209-231.

Beyth-Marom, R. How probable is probable? Numeric translation of verbal probability expressions. J. of Forecasting, 1982, 1, 257-269.

Beyth-Marom, R. Perception of correlation reexamined. Memory and Cognition, 1982, 10, 511-519.

Beyth-Marom, R., Dekel, S., Gombo, R. & Shaked, M. An elementary approach to thinking under uncertainty. Erlbaum, 1985.

Bisseret, A. Expert computer aided decision in supervisory control. Proc. of the IFAC'84 Conference, Budapest, 1984.

Blanche, R. Raison et discours. Paris, 1967.

Blaauw, G. J. Car driving as a supervisory control task. PhD Thesis, DUT / IZF, Delft / Soesterberg, 1984.

Bobrow, D. G. (Ed.), Special volume on qualitative reasoning about physical systems. Artificial Intelligence, 1984, 24.

Bobrow, D. G. & Norman, D. A. Some principles of memory schemata. In D. G. Bobrow & A. M. Collins (Eds.), Representation and understanding: Studies in cognitive science. Academic Press, 1975.

Bochenski, J. M. Formale Logik. Verlag Karl Alber GmbH, 1956.

Bolanos, M. J., Lamata, M. T. & Moral, S. A decision model under general information. BUSEFAL (L. S. I., University Paul Sabatier, Toulouse), 1985, 24, 56-63.

Bolker, E. A simultaneous axiomatization of utility and subjective probability. Philos. of Sci., 1967, 34, 333-340.

Borland International Inc., The TURBO Pascal tutor. 1984.

Bousfield, W. A. & Barclay, W. D. The relationship between order and frequency of occurrences of restricted associative responses. J. of Exp. Psych., 1950, 40, 643-647.

Bransford, J. D., Barclay, J. R. & Franks, J. J. Sentence memory: A constructive versus interpretive approach. Cognitive Sci., 1972, 3, 193-209.

Brehmer, B. In one word: Not from experience. Acta Psychologica, 1980, 45, 223-241.

Brembo, J. C. Thermodynamic condition monitoring of steam turbine plants. Proc. of ICMES Conf. on Condition Monitoring and Preventive Maintenance, Paris, 1977.

Breuker, J. & Wielinga, B. KADS: Structured knowledge acquisition for expert systems. Proc. of the 5th Int. Workshop on Expert Sys. and their Applications, Avignon, 1985.

Brooke, J. B. & Duncan, K. D. Effects of system display format on performance in a fault location task. Ergonomics, 1981, 24, 175-189.

Brooking, A. Expert systems and elicitation of expert knowledge. Unpublished Lecture at Technical Research Centre, Helsinki, Finland, 1985.

Brooks, H. M. Developing and using problem descriptions. Proc. IRFIS 6, Intelligent Inf. Sys. for the Inf. Society, Frascati, Italy, 1985.

473

Brooks, H. M., Daniels, P. J. & Belkin, N. J. Problem descriptions and user models: Developing an intelligent interface for document retrieval systems. Advances in intelligent retrieval. Proc. of Informatics 8, Aslib, London, 1985, in press.

Brown, R., Kahr, A. J. & Peterson, C. Decision analysis for the manager. Holt, Rinehart & Winston, 1974.

Brown, R. V. Acts or events? Heresy in decision modelling. Proc. Int. Conf. ORSA / TIMS, October 1974.

Brownbridge, G., Fitter, M. & Sime, M. The doctors' use of a computer in the consulting room: An analysis. Int. J. Man-Machine Studies, 1984, 21, 65-90.

Bruner, J. S., Goodnow, J. J. & Austin, G. A. A study of thinking. Wiley, 1956.

Buch, G. & Diehl, A. An investigation of the effectiveness of pilot judgment training. Human Factors, 1984, 26(5), 557-564.

Buchanan, B. G. & Duda, R. O. Principles of rule-based expert systems. In M. C. Yovits (Ed.), Advances in computers, II. Academic Press, 1983.

Buchanan, B. G. & Feigenbaum, E. A. DENDRAL and Meta-DENDRAL: Their applications dimension. Artificial Intelligence, 1978, 11, 5-24.

Bunn, M. & Tsipis, K. The uncertainties of preemptive nuclear attack. Scientific American, 1983, 249(5), 38-47.

Cacciabue, P., Lisanti, B. & Tozzi, A. ALMOD-JRC computer, program part I. EUR 9422EN, 1984.

Cacciabue, P., Amendola, A. & Mancini, G. Accident simulator development for probabilistic safety analysis. Proc. of the Int. ANS/ENS Top. Meet. on Probabilistic Safety Methods and Application, San Francisco, 1985.

Cacciabue, P. & Cojazzi, G. Analysis and design of a nuclear safety system versus the operator time constraints. Proc. 2nd IFAC Conf. on Analysis, Design, and Evaluation of Man-Machine Sys., Varese, Italy, 1985.

Campbell, D. T. Factors relevant to validity of experiments in social settings. Psych. Bull., 1957, 54, 297-312.

Carbonell, J. A queueing model for many-instrument visual sampling. IEEE Trans. on Human Factors in Electronics, 1966, HFE-7, 157-164.

Carbonell, N. et al. Acquisition et formalisation du raisonnement dans un systeme expert de lecture de spectrogrammes vocaux. Actes du Colloque 'Les modes de raisonnement', Orsay, ARC, 1984, 67-88.

Carnap, R. Meaning and necessity. University of Chicago Press, 1947.

Carnap, R. The logical foundations of probability. University of Chicago Press, 1950.

Carswell, C. M. & Wickens, C. D. Stimulus integrality in displays of system input-output relationships: A failure detection study. Proc. of the Human Factors Society, 28th Ann. Meeting, 1984.

Cavozzi, J. Contribution a une etude des representations des lois elementaires de la physique des gaz. 1er Congres de Psychologie du Travail, Paris, France, Fevrier 1980.

Chambers, J. M., Cleveland, W. S., Kleiner, P. A. & Tukey, P. A. Graphical methods for data analysis. Wadsworth, 1983.

Chandrasekaran, B. Special issue on natural and social metaphors for distributed problem solving. IEEE Trans. on Sys., Man, Cyb., 1981, SMC-11, 1-96.

Chandrasekaran, B. & Milne, R. (Eds.) Special section on reasoning about structure, behavior and function. ACM SIGART Newsletter, 1985, 93, 4-55.

Chase, W. & Simon, H. A. Perception in chess. Cognitive Psych., 1973, 4, 55-81.

Christensen-Szalanski, J. J. J. & Busyhead, J. B. Physicians' use of probabilistic information in a real clinical setting. J. of Exp. Psych.: Human Perception and Performance, 1981, 7, 928-935.

Cheyn, G. H.-L. & Levis, A. H. Analysis of preprocessors and decision aids in organisations. Proc. 2nd IFAC Conf. on Analysis, Design, and Evaluation of Man-Machine Sys., Varese, Italy, 1985.

Clancey, W. J. Methodology for building an intelligent tutoring system. PhD Thesis, Stanford University, CA, 1981.

Clancey, W. J. The epistemology of rule-based expert systems. A framework for explanation. Artificial Intelligence, 1983, 20, 215-252.

Clancey, W. J. Software tools for developing expert systems. In I. DeLotto & M. Stefanelli (Eds.), Artificial intelligence in medicine. North-Holland, 1985.

Cockerill-Sambre. Prise en compte des facteurs humains des la conception de la coulee continue de la S. A. Cockerill-Sambre-Chertal. Rapport de la recherche CECA No. 7.247-11-026, Liege, Belgium, 1984.

Colby, K. M. & Hilf, F. D. Multidimensional evaluation of a simulation of paranoid thought processes. In L. W. Gregg (Ed.), Knowledge and cognition. Erlbaum, 1974.

Conant, R. C. Laws of information which govern systems. IEEE Trans. on Sys., Man, Cyb., 1976, SMC-6, 240-250.

Coombs, M. J. & Alty, J. Face-to-face guidance of university computer users - II: Characterising advisory interactions. Int. J. Man-Machine Studies, 1980, 12, 407-429.

Coombs, M. J. & Alty, J. L. Expert systems: An alternative paradigm. Int. J. Man-Machine Studies, 1984, 20, 21-43. Also in M. J. Coombs (Ed.), Developments in expert systems. Academic Press, 1984.

Coombs, M. J. & Stell, J. G. A model for debugging PROLOG by symbolic execution: The separation of specification and procedure. University of Strathclyde, Department of Computer Science, 1984.

Coser, L. A. The social functions of conflict. The Free Press, 1954.

Council for Science and Society. New technology: Society, employment & skill. Blackrose Press, 1981.

Croft, W. B. An expert assistant for a document retrieval system. RIAO'85, IMAG, Grenoble, 1985, 131-149.

Crossman, E. R. F. W. A theory of acquisition of speed-skill. Ergonomics, 1959, 2, 153-166.

Cuny, X. Recherche sur l'apprentissage des outils-signes: L'apprentissage du schema developpe en electricite. 1er Congres de Psychologie du Travail, Paris, France, 1980.

Curry, R. E. & Gai, E. G. Detection of random process failures by human monitors. In T. B. Sheridan & G. Johannsen (Eds.), Monitoring behaviour and supervisory control. Plenum Press, 1976.

Daan, H. & Murphy, A. H. Subjective probability forecasting in the Netherlands: Some operational and experimental results. Meteorologische Rundschau, 1982, 35, 99-112.

Daniellou, F. Strategie de resolution d'incidents sur presse automatique: Le poids de la technologie et de l'organisation du travail. Congres de la Societe d'Ergonomie de Langue Francaise, Paris, France, 1982.

Daniels, P. J. The user modelling function of an intelligent interface for document retrieval systems. Proc. IRFIS 6, Intelligent Inf. Sys. for the Inf. Society, Frascati, Italy, 1985.

Daniels, P. J., Brooks, H. M. & Belkin, N. J. Using problem structures for driving human-computer dialogues. RIAO'85, IMAG, Grenoble, 1985, 646-660.

Davis, J. H. Group performance. Addison-Wesley, 1982.

Davis, R. TEIRESIAS: Experiments in communication with a knowledge-based system. In M. E. Sime & M. J. Coombs (Eds.), Designing for human-computer interaction. Academic Press, 1983.

Davis, R. Reasoning from first principles in electronic troubleshooting. In M. J. Coombs (Ed.), Developments in expert systems. Academic Press, 1984.

Davis, R. & King, J. An overview of production systems. In E. W. Elcock & D. Michie (Eds.), Machine Intelligence 8. Wiley, 1976.

Dawes, R. M. The role of the expert in constructing predictive systems. Proc. 1974 Int. Conference on Sys., Man and Cyb. of the IEEE SMC, 1974, 522-525.

Dawes, R. M. The robust beauty of improper linear models in decision making. American Psychologist, 1979, 34, 571-582.

Dawes, R. M. Confidence in intellectual judgments vs. confidence in perceptual judgments. In E. D. Lauterman & H. Feger (Eds.), Similarity and choice. Hans Huber, 1980.

Decreton, M. Supervising a complex measurement and control network using an expert system. Proc. Workshop and Conf. on Applied AI and Knowledge-Based Expert Sys., Stockholm, Sweden, 1984.

De Groot, A. Thought and choice in chess. Mouton, 1965.

De Jong, J. J. & Koster, E. P. The human operator in the computer-controlled refinery. In E. Edwards & F. P. Lees (Eds.), The human operator in process control. Taylor & Francis, 1974.

De Keyser, V. Fiabilite et securite. Fiabilite et securite, Etudes de Physiologie et de Psychologie du Travail No. 7, CECA, Luxembourg, 1972.

De Keyser, V. L'analyse des habilites mentales, mise en valeur ou confiscation de l'experience des travailleurs. Conditions de Travail, 1982, 1, 25-31.

De Keyser, V. Communications sociales et charge mentale dans les postes automatises. In M. de Montmollin (Ed.), Communication et travail (special issue), Psychologie Francaise, 1983, 28, 3-4, 239-241.

De Keyser, V. Structuring of knowledge of operators in continuous processes: Case study of a continuous casting plant startup. In J. Rasmussen, L. Leplat & K. Duncan (Eds.), New technology and human error. Wiley, 1985, in press.

De Keyser, V. Les communications dans les systemes automatises: Champs cognitifs et supports d'information chez les travailleurs. Actes des XIX Journees d'Etudes de l'A. P. S. L. F., La Communication, Presses universitaires de France, Paris, 1985.

De Keyser, V. & Decortis, F. Collective control in an automatized system as apprehended in verbal communications. Proc. 2nd IFAC Conf. on Analysis, Design and Evaluation of Man-Machine Systems, Varese, Italy, 1985.

De Keyser, V. & Piette, A. Analyse de l'activite des operateurs au tableau synoptique d'une chaine d'agglomeration en siderurgie. Le Travail Humain, 1970, 33, 341-352.

de Kleer, J. Qualitative and quantitative knowledge in classical mechanics. MSc Dissertation, MIT, MA, 1975.

Delattre, P. La notion de systeme dans les sciences contemporaines. In J. Lesourne (Ed.), Epistomologies, Vol. II. Librarie de l'Universite Aix-en-Provence, 1982.

Delattre, P. & Thellier, M. (Eds.), Elaboration et justification des modeles, Vol. 2. Maloine, 1979.

Dempster, A. P. Upper and lower probabilities induced by a multi-valued mapping. Annals of Math. Statistics, 1967, 38, 325-339.

DeSmet, A. A., Fryback, D. G. & Thornbury, J. R. A second look at the utility of radiographic skull examination for trauma. American J. of Radiology, 1979, 132, 95-99.

Domotor, Z. Qualitative information and entropy structures. In J. Hintikka & P. Suppes (Eds.), Information and inference. Reidel, 1970.

Domotor, Z. Axiomatization of Jeffrey utilities. Synthese, 1978, 39, 165-210.

Dorner, D. Heuristics and cognition in complex systems. In R. Groner, M. Groner & W. F. Bischof (Eds.), Methods of heuristics. Erlbaum, 1983.

Dorner, D. Of the difficulties people have in dealing with complexity. Simulation and Games, 1984, 11, 67-106.

Drazovich, R. J., Brooks, S. & Foster, S. Knowledge-based ship classification. Workshop on Application of Image Understanding and Spatial Processing to Radar Signals for Aut. Ship Classification, New Orleans, LO, 1979.

Dubois, D. Modeles mathematiques de l'imprecis et de l'incertain en vue d'applications aux techniques d'aide a la decision. These d'Etat, Universite de Grenoble, 1983.

Dubois, D. Steps to a theory of qualitative possibility. Proc. of the 6th Int. Conf. of Cyb. and Sys.. Paris, September 1984, 147-152.

Dubois, D. & Prade, H. Fuzzy sets and systems: Theory and applications. Academic Press, 1980.

Dubois, D. & Prade, H. Additions of interactive fuzzy numbers. IEEE Trans. on Aut. Contr., 1981, TAC-26, 926-936.

Dubois, D. & Prade, H. A class of fuzzy measures based on triangular norms. Int. J. of General Sys., 1982, 8, 43-61.

Dubois, D. & Prade, H. The use of fuzzy numbers in decision analysis. In M. M. Gupta & E. Sanchez (Eds.), Fuzzy information and decision processes. North-Holland, 1982.

Dubois, D. & Prade, H. On several representations of an uncertain body of evidence. In M. M. Gupta & E. Sanchez (Eds.), Fuzzy information and decision processes. North-Holland, 1982.

Dubois, D. & Prade, H. Criteria aggregation and ranking of alternatives in the framework of fuzzy set theory. TIMS Study in the Management Sciences, 1984, 20, 209-240.

Dubois, D. & Prade, H. Fuzzy sets and statistical data. European J. of Operational Res., 1985, in press.

Dubois, D. & Prade, H. Theorie des possibilites: Applications a la representation des connaissances en informatique. Masson, 1985.

Dubois, D. & Prade, H. Evidence measures based on fuzzy information. Automatica, 1985, 21, 547-562.

Dubois, D. & Prade, H. A review of fuzzy set aggregation connectives. Inf. Sciences, 1985, 35, in print.

Dubois, D. & Prade, H. Fuzzy numbers: An overview. In J. C. Bezdek (Ed.), The analysis of fuzzy information, Vol. I. CRC Press, 1985 (in print).

Dubois, D. & Prade, H. A note on measures of specificity for fuzzy sets. Int. J. of General Sys., 1985, 10, 289-293.

Dubois, D. & Prade, H. Recent models of uncertainty and imprecision as a basis for decision theory: Towards less normative frameworks. This volume.

Dubois, D. & Prade, H. A set theoretic view of belief functions: Logical operations and approximations by fuzzy sets. Int. J. of General Sys., 1986, in print.

Dubois, D. & Prade, H. Properties of information measures in Shafer's theory of evidence. Fuzzy Sets and Systems, 1986, in print.

Duda, R., Gaschnig, J. & Hart, P. Model design in the prospector consultant system for mineral exploration. In D. Michie (Ed.), Expert systems in the micro-electronic age. Edinburgh University Press, 1979.

Duncan, J. Selective attention and the organisation of visual information. J. of Exp. Psych.: General, 1984, 113, 501-517.

Duncan, K. D. Training for fault diagnosis in industrial process plants. In J. Rasmussen & W. B. Rouse (Eds.), Human detection and diagnosis of system failures. Plenum, 1981.

Edwards, W. The theory of decision making. Psych. Bull., 1954, 51, 201-214.

Edwards, W. Behavioral decision theory. Ann. Rev. of Psych., 1961, 12, 473-498.

Eilenber, S. Automata, languages, and machines, Vol. B. Academic Press, 1976.

Einhorn, H. J. & Hogarth, R. M. Confidence in judgments: Persistence in the illusion of validity. Psych. Rev., 1978, 85, 395-416.

Einhorn, H. J. & Hogarth, R. M. Behavioral decision theory: Processes of judgment and choice. Ann. Rev. of Psych., 1981, 32, 53-88.

Einhorn, H. J. & Hogarth, R. M. Ambiguity and uncertainty in probabilistic inference. Psych. Rev. 1985, 92, 433-461.

Ephrath, A. R. & Young, L. R. Monitoring vs. man-in-the-loop detection of aircraft control failures. In J. Rasmussen & W. B. Rouse (Eds.) Human detection and diagnosis of system failures. Plenum, 1981.

Ericsson, A. & Simon, H. A. Verbal reports as data. Psych. Rev., 1980, 87, 215-251.

Evans, J. St. B. T. Thinking and reasoning: Psychological approaches. Routledge & Kegan Paul, 1983.

Fagerland, H., Rothaug, T. & Tokle, P. Monitoring and diagnosis of process deviations in marine diesel engines. Proc. of the Institute of Marine Engineers, London, 1978.

Faverge, J. M. et al. L'ergonomie des processus industriels. Editions de l'Institut de Sociologie d l'Universite de Bruxelles, 1966.

Feather, N. (Ed.), Expectancy, incentive and action. Erlbaum, 1982.

Feehrer, C. E. & Baron, S. Artificial intelligence for cockpit aids. IFAC Proc. 2nd Conf. on Analysis, Design, and Evaluation of Man-Machine Sys., Varese, Italy, 1985.

Fichet-Clairfontaine, P. Y. Etude ergonomique de l'influence de la conception de la salle de commande, de la stabilite du processus et de la diversification de la production sur l'activite des operateurs dans les unites a processus continus. Unpublished thesis, Universite de Paris XIII, 1985.

Finegold, A. The engineer's apprentice. In P. H. Winston & K. A. Prendergast (Eds.), The AI business: The commercial use of artificial intelligence. MIT Press, 1984.

Fischhoff, B. Hindsight /= foresight: The effect of outcome knowledge on judgment under uncertainty. J. of Exp. Psych.: Human Perception and Performance, 1975, 1, 288-299.

Fischhoff, B. Perceived information of facts. J. Exp. Psych.: Human Perception and Performance, 1977, 3, 349-358.

Fischhoff, B. Clinical decision analysis. Operations Res., 1980, 28, 28-43.

Fischhoff, B. Debiasing. In D. Kahneman & A. Tversky (Eds.), Judgment under uncertainty: Heuristics and biases. Cambridge University Press, 1982.

Fischhoff, B. Setting standards: A systematic approach to managing public health and safety risks. Management Sci., 1984, 30, 834-843.

Fischhoff, B. Judgmental aspects of risk analysis. Office of Management and Budget Handbook of Risk Analysis. Plenum, in press.

Fischhoff, B. Decision making in complex systems. This volume.

Fischhoff, B. & Beyth-Marom, R. Hypothesis evaluation from a Bayesian perspective. Psych. Rev., 1983, 90, 239-260.

Fischhoff, B. & Cox, L. A. Jr. Conceptual framework for benefit assessment. Handbook of benefit assessment, National Science Foundation for Office of Management and Budget. In press.

Fischhoff, B., Goitein, B. & Shapira, Z. The experienced utility of expected utility approaches. In N. Feather (Ed.), Expectations and actions: Expectancy value models in psychology. Erlbaum, 1982.

Fischhoff, B., Slovic, P. & Lichtenstein, S. Fault trees: Sensitivity of estimated failure probabilities to problem representation. J. of Exp. Psych.: Human Perception and Performance, 1978, 4, 330-344.

Fischhoff, B., Slovic, P. & Lichtenstein, S. Knowing what you want: Measuring labile values. In T. Wallsten (Ed.), Cognitive processes in choice and decision behavior. Erlbaum, 1980.

Fischhoff, B., Svenson, O. & Slovic, P. Active responses to environmental hazards. In D. Stokols & I. Altman (Eds.), Handbook of environmental psychology. Wiley, in press.

Fischhoff, B., Watson, S. & Hope, C. Defining risk. Policy Sciences, 1984, 17, 123-139.

Fine, T. Theories of probability. New York: Academic Press, 1973.

Fiske, S. & Taylor, S. E. Social cognition. Addison Wesley, 1984.

Fitter, M. J. & Sime, M. E. Responsibility and shared decision making. In H. T. Smith & T. R. G. Green (Eds.), Human interaction with computers. Academic Press, 1980.

Forbus, K. D. Qualitative process theory. AI Memo 664, Tech. Rep. MIT AI Lab., 1982.

Forte, B. & Pintacuda, N. Information fournie par une experience. Comptes Rendues de l'Academie des Sciences de Paris, 1968, 266 A, 242-245.

Forte, B. & Pintacuda, N. Sull'informazione associata alle esperienze incomplete. Annali di Matematica Pura ed Applicata, 1968, 80, 215-234.

Fortin, D. A., Rooney, T. B. & Bristol, E. H. Of christmas trees and sweaty palms. Proc. 9th Ann. Advanced Contr. Conf., West Lafayette, IN, 1983.

Fox, M. S. Artificial intelligence in the factory of the future. Proc. ACM 12th Ann. Computer Sci. Conf., Philadelphia, PA, 1984.

Fox, M. S., Lowenfeld, S. & Kleinosky, P. Techniques for sensor-based diagnosis. Proc. 8th. IJCAI, Karlsruhe, BRD, 1983.

Fraisse, P. Psychologie du temps. Presses Universitaire de France, 1957.

Frank, M. J. On the simultaneous associativity of F(x,y) and x+y-F(x,y). Aequationes Mathematicae, 1979, 19, 194-226.

Fraser, J. T., Harber, F. C. & Muller, G. H. (Eds.) The study of time. Springer Verlag, 1972.

Freeling, A. N. S. Fuzzy sets and decision analysis. IEEE Trans. on Sys., Man, Cyb., 1980, SMC-10, 341-354.

Funk, K., Greitzer, F. L. & Hutchins, S. G. _Prototype intelligent tactical assistant (ITA): Conceptual design and initial implementation_. NPRDC 84-38, San Diego, CA, 1984.

Fussel, J. B. Fault-tree analysis. Concepts and techniques. In E. J. Henley & J. W. Lynn (Eds.), _Generic techniques in system reliability assessment_. Noordhoff, 1976.

Gallanti, M. & Guida, G. _Intelligent decision aids for process environments: An expert system approach_. This volume.

Gallanti, M., Guida, G., Spampinato, L. & Stefanini, A. Representing procedural knowledge in expert systems: An application to process control. _Proc. 9th IJCAI_, Los Angeles, CA, 1985.

Galperine, P. Essai sur la formation par etages des actions et des concepts. _Recherches psychologiques en URSS_. Moscou: Editions du Progres, 1966.

Gammack, J. G. & Young, R. M. Psychological techniques for eliciting expert knowledge. In M. Bramer (Ed.), _Research and development in expert systems_. Cambridge University Press, 1985.

Garbolino, P. _Decision complexity and information measures_. This volume.

Gardies, J. L. _La logique du temps_. Presses Universitaires de France, 1975.

Gaschnig, J., Klahr, P., Pople H., Shortliffe, E. H. & Terry, A. Evaluation of expert systems: Issues and case studies. In D. A. Waterman, R. Hays-Roth & D. Lenat (Eds.), _Building expert systems_. Addison-Wesley, 1984.

Gauthier, D. P. _Practical reasoning_. Clarendon Press, 1963.

Gentner, D. & Stevens, A. L. _Mental models_. Erlbaum, 1983.

George, C. _Apprendre par l'action_. Paris: PUF, 1983.

Georgeff, M. P. & Bonollo, U. Procedural expert systems. _Proc. 8th IJCAI_, Karlsruhe, BRD, 1983.

Gibson, J. J. _The senses considered as perceptual systems_. Houghton Mifflin, 1969.

Gibson, J. J. _The ecological approach to visual perception_. Houghton Mifflin, 1979.

Giles, R. Foundations for a theory of possibility. In M. M. Gupta & E. Sanchez (Eds.), _Decision processes_. North-Holland, 1982.

Goldberg, L. R. Simple models or simple processes? Some research on clinical judgments. _American Psychologist_, 1968, 23, 483-496.

Goldsmith, T. E. & Schvaneveldt, R. W. Facilitating multiple-cue judgments with integral information displays. In J. C. Thomas & M. L. Schneider (Eds.) Human factors in computer systems. Ablex Publishing, 1984.

Gonzales, R. C. & Houngton, L. C. Machine recognition of abnormal behavior in nuclear reactors. IEEE Trans. on Sys., Man, Cyb., 1977, SMC-7, 717-728.

Good, I. J. Subjective probability as the measure of a non measurable set. In E. Nagel, P. Suppes & A. Tarski (Eds.), Logic methodology and philosophy of science. Stanford University Press, 1962.

Goodstein, L. P. Display support for detection and identification of disturbances in industrial process systems. In J. Rasmussen & W. B. Rouse (Eds.), Human detection and diagnosis of system failures. Plenum, 1981.

Goodstein, L. P. An integrated display set for process operators. In G. Johannsen & J. E. Rijnsdorp (Eds.), Analysis, design and evaluation of man-machine systems. Pergamon Press, 1983.

Gottinger, H. Qualitative information and comparative informativeness. Kybernetik, 1973, 13, 81-94.

Gottinger, H. Subjective qualitative information structures based on orderings. Theory and Decision, 1974, 5, 69-97.

Gottinger, H. Coping with complexity. Reidel, 1983.

Graf, P. & Schacter, D. L. Implicit and explicit memory for new associations in normal and amnesic patients. J. of Exp. Psych.: Learning, Memory, and Cognition, 1985, 11, 510-518.

Green, T. Human interaction with computers. Academic Press, 1980.

Grether, D. M. & Plott, C. R. Economic theory of choice and the preference reversal phenomenon. American Economic Rev., 1979, 69, 623-638.

Grosjean, V. Etude de l'erreur a travers un cadre explicatif complexe. Le Travail Humain, 1986, in print.

Haber, R. N. The power of visual perceiving. J. of Mental Imagery, 1981, 5, 1-40.

Hacking, I. Slightly more realistic personal probability. Philos. of Sci., 1967, 34, 311-325.

Hamblin, C. L. The modal "probably". Mind, 1959, 68, 234-240.

Hammer, W. Product safety and management engineering. Prentice-Hall, 1980.

Harman, G. Reasoning and explanatory coherence. American Philosophical Quart., 1980, 17, 151-157.

Harmon, P. & King, D. Expert systems: Artificial intelligence in business. Wiley, 1985.

Hart, P. G. Directions for AI in the eighties. ACM SIGART Newsletter, 1982, 79, 11-16.

Hart, S. G. & Sheridan, T. B. Pilot workload, performance, and aircraft control automation. Proc. AGARD Symposium on Human Factors Considerations in High Performance Aircraft. NATO, Brussels, 1984.

Hasher, L. & Zacks, R. T. Automatic processing of fundamental information: The case of frequency of occurrence. American Psychologist, 1984, 39, 1372-1388.

Hawkins, D. An analysis of expert thinking. Int. J. of Man-Machine Studies, 1983, 18, 1-48.

Hayes-Roth, F., Waterman, D. A. & Lenat, D. Principles of pattern-directed inference systems. In D. A. Waterman & F. Hayes-Roth (Eds.), Pattern-directed inference systems. Academic Press, 1978.

Hayes-Roth, F., Waterman, D. A. & Lenat, D. (Eds.), Building expert systems. Addison-Wesley, 1983.

Herriot, P. Attributes of memory. Methuen & Co., 1974.

Hery, J. F. A prototype expert system in PWR power plant conducting. In T. Bernold & G. Albers (Eds.), Artificial intelligence: Towards practical application. North-Holland, 1985.

Hershey, J. C., Kuhnreuter, H. C. & Schoemaker, P. J. H. Sources of bias in assessment procedures for utility functions. Management Sci., 1982, 28, 936-954.

Higashi, M. & Klir, G. Measures of uncertainty and information based on possibility distributions. Int. J. of General Sys., 1983, 9, 43-58.

Hoc, J. M. & Leplat, J. Evaluation of different modalities of verbalization in a sorting task. Int. J. of Man-Machine Studies, 1983, 18, 283-306.

Hogarth, R. M. Beyond discrete biases: Functional and dysfunctional aspects of judgmental heuristics. Psych. Bull., 1982, 90, 197-217.

Hollnagel, E. The paradigm for understanding in hermeneutics and cognition. J. of Phenomenological Psych., 1978, 9, 188-217.

Hollnagel, E. What we do not know about man-machine systems. Int. J. of Man-Machine Studies, 1983, 18, 135-143.

Hollnagel, E. Inductive and deductive approaches to modelling of human decision making. Psyke & Logos, 1984, 5, 288-301.

Hollnagel, E. A survey of man-machine system evaluation methods. HWR-148, OECD Halden Reactor Project, Halden, Norway, 1985.

Hollnagel, E. Cognitive system performance analysis. This volume.

Hollnagel, E., Hunt, G. & Marshall, E. The experimental validation of the critical function monitoring system. Preliminary results of analysis. HWR-111, OECD Halden Reactor Project, Halden, Norway, 1983.

Hollnagel, E. & Lind, M. Self-reference as a problem in the control of complex systems. Psyke & Logos, 1982, 2, 323-332.

Hollnagel, E., Pedersen, O. M. & Rasmussen, J. Notes on human performance analysis. Riso-M-2285, Riso National Laboratory, Roskilde, Denmark, 1981.

Hollnagel, E. & Woods, D. Cognitive systems engineering: New wine in new bottles. Int. J. of Man-Machine Studies, 1983, 18, 583-306.

Hollo, E. & Taylor, J. R. Algorithm and program for consequence diagram and fault-tree construction. Riso-M-1907, Riso National Laboratory, Roskilde, Denmark, 1976.

Hoogovens report. Human factors evaluation: Hoogovens No. 2 hot strip mill. Tech. Rep. FR251, British Steel Corporation /Hoogovens, 1976.

Hovland, C. I., Janis, I. L. & Kelley, H. H. Communication and persuasion. Yale University Press, 1976.

Humphreys, P., Svenson, O. & Vari A. (Eds.) Analysing and aiding decision processes. North-Holland, 1983.

Hunt, R. M. & Rouse, W. B. A fuzzy rule-based model of human problem solving. IEEE Trans. on Sys., Man, Cyb., 1984, SMC-14, 112-120.

Hutchins, E., Hollan, J. D. & Norman, D. A. Direct manipulation interfaces. In Norman, D. A. & Draper, S. W. (Eds.), User centered system design: New perspectives on human-computer interaction. Erlbaum, 1985.

Iosif, G. Functio de supraveghere a tablourila de commanda. Editera Academici Republicii Socialiste Romania, Bukarest, 1970.

Iosif, G. & Ene P. On operator's mental proceedings in diagnosis. Rev. Roum. Sci. Sociales, 1978, 22, 173-183.

Jackson, P. & Lefrere, P. On the application of rule-based techniques to the design of advice giving systems. Int. J. of Man-Machine Studies, 1984, 20, 63-86.

Janis, I. L. Victims of groupthink. Houghton Mifflin, 1972.

Janis, I. L. Counselling on personal decisions. Yale University Press, 1982.

Janis, I. L. & Mann, L. Decision making. The Free Press, 1980.

Jeffrey, R. C. Axiomatizing the logic of decision. In C. A. Hooker (Ed.), Foundations and applications of decision theory, Vol. 1. Reidel, 1978.

Jeffrey, R. C. The logic of decision, 2nd ed. University of Chicago Press, 1983.

Jeffrey, R. C. De Finetti's probabilism. Synthese, 1984, 60, 73-90.

Johannsen, G. Fault management and supervisory control of decentralized systems. In J. Rasmussen & W. B. Rouse (Eds.), Human detection and diagnosis of system failure. Plenum Press, 1981.

Johannsen, G. & Borys, B.-B. Investigation of display contents and decision support in a rule=based fault correction task. Proc. 2nd IFAC Conf. on Analysis, Design, and Evaluation of Man-Machine Sys., Varese, Italy, 1985.

Johannsen, G., Rijnsdorp, J. E. & Sage, A. P. Human system interface concerns in support system design. Automatica, 1983, 19, 595-603.

Johannsen, G. & Rouse, W. B. Studies of planning behavior of aircraft pilots in normal, abnormal, and emergency situations. IEEE Trans. Sys., Man, Cyb., 1983, SMC-13, 267-278.

Johnson-Laird, P. N. Mental models: Towards a cognitive science of language, inference, and consciousness. Cambridge University Press, 1983.

Joyce, J. P. & Lapinsky, G. W. A history and overview of the safety parameter display system concept. IEEE Trans. on Nuclear Sci., 1983, NS-30.

Jungermann, H. The two camps on rationality. In R. W. Scholz (Ed.), Decision making under uncertainty. Elsevier, 1984.

Kahneman, D. & Miller, D. T. Norm theory: Comparing reality to its alternatives. Manuscript submitted for publication, 1984.

Kahneman, D., Slovic, P. & Tversky, A. (Eds.) Judgment under uncertainty: Heuristics and biases. Cambridge University Press, 1982.

Kahneman, D. & Treisman, A. Changing views of attention and automaticity. In R. Parasuraman & D. R. Davies (Eds.), Varieties of attention. Academic Press, 1984.

Kahneman, D. & Tversky, A. Subjective probability: A judgment of representativeness. Cognitive Psych., 1972, 3, 430-454.

Kahneman, D. & Tversky, A. Prospect theory: An analysis of decision under risk. Econometrica, 1979, 47, 263-292.

Kappos, D. A. Strukturtheorie der Wahrscheinlighkeitsfelder u.- raume. Springer Verlag, 1960.

Kappos, D. A. Probability algebras and stochastic spaces. Academic Press, 1969.

Karmiloff-Smith, A. & Inhelder, B. If you want to get ahead, get a theory. Cognition, 1975, 3, 195-212.

Kautto, A. Information presentation in power plant control rooms. Research Report 320, Technical Research Centre of Finland, Helsinki, Finland, 1984.

Keeney, R. Multiplicative utility functions. _Operations Res._, 1974, _22_, 22-34.

Keeney, R. L. & Raiffa, H. _Decision with multiple objectives._ Wiley, 1976.

Kemeny, J. et al. _The need for change: The legacy of TMI. Report of the President's Commission on the Accident at Three Mile Island._ Washington, Government Printing Office, 1979.

Kempf, K. G. Manufacturing and artificial intelligence. In T. Bernold & G. Albers (Eds.), _Artificial intelligence: Towards practical application._ North-Holland, 1985.

Kepner, C. H. & Tregoe, B. B. _The rational manager: A systematic approach to problem solving and decision making._ McGraw Hill, 1960.

Keren, G. B. _On the calibration of experts and lay people._ Manuscript submitted for publication, 1984.

Keren, G. B. _How much do we know about what we perceive?_ Manuscript submitted for publication, 1985.

Kidd, J. B. The utilization of subjective probabilities in production planning. _Acta Psychologica_, 1970, _34_, 338-347.

Klebau, J., Baldeweg, F., Fiedler, U. & Lindner, A. An interactive approach to disturbance analysis in nuclear power plants. _Proc. 2nd IFAC Conf. on Analysis, Design, and Evaluation of Man-Machine Sys._, Varese, Italy, 1985.

Kleinman, D., Baron, S. & Levison, W. A control theoretic approach to manned-vehicle analysis. _IEEE Trans. on Aut. Contr._, 1971, _AC-16_, 824-832.

Kleinman, D., Baron, S. & Levison, W. An optimal control model of human response, Part I: Theory and validation. _Automatica_, 1974, _6_, 357-369.

Kok, J. J. & van Wijk, R. A. _Evaluation of models describing human operator control of slowly responding complex systems._ PhD Thesis, DUT, Delft,, 1978.

Kok, J. J. & stassen, H. G. Human operator control of slowly responding systems: Supervisory control. _J. of Cyb. and Inf. Sci._, 1980, _3_, 123-174.

Kolers, P. A. & Smythe, W. E. Images, symbols and skills. _Canadian J. of Psych._, 1979, _33_, 158-183.

Koriat, A., Lichtenstein, S. & Fischhoff, B. Reasons for confidence. _J. of Exp. Psych.: Human Learning and Motivation_, 1980, _6_, 107-118.

Koslowski, B. & Bruner, J. S. Learning to use a lever. _Child Development_, 1972, _43_, 790-799.

Kotovsky, K., Hayes, J. R. & Simon, H. A. Why are some problems hard? Evidence from the tower of hanoi. _Cognitive Psychology_, 1985, _17_, 248-294.

487

Kraiss, K.-F. Fahrzeug-und Processfuhrung. Kognitives Verhalten des Menschen und Entscheidungshilfen. Springer Verlag, 1985.

Krantz, D. H. & Luce, R. D. Conditional expected utility. Econometrica, 1971, 39, 253-271.

Krantz, D. H., Luce, R. D., Suppes, P. & Tversky, A. Foundations of measurement, Vol. 1. Academic Press, 1971.

Kripke, S. Semantical considerations on modal logic. Acta Philosophica Fennica, 1963, 16, 83-94.

Kripke, S. Naming and necessity. In D. Davidson & G. Harman (Eds.), Semantics of natural languages. Reidel, 1972.

Kruskal, J. B. Nonmetric multidimensional scaling: A numerical method. Psychometrika, 1964, 29, 115-129.

Landa, L. N. Institutional regulation and control. Cybernetics, algorithmization and heuristics in education. Educational Technology Publications, 1976.

Landa, L. N. Algo-heuristic theory of performance, learning and instruction: Subject, problems, principles. Contemporary Educational Psych., 1984, 9, 235-245.

Langlotz, C. P. & Shortliffe, E. H. Adapting a consultation system to critique user plans. Int. J. of Man-Machine Studies, 1983, 19, 479-496.

Lanir, T. Strategic surprises. Tel Aviv University Press, 1982.

Latane, B., Williams, K. & Harkins, S. Many hands make light the work: The causes and consequences of social loafing. J. of Personality and Social Psych., 1979, 37, 822-832.

Lee, S. M. Goal programming for decision analysis. Auerbach, 1972.

Lees, F. P. Computer support for diagnostic tasks in the process industries. In J. Rasmussen & W. B. Rouse (Eds.), Human detection and diagnosis of system failures. Plenum Press, 1981.

Lees, F. P. Process computer alarm and disturbance analysis: Review of the state of the art. Computers and Chem. Eng., 1983, 7, 669-694.

Lehnert, W. The process of question answering. Erlbaum, 1978.

Lenat, D. B. The nature of heuristics. Artificial Intelligence, 1982, 19, 189-249.

Lenat, D. B. Theory formation by heuristic search. The nature of heuristics II: Background and examples. Artificial Intelligence, 1983, 21, 31-60.

Lenat, D. B. & Brown, J. S. Why AM and Eurisko appear to work. Artificial Intelligence, 1984, 23, 269-294.

Leontyev, A. N. Problems of the development of the mind. Moscow: Progress, 1981 (org. 1959).

Leplat, J. Attention et incertitude dans les travaux de surveillance et d'inspection. Dunod, 1968.

Leplat, J. Task analysis and activity analysis in situations of field diagnosis. In J. Rasmussen & W. B. Rouse (Eds.), Human detection and diagnosis of system failures. Plenum Press, 1981.

Leplat, J. The elicitation of expert knowledge. This volume.

Leplat, J. & Cuny, X. Introduction a l'analyse du travail (2nd ed.) Paris, PUF, 1984.

Leplat, J. & Hoc, J. M. Subsequent verbalization in the study of cognitive processes. Ergonomics, 1981, 743-755.

Leplat, J. & Pailhous, J. L'acquisition des habiletes mentales: La place des techniques. Le Travail Humain, 1981, 44, 275-282.

Leplat, J. & Savoyant, P. Ordonanncement et coordination des activites dans les travaux individuels et collectifs. Bull. de Psych., 1984, 36-37, 270-278.

Levy-Schoen, A. Mesurer les mouvements des yeux: Pour quoi faire? Le Travail Humain, 1983, 46, 3-10.

Lewis, C. & Mack, R. The role of abduction in learning to use a computer system. Research Report RC9433 (#41620), IBM Thomas J. Watson Research Centre, Yorktown Heights, NY, 1982.

Lichtenstein, S. & Fischhoff, B. Do those who know more also know more about how much they know? The calibration of probability judgments. Organizational Behavior and Human Performance, 1977, 20, 159-183.

Lichtenstein, S. & Fischhoff, B. Training for calibration. Organizational Behavior and Human Performance, 1980, 26, 149-171.

Lichtenstein, S., Fischhoff, B. & Phillips, L. Calibration of probabilities: State of the art to 1980. In D. Kahneman, P. Slovic & A. Tversky (Eds.), Judgment under uncertainty: Heuristics and biases. Cambridge University Press, 1982.

Lichtenstein, S., Slovic, P. & Zink, D. Effect of instruction in expected value on optimality of gambling decisions. J. of Exp. Psych., 1969, 79, 236-240.

Lieberman, H. Seeing what your programs are doing. Int. J. of Man-Machine Studies, 1984, 21, 311-331.

Lihou, D. A. Aiding process plant in fault finding and corrective action. In J. Rasmussen & W. B. Rouse (Eds.), Human detection and diagnosis of system failure. Plenum Press, 1981.

Lind, M. The use of flow models for automated plant diagnosis. In J. Rasmussen & W. B. Rouse (Eds.), Human detection and diagnosis of system failures. Plenum Press, 1981.

Lind, M. Multilevel flow modelling for process plant diagnosis and control. Riso-M-2357, Riso National Laboratories, Roskilde, Denmark, 1982.

Lind, M. Information interfaces for process plant diagnosis. Riso-M-2417, Riso National Laboratories, Roskilde, Denmark, 1984.

Lind, M. Representing goals and functions of complex systems: An introduction to multilevel flow modelling. In preparation, 1985.

Lind, R. & Greeburger, M. (Eds.) Rate of discount; Its meaning and appropriateness in energy investment and R&R decision making. Resources for the Future, Washington, 1980.

Loftus, E. F. Eyewitness testimony. Harvard University Press, 1979.

Loftus, E. F., Miller, D. G. & Burns, H. J. Semantic integration of verbal information in a visual memory. J. of Exp. Psych.: Human Learning and Memory, 1978, 4, 19-31.

Luhmann, N. Macht. Ferdinand Enke Verlag, 1975.

Luhmann, N. Gesellschaftsstruktur und Semantik. Suhrkamp Verlag, 1980.

Lukau, A. L'activite de dispatcher de chemin de fer a differents niveaux de contraintes. Laboratory of Industrial Psychology, IPSE, University of Liege, Belgium, 1985. (Unpublished thesis.)

Lunteren, A. van Identification of human operator describing function models with one or two inputs in closed loop systems. PhD Thesis, DUT, Delft, 1979.

Lusk, E. & Stratton, R. Automated reasoning in man-machine control systems. Proc. 9th Ann. Advanced Contr. Conf., West Lafayette, IN, 1983.

Lusted, L. B. A study of the efficacy of diagnostic radiologic procedures: Final report on diagnostic efficacy. Efficacy Study Committee of the American College of Radiology, 1977.

Madni, A. M., Samet, M. G. & Freedy, A. A trainable on-line model of the human operator in information acquisition tasks. IEEE Trans. Sys., Man, Cyb., 1982, SMC-12, 504-511.

Mandler, G. Cognitive psychology. An essay in cognitive science. Erlbaum, 1985.

March, J. G. Bounded rationality, ambiguity, and the engineering of choice. The Bell J. of Economics, 1978, 9, 587-608.

Marczewski, E. Independance d'ensembles et prolongement de mesures. Colloquium Mathematicum, 1948, 1, 122-132.

Marine, C. & Navarro, C. Role de l'organisation informelle du travail en equipe lors d'un dysfonctionnement technique. Bull. de Psych., 1980, 33-34, 311-316.

Marr, D. Vision. Freeman, 1982.

Martin-Solis, G. A., Andow, P. K. & Lees, F. P. An approach to fault-tree synthesis for process plants. Loss prevention and safety promotion in the process industries, Vol. DECHEMA, Frankfurt, 1977.

McClelland, J. L. & Rumelhart, D. E. (Eds.) Parallel distributed processing: Explorations in the microstructure of cognition. Vol. II: Applications. Bradford Books & MIT Press, 1986.

McCormick, N. J. Reliability and risk analysis. Academic Press, 1981.

McDermott, D. A temporal logic for reasoning about processes and plans. Cognitive Sci., 1982, 6, 101-155.

McRuer, D. & Jex, H. R. A review of quasi-linear pilot models. IEEE Trans. on Human Factors in Electronics, 1967, HFE-8, 231-249.

McRuer, D. & Krendel, E. Mathematical models of pilot behavior. NATO Agardograph #188, 1974.

Medin, D. L. & Schaffer, M. M. Context theory of classification learning. Psych. Rev., 1978, 85, 207-238.

Meehl, P. E. Clinical vs. statistical prediction: A theoretical analysis and a review of evidence. University of Minnesota Press, 1954.

Meijer, C. H. A critical function expert system for nuclear power plants. Proc. Enlarged Halden Programme Group Meeting on Fuel Performance Experiments and Analysis and Computerised Man-Machine Communication, Loen, Norway, 1983.

Meijer, C. H. Potential applications of knowledge based expert systems to industrial environments. Proc. 4th C-E Corporate Technology Awareness Conf., Atlanta, GA, 1983.

Miller, G. A. The magical number seven, plus or minus two: Some limits on our capacity for processing information. Psych. Rev., 1956, 63, 81-97.

Miller, P. L Attending: Critiquing a physician's management plan. IEEE Trans. on Pattern Analysis and Machine Intelligence, 1983, PAMI-5, 449-461.

Miller, R. B. Task description and analysis. In R. M. Gagne (Ed.), Psychological principles in system development. Holt, Rinehart and Winston, 1963.

Millot, P. & Willaeys, D. An approach of dynamical allocation of supervision tasks between man and computer in control rooms of automatized production systems. Proc. 2nd IFAC Conf. on Analysis, Design, and Evaluation of Man-Machine Sys., Varese, Italy, 1985.

Minsky, M. A framework for representing knowledge. In P. H. Winston (Ed.), The psychology of computer vision. McGraw-Hill, 1975.

Mischel, W. Personality and assessment. Wiley, 1968.

Mitroff, I. & Featheringham, T. R. On systemic problem solving and the error of the third kind. Behavioral Sci., 1974, 19, 383-393.

Mondadori, M. On a well-known difficulty for the subjectivistic research programme. Epistemologia, 1984, 7, 33-42.

Montague, R. Formal philosophy. Yale University Press, 1974.

Montgomery, H. Decision rules and the search for a dominance structure: Towards a process model of decision making. In P. Humphreys, O. Svenson & A. Vari (Eds.), Analyzing and aiding decision processes. North-Holland, 1983.

Montmollin, M. L'intelligence de la tache. Elements d'ergonomie cognitive. Peter Lang, 1984.

Montmollin, M. & De Keyser, V. Expert logic vs. operator logic. In G. Johannsen, G. Mancini & L. Martenson (Eds.), Analysis, design and evaluation of man-machine systems. CEC-JRC Ispra, Italy: IFAC, 1985.

Moore, R. L., Hawkinson, L. B., Knickerbocker, C. G. & Churchman, L. M. A real-time expert system for process control. Proc. 1st Conf. on Artificial Intelligence Applications, Denver, CO, 1984.

Moray, N. The role of attention in the detection of errors and the diagnosis of failures in man-machine systems. In J. Rasmussen & W. B. Rouse (Eds.), Human detection and diagnosis of system failures. Plenum Press, 1981.

Moray, N. Monitoring behavior and supervisory control. In K. Boff, J. Beatty & L. Kaufmann (Eds.), Handbook of human perception and performance, Wiley, 1985, in press.

Moray, N. Modelling cognitive activities: Human limitations in relation to computer aids. This volume.

Morick, H. Cartesian privilege and the strictly mental. Philos. and Phenomenological Res., 1971, 31, 546-551.

Motoda, H., Yamada, N. & Yoshida, K. A knowledge based system for plant diagnosis. Proc. Int. Conf. on 5th Generation Computer System, ICOT, Tokyo, Japan, 1984.

Muralidharan, R. & Baron, S. DEMON: A human operator model for decision making, monitoring and control. J. of Cyb. and Inf., 1980, 3.

Mulsant, B. & Servan-Schreiber, D. Knowledge engineering: A daily activity on a hospital ward. Technical Report STAN-CS-82-998, Stanford University, 1983.

Murphy, A. H. & Winkler, R. L. Probability of precipitation forecasts. J. of the American Statistical Association, in press.

Myers, D. G. & Lamm, H. The group polarization phenomenon. Psych. Bull., 1976, 83(4), 602-627.

National Interagency Management System. The what, why, and how of NIIMS. Us Dept. of Agriculture, 1982.

National Research Council. Survey measure of subjective phenomena. The Council, 1982.

National Research Council. Research needs in human factors. The Council, 1983.

National Science Foundation. Office of Management and Budget handbook of risk assessment. Plenum, in press.

Negotia, C. V. Expert systems and fuzzy systems. The Benjamin Cummings Publishing Company, 1985.

Neisser, U. Cognition and reality. Freeman, 1976.

Nelson, W. REACTOR: An expert system for diagnosis and treatment of nuclear reactor accidents. Proc. 2nd Nat. Conf. on Artificial Intelligence, Pittsburgh, PA, 1982.

Newell, A. Artificial intelligence and the concept of mind. In R. C. Schank & K. M. Colby (Eds.), Computer models of thought and language. Freeman, 1973.

Newell, A. & Simon, H. A. Human problem solving. Prentice-Hall, 1972.

Nielsen, D. S. Use of cause-consequence charts in practical systems analysis. In R. E. Barlow (Ed.), Proc. conf. on reliability and Fault-tree analysis. University of California at Berkeley, 1974.

Nii, H. P. & Feigenbaum, E. A. Rule based understanding of signals. In D. A. Waterman & R. Hayes-Roth (Eds.), Pattern directed inference systems. New York: 1978.

Nilsson, N. J. Principles of artificial intelligence. Tioga Publishing, 1980.

Nisbett, R. E. & Ross, L. Human inference: Strategies and shortcomings of social judgment. Prentice-Hall, 1980.

Nisbett, R. E. & Wilson, T. D. Telling more than we can know: Verbal reports on mental processes. Psych. Rev., 1977, 74, 231-259.

Norman, D. A. Categorization of action slips. Psych. Rev., 1981, 88, 1-15.

Norman, D. A. Position paper on human error. NATO Conf. on Human Error, Bellagio, Italy, 1983.

Norman, D. A. Cognitive engineering. In D. A. Norman & S. W. Draper (Eds.), User centered system design: New perspectives in human-machine interaction. Erlbaum, 1986.

493

Norman, D. A. New views of information processing: Implications for intelligent decision support systems. This volume.

Norman, D. A. & Bobrow, D. G. On the role of active memory processes in perception and cognition. In C. N. Cofer (Ed.), The structure of human memory. Freeman, 1976.

Norman, D. A. & Bobrow, D. G. Descriptions: An intermediate stage in memory retrieval. Cognitive Psych., 1979, 11, 107-123.

Norman, D. A. & Draper, S. W. (Eds.), User centered system design: New perspectives in human-machine interaction. Erlbaum, 1986.

Norwich, A. M. & Turksen, I. B. The fundamental measurement of fuzziness. In R. Yager (Ed.), Fuzzy set and possibility theory: Recent developments. Pergamon Press, 1982.

Nowell-Smith, P. H. Ethics. Penguin, 1954.

Oblow, E. A hybrid uncertainty theory. Proc. of the 5th Int. Workshop on Expert Sys. and their Applications, Avignon, May 13-15. ADI, 1985.

Ochanine, D. French translation of nine unpublished papers. L'image operative. Universite de Paris I, 1981.

Olson, D. R. The languages of experience: On natural language and formal education. Bull. of the Brit. Psych. Society, 1975, 28, 363-373.

Ornstein, R. E. On the experience of time. Penguin, 1969.

Pailhous, J. La representation de l'espace urbain. Presse Universitaires de France, 1970.

Papenhuijzen R. On the modelling of the navigator's behavior on board ships. MSc Thesis, A324, DUT, Delft, 1985.

Parinello, S. La teoria delle scelte: Anzioni presenti a gradi di liberta di azioni future. Universita di Roma, Istituto di Automatica, R. B. 0. 5, 1981.

Pattipati, K. R., Ephrath, A. R. & Kleinman, D. L. Analysis of human decision making in multi-task environments. Technical Report EECS-TR-79-15, University of Connecticutt Proc. Int. Conf. on Cyb. and Soc., 1980.

Pattipati, K. R., Kleinman, D. L. & Ephrath, A. E. A dynamic decision model of human task selection performance. IEEE Trans. Sys., Man, Cyb., 1983, SMC-13, 145-166.

Payne, J. W. Contingent decision behavior. Psych. Bull., 1982, 92, 382-401.

Payne, S. J., Sime, M. E. & Green, T. R. G. Perceptual structure cueing in a simple command language. Int. J. of Man-Machine Studies, 1984, 21, 19-29.

494

Pearl, J., Leal, A. & Saleh, J. GODDESS: A goal directed decision structuring system. _IEEE Trans. on Pattern Analysis and Machine Intelligence_, 1982, _PAMI-4_, 250-262.

Pejtersen, A. M. Investigation of search strategies in fiction based on an analysis of 134 user-librarian conversations. _IRFIS 3, Proc. of the Third Int. Res. Forum in Inf. Sci._, Oslo, 1979.

Perrow, C. _Normal accidents._ Basic Books, 1984.

Peterson, C. R. (Ed.) Special issue: Cascaded inference. _Organizational Behavior and Human Performance_, 1973, _10_, 310-432.

Peterson, C. R. & Beach, L. R. Man as an intuitive statistician. _Psych. Bull._, 1967, _63_, 29-46.

Pew, R. W. & Baron, S. Perspectives on human performance modelling. _Proc. IFAC Conf. on Analysis, Design, and Evaluation of Man-Machine Sys._, Baden-Baden, BRD, 1982.

Pew, R. W., Miller, D. C. & Fehrer, C. E. _Evaluation of control room improvements through analysis of critical operator decisions._ EPRI NP-1982, Palo Alto, CA, 1981.

Piaget, J. _Le developpement de la notion de temps chez l'enfant._ Presses Universitaire de France, 1946.

Pitz, G. S. & Sachs, N. J. Behavioral decision theory. _Ann. Rev. of Psych._, 1984, _35_.

Pitz, G. S., Sachs, N. J. & Heerboth, J. Procedures for eliciting choices in the analysis of individual decisions. _Organizational Behavior and Human Performance_, 1980, _26_, 396-408.

Polanyi, M. _Personal knowledge._ Routledge & Kegan Paul, 1962.

Polanyi, M. _The tacit dimension._ Doubleday, 1966.

Pollack, M. E., Hirschberg, J. & Webber, B. User participation in the reasoning processes of expert systems. _Proc. of the National Conf. on Artificial Intelligence_, 1982.

Polya, G. _Les mathematiques et le raisonnement plausible._ Gauthier-Villars, 1958.

Pople, H. Heuristic methods for imposing structure on ill-structured problems: The structuring of medical diagnosis. In P. Szolovitz (Ed.), _Artificial intelligence in medicine._ Westview Press, 1982.

Powers, G. J. & Tomkins, F. C. Fault-tree synthesis for chemical processes. _A. I. Chem. Eng. Journ._, 1974, _20_, 376.

Prade, H. A computational approach to approximate and plausible reasoning with applications to expert systems. _IEEE Trans. on Pattern Analysis and Machine Intelligence_, 1985, _PAMI-7_(3), 260-283.

Prior, A. N. _Time and modality._ Clarendon Press, 1957.

Pushkin, V. N. (Ed.) Problems of heuristics. Jerusalem: Program for scientific translation, 1972.

Putnam, H. The mental life of some machines. In H. N. Castaneda (Ed.), Intentionality, minds and perception. Wayne State University Press, 1967.

Quinlan, J. R. Internal consistency in plausible reasoning systems. New Generation, 1985, 3(2), 157-180.

Rabardel, J. Influence des representations preexistantes sur la lecture du dessin technique. Le Travail Humain, 1982, 41(2), 251-266.

Rabbitt, P. The control of attention in visual search. In R. Parasuraman & D. R. Davies (Eds.), Varieties of attention. Academic Press, 1984.

Raiffa, H. Decision analysis. Addison-Wesley, 1968.

Rasmussen, J. The human data processor as a system components: Bits and pieces of a model. Riso-M-1722, Riso National Laboratory, Roskilde, Denmark, 1974.

Rasmussen, J. Outlines of a hybrid model of the process operator. In T. B. Sheridan & G. Johannsen (Eds.), Monitoring behavior and supervisory control. Plenum Press, 1976.

Rasmussen, J. What can be learned from human error reports? In K. D. Duncan, M. Gruneberg & D. Wallis (Eds.), Changes in working life. Wiley, 1980.

Rasmussen, J. Models of mental strategies in process plant environments. In J. Rasmussen & W. B. Rouse (Eds.), Human detection and diagnosis of system failure. Plenum Press, 1981.

Rasmussen, J. Skills, rules, and knowledge: Signals, signs,and symbols, and other distinctions in human performance. IEEE Trans. on Sys., Man, Cyb., 1983.

Rasmussen, J. Human error data: Facts or fiction. Riso-M-2499, 4th Nordic Accident Seminar, Rovaniemi, Finland, 1984.

Rasmussen, J. Strategies for state identification and diagnosis in supervisory control tasks, and design of computer-based support systems. Advances in Man-Machine Sys. Res. (Vol. I), J. A. I Press Inc., 1984.

Rasmussen, J. On information systems and man-machine interaction. An approach to cognitive engineering. Elsevier, in print, 1985.

Rasmussen, J. & Jensen, A. Mental procedures in real life tasks: A case study of electronic trouble shooting. Ergonomics, 1974, 17, 293-307.

Rasmussen, J., Pedersen, O. M., Mancini, G. et al. Classification system for reporting events involving human malfunctions. EUR 7444EN, 1981.

Rasmussen, J. & Rouse, W. B. (Eds.) Human detection and diagnosis of system failure. Plenum, 1981.

Rawls, J. A theory of justice. The Belknap Press, 1971.

Reason, J. T. Metacognitive ratings and order of output in category generation: Evidence for a salience gradient. Paper given to the Cognitive Section of the British Psychological Society, St. Peter's College, Oxford, 1984.

Reason, J. General error modelling system (GEMS). In J. Rasmussen, K. D. Duncan, & J. Leplat (Eds.), New technology and human error. Wiley, 1985, in print.

Reason, J. Recurrent errors in process environments: Some implications for the design of intelligent decision support systems. This volume.

Reason, J. & Mycielska, K. Absent minded? The psychology of mental lapses and everyday errors. Prentice-Hall, 1982.

Reichenbach, H. The philosophy of space and time. University of California Press, 1958.

Reina, G. & Squellati, G. LAM techniques: Systematic generation of logical structures in system reliability studies. Proc. of NATO ASI on Synthesis and Analysis Methods for Safety and Reliability Assessment. Plenum Press, 1978.

Reiser, B. J., Black, J. B. & Abelson, R. P. Knowledge structures in the organisation and retrieval of autobiographical memories. Cog. Psych., 1985, 17, 89-137.

Rescher, N. Plausible reasoning. Van Gorcum, 1976.

Rokeach, M. The nature of human values. The Free Press, 1973

Rosch, E., Mervis, C. B., Gray, W., Johnson, D. & Boyes-Bream, P. Basic objects in natural categories. Cog. Psych., 1976, 8, 382-439.

Roscoe, S. N. & Eisele, J. E. Integrated computer-generated cockpit displays. In T. B. Sheridan & G. Johannsen (Eds.), Monitoring behavior and supervisory control. Plenum Press, 1976.

Ross, L. The intuitive psychologist and his shortcomings: Distortions in the attribution process. In L. Berkowitz (Ed.), Advances in experimental social psychology, Vol. 10. Academic Press, 1977.

Rouse, S. H., Rouse, W. B. & Hammer, J. M. Design and evaluation of an on-board computer-based information system for aircraft. IEEE Trans. Sys., Man, Cyb., 1982, SMC-12, 451-463.

Rouse, W. B. Experimental studies and mathematical models of human problem solving performance in fault diagnosis tasks. In J. Rasmussen & W. B. Rouse (Eds.), Human detection and diagnosis of system failure. Plenum Press, 1981.

Rouse, W. B. Models of human problem solving: Detection, diagnosis, and compensation for system failures. Proc. of the IFAC Conf. on Analysis, Design and Evaluation of Man-Machine Sys.. Baden-Baden, BRD, 1982.

Rouse, W. B. & Rouse, S. H. A framework for research on adaptive decision aids. AFAMRL-TR-83-082, Wright-Patterson AFB, 1983.

Ruitenbeek, J. C. Visual and proprioceptive information in goal directed movements: A system theoretical approach. PhD Thesis, DUT, Delft, 1985.

Rumelhart, D. E. & McClelland, J. L. (Eds.) Parallel distributed processing: Explorations in the microstructure of cognition. Vol. I: Foundations. Bradford Books & MIT Press, 1986.

Sage, A. P. Sensitivity analysis in systems for planning and decision support. J. Franklin Institute, 1981, 312, 265-291.

Sage, A. P. Behavioral and organisational considerations in the design of information systems and processes for planning and decision support. IEEE Trans. Sys., Man, Cyb., 1981, SMC-11, 640-678.

Samet, M. G. Quantitative interpretation of two qualitative scales used to rate military intelligence. Human Factors, 1975, 17, 192-202.

Sage, A. P. Behavioral and organisational considerations in the design of information systems and processes for planning and decision support. IEEE Trans. on Sys., Man, Cyb., 1981, SMC-11, 640-678.

Sainsaulieu, R. L'identite au travail. Presses de la Fondation nationale des Sciences politiques, 1977.

Sakuguchi, T. & Matsumoto, K. Development of a knowledge based system for power system restoration. IEEE Trans. on Power Apparatus and Sys., 1983, PAS-102, 320-329.

Sanderson, P. Reasoning about links between logical gates. Unpublished manuscript, University of Illinois, IL, 1985.

Sanderson, P., Thornton, C. & Vicente, K. Personal communication. Research in progress. Dept. of Industrial Engineering, University of Toronto, 1985.

Sauers, R. Controlling expert systems. In L. Bolc & M. J. Coombs (Eds.), Computer expert systems. Springer Verlag, in press.

Sauers, R. & Walsh, R. On the requirements of future expert systems. Proc. 8th IJCAI, Karlsruhe, BRD, 1983.

Savage, L. J. The foundations of statistics. Wiley, 1954 & Dover, 1972.

Scarl, E. A., Jamison, J. R. & Delaune, C. I. A fault detection and isolation method applied to liquid oxygen loading for the space shuttle. Proc. 9th IJCAI, Los Angeles, 1985.

Schank, R. C. & Abelson, R. P. Scripts, plans, goals, and understanding. Erlbaum, 1977.

Schoemaker, P. J. The expected utility model: Its variants, purposes, evidence and limitations. J. of Economic Literature, 1983, 20, 528-563.

Schon, D. A. The reflective practitioner. How professionals think in action. Basic Books, 1983.

Schum, D. A. Current developments in research on cascaded inference. In T. D. Wallstein (Ed.), Cognitive processes in decision and choice behavior. Erlbaum, 1980.

Schvaneveldt, R. W., Durso, F. T. & Dearholt, D. W. Pathfinder: Scaling with network structures. Memo in Computer and Cognitive Science, MCCS-85-9, Computing Research Laboratory, New Mexico State University, NM, 1985.

Schvaneveldt, R. W., Durso, F. T., Goldsmith, T. E., Breen, T. K., Cooke, N. M., Tucker, R. G. & DeMaio, J. C. Measuring the structure of expertise. Int. J. of Man-Machine Studies, 1986, in press.

Schvaneveldt, R. W., Goldsmith, T. E., Durso, F. T., Maxwell, K., Acosta, H. M. & Tucker, R. G. Structures of memory for critical flight information. Technical Paper No. AFHRL-TP-81-46, Williams AFB, AZ, 1982.

Schweizer, B. & Sklar, A. Associative functions and abstract semi-groups. Publ. Math. Debrecen., 1963, 10, 69-81.

Schwyhla, W. About the isomorphism between some Sugeno-measures and classical measures. In E. P. Klement (Ed.), Proc. 2nd. Int. Seminar on Fuzzy Set Theory, J. Keppler University, Linz, Austria, 1980.

Scott, D. Measurement models and linear inequalities. J. of Math. Psych., 1964, 1, 233-247.

Scott, D. Domains for denotational semantics. In G. Goos & J. Hartmanis (Eds.), Automata, languages, and programming. Lecture Notes in Computer Sci., 1982, 140, 577-613.

Selfridge, O. G., Rissland, E. L. & Arbib, M. A. Adaptive control of ill-defined systems. Plenum Press, 1984.

Senders, J. W. Visual scanning processes. University of Tilburg Press, 1984.

Seo, F. & Sakawa, M. Fuzzy multiattribute utility analysis for collective choice. IEEE Trans. on Sys., Man., Cyb., 1985, SMC-15, 45-53.

Shackle, G. L. S. Decision, order and time in human affairs. Cambridge University Press, 1961.

Shafer, G. A mathematical theory of evidence. Princeton University Press, 1976.

Shafer, G. Non-additive probabilities in the works of Bernoulli and Lambert. Archives for the History of Exact Sciences, 1978, 19, 309-370.

Shafer, G. Constructive probability. Synthese, 1981, 48, 1-60.

Shafer, G. Belief functions and possibility measures. In J. C. Bezdek (Ed.), The analysis of fuzzy information, Vol. I. CRC Press, 1985.

Shaklee, H. & Tucker, D. A rule analysis of judgments of covariation events. Memory and Cognition, 1980, 8, 459-467.

Shannon, C. E. & Weaver, W. The mathematical theory of communication. University of Illinois Press, 1949.

Sheil, B. A. Teaching procedural literacy. Proc. of the 1980 ACM National Conf., Nashville, TN, 1980.

Sheil, B. A. Coping with complexity. Xerox PARC Report No. CIS-15, Palo Alto, CA, 1981.

Shepard, R. N. Analysis of proximities: Multidimensional scaling with an unknown distance function. I. Psychometrika, 1962, 27, 125-140.

Shepard, R. N. Analysis of proximities: Multidimensional scaling with an unknown distance function. II. Psychometrika, 1962, 27, 219-246.

Shepard, R. N. On subjective optimum selections among multiattribute alternatives. In M. Shelley & G. Bryan (Eds.), Human judgments and optimality. Wiley, 1964.

Sheperd, A., Duncan, K. D., Marshall, E. & Turner, A. Diagnosis of plant failures from a control panel: A comparison of three training methods. Ergonomics, 1977, 20, 347-367.

Sheridan, T. B. Interaction of human cognitive models and computer-based models in supervisory control. Report, MIT, MMS Lab., Cambridge, MA, 1984.

Sheridan, T. B. Monitoring behavior and supervisory control. In G. Salvendy (Ed.), Handbook of human factors / ergonomics. Wiley, 1985, in press.

Sheridan, T. B. & Hennessy R. (Eds.) Research and modeling of supervisory control behavior. National Academy Press, 1984.

Sheridan, T. B. & Johannsen, G. (Eds.) Monitoring behaviour and supervisory control. Plenum, 1976.

Shirley, R. S. & Fortin, D. A. Status report: An expert system to be used for process control. Intern Report, The Foxboro Company, Foxboro, MA, 1985.

Shlaim, A. Failures in national intelligence estimates: The case of the Yom Kippur war. World Politics, 1976, 28, 348-380.

Shortliffe, E. H. Computer-based medical consultations: MYCIN. Elsevier, 1976.

Shortliffe, E. H. The computer and medical decision making: Good advice is not enough. IEEE Eng. in Medicine and Biology Magazine, 1982, 1, 16-18.

Shortliffe, E. H. & Buchanan, B. G. A model of inexact reasoning in medicine. Math. Biosciences, 1975, 23, 351-379.

Sikorski, R. Independent fields and Cartesian products. Studia Mathematica, 1950, 37, 25-54.

Simon, H. A. Models of man: Social and rational. Wiley, 1957.

Simon, H. A. The sciences of the artificial. MIT Press, 1981.

Simon, H. A. Models of bounded rationality, Vol. I. M.I.T. Press, 1982.

Skorstad, E. Technology and overall control. In J. Rasmussen, L. Leplat & K. Duncan (Eds.), New technology and human error. Wiley, 1985, in press.

Sloman, A. The computer revolution in philosophy: Philosophy, science and models of mind. Harvester Press, 1978.

Slovic, P. & Fischhoff, B. On the psychology of experimental surprises. J. of Exp. Psych.: Human perception and Performance, 1977, 3, 1-39.

Slovic, P. & Lichtenstein, S. Preference reversals: A broader perspective. American Economic Rev., 1983, 73, 596-605.

Slovic, P., Lichtenstein, S. & Fischhoff, B. Decision making. In R. C. Atkinson, R. J. Hernnstein, G. Lindzey & R. D. Luce (Eds.), Steven's handbook of experimental psychology (2nd ed.). Wiley, in press.

Slovic, P. & Tversky, A. Who accepts Savage's axiom? Behavioral Sci., 1974, 19, 368-373.

Smith, C. A. B. Consistency in statistical inference and decision. J. of the Royal Statistical Society B, 1961, 23, 1-25.

Smith, M. J., Cohen, B. G., Stammerjohn, L. W. & Happ, A. An investigation of health complaints and job stress in video display operations. Human Factors, 1981, 23, 387-400.

Smith, S. L. Man-computer information transfer. In J. H. Howard (Ed.), Electronic information display systems. Spartan Books, 1963.

Sorkin, R. D. & Woods, D. D. Systems with human monitors: A signal detection analysis. Human-Computer Interaction, 1985, 1, 49-75.

Sperandio, J. C. La psychologie du travail en ergonomie. Presses universitaires de France, 1980.

Sprague, R. H. Jr. & Carlson, E. D. Building effective decision support systems. Prentice-Hall, 1982.

Stassen, H. Decision demands and task requirements in work environments. What can be learned from human operator modelling. This volume.

Stassen, H. G., Kok, J. J. van der Veldt, R. & Heslinga, G. Modelling human operator performance: Possibilities and limitations. Proc. 2nd IFAC Conf. on Analysis, Design, and Evaluation of Man-Machine Sys., Varese, Italy, 1985.

Stassen, H. G., Lunteren, A. van, Hoogendoorn, R., Kolk, G. J. van der, Balk, P., Morsink, G. & Schuurman, J. C. A computer model as an aid in the treatment of patients with injuries of the spinal cord. Proc. of the Int. Conf. on Cyb. and Soc., Cambridge, MA, 1980.

Stefik, M., Aikins, J., Balzer, R., Beniot, J., Birnbaum, L., Hayes-Roth, F. & Sacerdoti, E. The organization of expert systems: A prescriptive tutorial. Xerox Reports VLSI-82-1, Palo Alto, CA, 1982.

Stokey, E. & Zeckhauser, R. A primer for policy analysis. Norton, 1978.

Stoy, J. Denotational semantics: The Scott-Strachey approach to programming language theory. MIT Press, 1977.

Sugeno, M. Theory of fuzzy integral and its applications. Ph. D. Thesis, Tokyo Institute of Technology, Japan, 1974.

Suppes, P. The measurement of belief. J. of the Royal Statistical Society B, 1974, 36, 160-175.

Swain, A. & Guttman, H. C. Handbook of human reliability analysis with emphasis on nuclear power plant applications. NUREG/CR-1278, USNRC, Washington, 1983.

Sweller, J., Mawer, R. F. & Ward, M. R. Development of expertise in mathematical problem solving. J. of Exp. Psych., 1983, 112, 639-661.

Terssac, G. de Activite mentale et regulation des conduites: Etude de quelques postes dans l'industrie chimique. In C. Benayoun, G. de Terssac, Y. Lucas, M. Membrado & E. Soula (Eds.), Les composantes mentales du travail ouvrier. ANACI, Paris, 1980.

Thijs, W. L. Th., Stassen, H. G., Kok, J. J. & van der Veldt, R. J. Supervisory control and fault management. Voyage of discovery through the estate of theory and model creation. Proc. of the 9th World Congress of IFAC, Budapest, 1984.

Thijs, W. L. Th. & Mendel, M. B. Fault management and subjective probability. Proc. of the 3rd Europ. Conf. on Man. Contr.,Roskilde, Denmark, 1983.

Tichomirow, O. K. Struktur der Denktatigkeit der Menschen. In A. W. Bruschlinski & O. K. Tichomirow (Eds.), Zur Psychologie des Denkens. Deutscher Verlag der Wissenschaften, 1975.

Tihansky, D. Confidence assessment of military air frame cost predictions. Operations Res., 1976, 24, 26-43.

Toikka, K., Engestrom, Y. & Norros, L. Entwickelnde Arbeitsforschung. Theoretische und methodologische Elemente. Forum kritische Psychologie, 1985, 15, Argument Sonderband 121, 5-41.

Towill, D. R. A model for describing process operator performance. In E. Edwards & F. P. Lees (Eds.), The human operator in process control. Taylor & Francis, 1974.

Tulving, E. Precis of 'Elements of episodic memory.' Behavioral and Brain Sciences, 1984, 7, 223-268.

502

Turing, A. M. Computing machinery and intelligence. Mind, 1950, 433-460.

Turner, J. A. Software ergonomics: Effects of computer application design parameters on operator task performance and health. Ergonomics, 1984, 27(6), 663-690.

Tversky, A. Intransitivity of preferences. Psych. Rev., 1969, 76, 31-48.

Tversky, A. & Kahneman, D. Availability: A heuristic for judging frequency and probability. Cognitive Psych., 1973, 4, 207-232.

Tversky, A. & Kahnemand, D. The framing of decisions and the psychology of choice. Sci., 1981, 211, 453-458.

Underwood, W. E. (A CSA model-based nuclear power plant consultant. Proc. 2nd Nat. Conf. on Artificial Intelligence, Pittsburgh, PA, 1982.

US Nuclear Regulatory Commission. PRA procedure guide (NUREG/CR-2300). The Commission, 1983.

Veldhuyzen, W. Ship maneuvering under human control. Analysis of the helmsman's control behavior. PhD Thesis, DUT, Delft, 1976.

Veldhuyzen, W. & Stassen, H. G. The internal model concept: An application to modelling human control of large ships. Human Factors, 1977, 19, 367-380.

Veldt, R. J. van der Looking ahead in supervisory control. A study of prediction as an element of human information processing. Proc. of the 4th Ann. Conf. on Man. Contr., Soesterberg, Holland, 1984.

Veldt, R. J. van der Predictive information in the control room. A forecast of the future. Proc. of the 5th Ann. Conf. on Man. Contr., Berlin, 1985.1985

Vermesch, P. Une problematique theoretique en psychologie du travail. Essais d'application des theories de J. Piaget a l'analyse du fonctionnement cognitif de l'adulte. Le Travail Humain, 1978, 41, 265-278.

Volta, G. Modelli per la stima e la gestione del rischio tecnologico. ENEA Direzione Centrale Studi, Serie Seminari, Roma, 1983.

Volta, G. Time and decision. This volume.

Von Neumann, J. & Morgenstern, O. Theory of games and economic behavior. Princeton University Press, 1944 & 1953.

Von Winterfeldt, D. & Edwards, W. Costs and payoffs in perceptual research. Psych. Bull., 1982, 93, 609-622.

Von Winterfeldt, D. & Edwards, W. Decision making and behavioral research. Cambridge University Press, in press.

Wagenaar, W. A. My memory. Manuscript submitted for publication, 1985.

Wagenaar, W. A. & Keren, G. B. Calibration of probabilistic assessments by professional blackjack dealers, statistical experts, and lay people. Organizational Behavior and Human Performance, 1985, 36, in press.

Wagenaar, W. A. & Keren, B. Does the expert know? The reliability of predictions and confidence ratings of experts. This volume.

Wagenaar, W. A. & Sagaria, S. Misperception of experimental growth. Perception and Psychophysics, 1976.

Wallsten, T. & Budescu, D. Encoding subjective probabilities: A psychological and psychometric review. Management Sci., 1983, 29, 151-173.

Waltz, D. Artificial intelligence: An assessment of the state-of-the art and recommendations for future directions. The AI Magazine, 1983, 4, 55-67.

Wang, M. S. Y. & Courtney, J. F. Jr. A conceptual architecture for generalised decision support. IEEE Trans. Sys., Man, Cyb., 1984, SMC-14, 701-711.

Warfield, J. N. Intent structures. IEEE Trans. on Sys., Man, Cyb., 1973, SMC-3, 133-140.

Wason, P. C. On the failure to eliminate hypotheses in a conceptual task. Quart. J. of Exp. Psych., 1960, 12, 129-140.

Wason, P. C. Reasoning about a rule. Quart. J. of Exp. Psych., 1968, 23, 273-281.

Waterman, D. A. A guide to expert systems. Addison-Wesley, 1986.

Watson, S. R., Weiss, J. J. & Donnell, M. Fuzzy decision analysis. IEEE Trans. on Sys., Man, Cyb., 1979, CMS-9, 1-9.

Weber, S. I-decomposable measures and integrals for Archimedean t-conorms I. J. of Math. Analysis and Applications, 1984, 101, 114-138.

Weinstein, N. D. Unrealistic optimism about future events. J. of Personality and Social Psych., 1980, 39, 806-820.

Weiss, S. M. & Kulikowski, C. A. A practical guide to designing expert systems. Chapman & Hall, 1983.

Weizenbaum, J. ELIZA - A computer program for the study of natural language communication between man and machine. Communications of the ACM, 1966, 9, 36-45.

Wewerinke, P. A model of the human decision maker observing a dynamic system. Tech. Rep. NLR TR 81062 L, National Lucht - en Ruimtevaart Laboratorium, Holland, 1981.

Wheeler, D. D. & Janis, I. L. A practical guide for making decisions. The Free Press, 1980.

White, M. F. An expert system for condition monitoring and fault diagnosis of diesel engines. _Proc. of the Institute of Marine Engineers_, London, 1985, submitted for publication.

White, M. F. & Rasmussen, M. Vibration diagnosis data bank for machinery maintenance. _Condition Monitoring '84_, Pineridge Press, 1984.

White, T. N. _Human supervisory control behavior: Verification of a cybernetic model._ PhD Thesis, DUT, Delft, 1983.

Wickens, C. D. _Engineering psychology and human performance._ Charles E. Merril, 1984.

Wickens, C. D. & Kessel, C. Failure detection in dynamic systems. In J. Rasmussen & W. B. Rouse (Eds.), _Human detection and diagnosis of system failure_. Plenum Press, 1981.

Wickens, C. D., Boles, D., Tsang, P. & Carswell, M. _The limits of multiple resource theory in display formatting: Effects of task integration._ AD-P003 321, National Technical Inf. Service, 1984.

Wiener, E. L. Beyond the sterile cockpit. _Human Factors_, 1985, _27_, 75-90.

Wiener, E. L. & Curry, R. E. Flight-deck automation: Promises and pitfalls. _Ergonomics_, 1980, _23_, 955-1011.

Wierbicki, A. P. A mathematical basis for satisfying decision making. _Math. Modelling_, 1982, _3_(5), 391-405.

Wierzschon, S. T. An inference rule based on Sugeno measure. In J. C. Bezdek (Ed.), _The analysis of fuzzy information_. CRC Press, 1985, in print.

Wilcks, Y. & Bien, J. Beliefs, points of view, and multiple environments. _Cog. Sci._, 1983, _7_, 95-119.

Williams, M. D. What makes RABBIT run? _Int. J. of Man-Machine Studies_, 1984, _21_, 333-352.

Winograd, T. _Understanding natural language._ Academic Press, 1972.

Winograd, T. Frame representations and the declarative-procedural controversy. In D. G. Bobrow & A. Collins (Eds.), _Representation and understanding_. Academic Press, 1975.

Winston, P. H. _Artificial intelligence._ Addison-Wesley, 1979.

Wohlstetter, R. _Pearl Harbor: Warning and decision._ Stanford University Press, 1962.

Wood, D. J., Shotter, J. D. & Godden, D. An investigation of the relationships between problem-solving strategies, representation and memory. _Quart. J. of Exp. Psych._, 1974, _26_, 252-257.

Woods, D. D. _Operator decision behavior during the steam generator tube rupture at the Ginna nuclear power plant station._ Research Report 82-1C57-CONRM-R2, Westinghouse, R&D Pittsburgh, PA.

Woods, D. D. Visual momentum: A concept to improve the cognitive coupling of person and computer. Int. J. of Man-Machine Studies, 1984, 21, 229-244.

Woods, D. D. Some results on operator performance in emergency events. In D. Whitfield (Ed.), Ergonomic problems in process operations. Inst. Chem Eng. Symp. Ser. 90, 1984.

Woods, D. D. Knowledge based development of graphic display systems. Proc. of the Human Factors Society, 29th Ann. Meeting, 1985.

Woods, D. D. On the significance of data: The display of data in context. Manuscript in preparation, 1986.

Woods, D. D. Fault management. Manuscript in preparation, 1986.

Woods, D. D. Paradigms for intelligent decision support. This volume.

Woods, D. D. & Hollnagel, E. Cognitive task analysis. Manuscript in preparation, 1986.

Woods, D. D., Wise, J. & Hanes, L. An evaluation of nuclear power plant safety parameter display systems. Proc. of the 24th Ann. Meeting of the Human Factors Society, 1981.

Wright, G. Behavioural decision theory. Penguin, 1984.

Yager, R. R. Entropy and specificity in a mathematical theory of evidence. Int. J. of General Sys., 1983, 9, 249-260.

Yager, R. R. Arithmetic and other operations in Dempster-Shafer structures. Tech. Rep. MII-508, Iona College, New Rochelle, NY, 1985a.

Yager, R. R. Optimal alternative selection in the face of evidential knowledge. Tech. Rep. MII-513B, Iona College, New Rochelle, NY, 1985b.

Yamada, N. & Motoda, H. A diagnosis method of dynamic system using the knowledge on system description. Proc. 8th IJCAI, Karlsruhe, BRD, 1983.

Yates, J. The content of awareness is a model of the world. Psych. Rev., 1985, 92, 249-284.

Young, R. M. Production systems for modelling human cognition. In D. Michie (Ed.), Expert systems in the micro-electronic age. Edinburgh University Press, 1979.

Yu, V. et al. Antimicrobal selection by a computer. J. of the American Medical Association, 1979, 242, 1279-1282.

Zadeh, L. A. Fuzzy sets. Inf. and Contr., 1965, 8, 338-353.

Zadeh, L. A. Probability measures of fuzzy events. J. of Math. Analysis and Applications, 1968, 23, 421-427.

Zadeh, L. A. Outline of a new approach to the analysis of complex systems and decision processes. IEEE Trans. on Sys., Man, Cyb., 1973, SMC-3, 28-44.

Zadeh, L. A. Fuzzy logic and its application to approximate reasoning. Information processing, North-Holland, 1974.

Zadeh, L. A. Fuzzy set as a basis for a theory of possibility. Fuzzy Sets and Sys., 1978, 1, 3-28.

Zadeh, L. A. PRUF: A meaning representation language for natural languages. Int. J. of Man-Machine Studies, 1978, 10, 395-460.

Zadeh, L. A. Fuzzy sets and information granularity. In. M. M. Gupta, R. K. Ragade & R. R. Yager (Eds.), Advances in fuzzy set theory and applications. North-Holland, 1979.

Zadeh, L. A. On the validity of Dempster rule of combination of evidence. Memo, UCB-ERL 79/24, University of California at Berkeley, USA. 1979.

Zadeh, L. A. A simple view of the Dempster-Shafer theory of evidence. Berkeley Cognitive Science Rep. 27, University of California at Berkeley, 1984.

Zakay, D. & Wooler, S. Time pressure, training and decision effectiveness. Ergonomics, 1984, 27, 273-284.

Zheng, Y. P., Basar, T. & Cruz, J. B. Jr. Stackelberg strategies and incentives in multiperson deterministic decision problems. IEEE Trans. Sys., Man, Cyb., 1984, SMC-14, 10-24.

Zimmermann, H. J. & Zysno, P. Decision and evaluations by hierarchical aggregation of information. Fuzzy Sets and Sys., 1983, 10, 243-260.

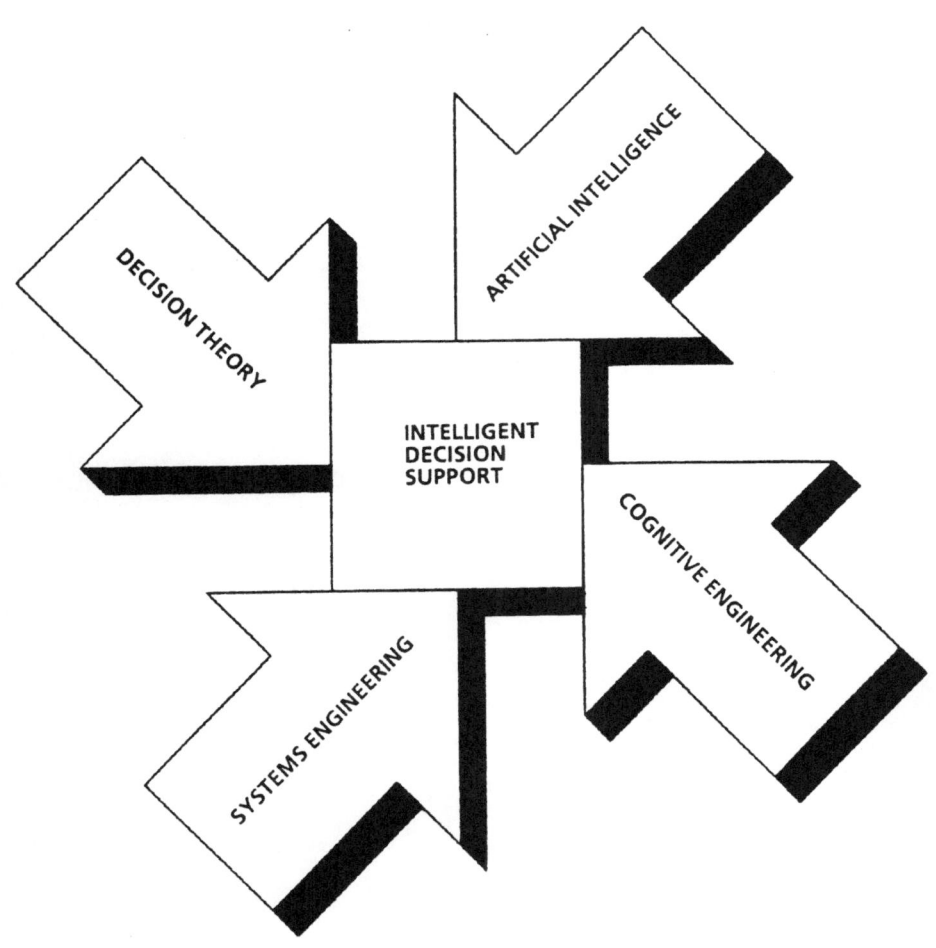

ABSTRACTS OF SHORT PAPERS

TOWARDS COGNITIVE METHODS ENGINEERING IN SUPERVISORY CONTROL

J.R. Buck and V.Kachitvichyanukul
Industrial & Management Engineering
University of Iowa

Several experiments have been conducted to test fundamental questions of primitive cognitive tasks by process
controllers. A partial taxonomy of cognitive tasks was created
and performance time statistics within- and between-persons
were investigated along with performance accuracies. One test
of these tasks has shown that stable time statistics exist for
tasks of a given type and complexity level. Also, the time
distribution within people is approximately normal or lognormal
and between subject means are normally distributed. All of the
tasks require information processing operations where the different types are defined by the nature of the operation.
Complexities within each type are specified by the size of an
operations reference and the field size for applying the
operation. Another test has established empirical evidence
that performance time statistics can be well described by variables of the reference and field within these tasks. Experiments addressed to concurrent cognitive tasks with tracking
have shown that performance times are insensitive to variations
in tracking frequencies. Another test was made on the arrival
rate and lumpiness of the task stream coming to an operator
but there appeared to be little change in performance time
statistics due to moderate changes in the stream variables.
Performance accuracies in these tasks appeared to change with
the type of task but not with any of the other variables
examined. While time-accuracy trade-offs were not observed,
the performance time statistics on correctly performed tasks
were different from those inaccurately performed.

The authors acknowledge support from the U.S.A.F. and
N.S.F. for this research.

INDIVIDUALS' ABILITY TO ASSESS THE IMPORTANCE OF PREDICTORS

V. Grayson Cuqlock
U.S. Army Human Engineering Laboratory
Aberdeen Proving Ground, Maryland   21005-5001   USA
Warren S. Torgerson
Johns Hopkins University
Baltimore, Maryland   21218   USA

Interest in allowing supervisors of automated systems to manually select displays to monitor has arisen, in part, from a need to decrease the human error caused by display-panel clutter. Unfortunately, humans can also be expected to make errors in sampling information; people are limited in their ability to detect the relative validities of information.

In this study we examined the ability of subjects to select the better of two variables to use to make evaluative judgments. Individuals first learned to use values on two predictor variables to estimate the values of a given continuous criterion; they were then required to choose only one of the predictors to use to estimate the value of the criterion. Optimal choice, in the least squares sense, requires that individuals choose the variable with the higher validity. The differences in the levels of variance accounted for by each predictor in the set were selected so that a function relating differences in predictability to detectability could be determined. Results indicate that a large difference in the squared validity of two sources of information is required before people consistently choose the better source.

A Model for Information System Analysis and Design

Anthony Debons
University of Pittsburgh

The purpose of an information system is to augment those human capabilities that are important to the ability of the individual to take action on event contingencies that are related to his/her development and survival.

As often discussed in the literature, all organisms are information systems - although through evolution and other factors - vary in degree of complexity and capability. Thus, the basic framework of information system can be readily appreciated, namely, that they include sensors to detect states of events, processing elements that gather, store, manipulate the symbols representing these events through which interpretation (significance) of the states of events can be achieved and some means through which actions can be taken. These components of an information system are represented in the EATPUT model presented in Figure 1.

The integration of these components is achieved through an acknowledgment of the various theories that govern the respective functioning of the components (Table 2).

An intelligent decision support system can be understood within the framework of such a model. The important point to stress is that each of the components represent the necessary and sufficient conditions for the representation of an intelligent decision support system. Failure of any one of the components leads to some consequence to the other components of the system just as depriving the individual of sight (unless augmented) will influence the ability of the individual to respond adequately to the states of the event world.

There are numerous instances where component failure has resulted in serious malfunctions of information systems. The most recent accident at the Union Carbide Plant in Institute, West Virginia is almost a textbook case. Without presenting the detailed account of this accident, one can appreciate that acknowledgment (acquisition component) of the chemical leak was inaccurate and tardy; the data transmitted to the computer was processed via a model that did not account for the major physical parameters of the leak and circumstances surrounding it. The local community was not adequately alerted to the serious dimensions of the leak some time after the accident (transfer).

Information systems are viable, important environments that compliment human ability to deal with life's contingencies. The EATPUT model provides one basis for understanding the structure and functioning of such systems. The model can be used, in addition, to estimate component and ultimately, overall system efficiency and effectiveness.

IMAS: A systematic technique for eliciting process operators
diagnostic knowledge structures.

David Embrey B.Sc., Ph.D.
HUMAN RELIABILITY ASSOCIATES
1 School House, Higher Lane, Dalton
Parbold, Lancashire, WN8 7RP, England

The Influence Modelling and Assessment System (IMAS) is a computer based technique for eliciting the structural knowledge possessed by the operator which is used during the diagnosis of abnormal events in process plants. This knowledge base can be expressed in the form of a network connecting causes (events which could occur), consequences (the various alternative outcomes of these events), and indicators (externally observable information regarding the status of these events, e.g. via displays in a control room).

This information is elicited by asking the operator to describe a scenario in terms of the various alternative events that could occur as the situation develops. He identifies the status of the observable indicators associated with the events. This information is used to generate a relational database which represents the structure of the operator's mental model of the situation.

This Subjective Cause-Consequence Model has several applications. It can be used to identify the information sources used by the operator for diagnosis, and can thus provide a valuable input to the design of the interface. It can also be used to specify the procedural support required to facilitate diagnosis. A normative model can be developed using a group of experienced operators and designers. This can be used as a reference to assess the adequacy of the mental model of an operator at different stages of training.

In the context of on-line decision support during an abnormal event, it is possible to input the current state of the observable indicators into a program called Search. This traverses the database via the cause-consequence nodes, and simulates the diagnostic thought processes that would occur if the operator were considering the event in an unstressed environment. It therefore allows the operator access to the diagnostic knowledge which is normally suppressed in highly stressed situations. Applications of IMAS as a more general tool for knowledge elicitation in AI applications are also being investigated.

A CYBERNETIC META-MODEL
OF DYNAMIC MIND IN MAKING DECISION

Halim Ergunalp
Telecam A.Ş., Lüleburgaz, Turkey

Decision is an act of selection among alternatives. It is a product of human mind, or a product of a group of interacting minds. Mind has been a disputable philosophical issue, but there are also scientific convictions about its nature.

Categorically differentiated, mind entails interactively functioning sub-units within itself either operating in harmony or in conflict with each other. Intelligence, reason, feeling, consciousness are the categories as such, making up the integral whole that we denote as "mind".

Selection is a simple seeming complex process. Up to the spatio-temporal point where this process ends with the act of "decision being taken", lots of things happen. 'Intelligence' perceives , 'Reason' lines-and weighs up, 'Feeling' quenches, burns, mixes up and distils, 'Consciousness' directs and finally 'the mind' decides. But how ?

This paper attempts to formulate an answer to the question. In line with the fundamental object of science to understand and explain, cybernetics has developed a meaningful clarity upon the structure and nature of basic action 'to select', and hence drawn the relevant discussions into the cross-sectional area of philosophy and science.

Cybernetics, with almost four decades of accumulation and as 'Science of effective organizational action' or as 'Science of defendable metaphores' provides a rich conceptual framework for a search into the question: "How is it that mind decides?" In an attempt of doing cybernetics, the paper deploys cybernetic resources and constructs a meta-model of dynamic mind within the process of making decision.

OPERATOR ACCEPTANCE ISSUES IN INTERACTIVE DECISION-MAKING
SYSTEMS

L.P.Goodstein
Risø National Laboratory
DK 4000 Roskilde, Denmark

Truly interactive decision making in a human-computer context
calls for a human-human like communication because the user in
reality has to cooperate with the "ghost" of the designers as
embodied in the form of a computer intermediary.

A recognized issue in connection with the introduction of
these kinds of advanced aids (or indeed any kind of "improvement"), especially in an existing work situation, is the "user
acceptance" problem. Designers are (too) often confronted with
the fact that users do not react in a positive manner to the
systems with which they are provided. In effect, a mismatch
occurs between the designers' intentions and actual user
utilisation.

Much has already been written on the importance in the design
process of including user-relevant information (and even
active user participation) to ensure that proper attention is
paid to the actual operational environment, the actual operational problems (and the users' resources for dealing with
these) and, of course, training factors.

This paper emphasizes the need to treat human-computer communication in a human-human-like manner in order to enhance
acceptance. Various possible relationships are described
briefly. The need for error-tolerant designs is also emphasized.

The peer relationship is discussed with respect to the potential problems connected with the user(s) accepting advice (or
even criticism) from the computer intermediary. An alternate
designation is suggested in which the information transferred
is classified (and identified) as either being objective -
i.e., based on engineering analyses of valid data - or subjective (including the assumptions and conditions) and an
example is given for a diagnosis and planning situation.

ALARM SYSTEM DESIGN FOR FAULT DIAGNOSIS ON PROCESS PLANT

Robert Pashby
Loughborough University of Technology
Department of Chemical Engineering
Loughborough
Leicestershire
United Kingdom

Process Plants are generally equipped with alarm systems and many use process control computers to report alarm conditions to the process operator. When a fault occurs the operator must make a diagnosis to determine the fault and hence formulate his corrective action. However, if an unfamiliar fault (or several unrelated faults) occur then the operator may have difficulty in fulfilling his diagnostic task.

A Knowledge-based System performing Alarm Handling and Disturbance Analysis could then be used to filter the data and present higher-order information and advice to the operator. The Knowledge-based System would also be able to justify why an alarm was displayed.

Some of the development work to produce such an alarm system is discussed.

Existing alarm systems have many short-comings which would have to be amended before any alarm Handling and Disturbance Analysis could be usefully installed. Some of these problems are briefly discussed.

A HUMAN INTERFACE FOR AN ADVANCED MANUFACTURING PROCESS

John Wilson,
Department of Production Engineering & Production Management,
University of Nottingham
University Park,
Nottingham,
NG7 2RD

It is contended that advanced manufacturing processes, automated to some degree, do not necessarily imply the absence of a human operator, nor should they inevitably lead to de-skilled fragmentary human work. Two relevant programmes of work are underway; one to produce a predictor display for sophisticated robots to be used as an aid to safety, reliability and planning; the second is to develop an effective human interface for an individual unit of advanced manufacturing technology. This latter, the subject of this paper, is a rotary forging machine (RFM), programmable on- or off-line. The machine motions available now and envisaged are numerous and complex; even more so, the deformations set up in the workpiece are extremely complex and "invisible".

The sophistication provided by the technology necessitates an understanding of the process - for pre-programming - for which knowledge is not now available in industry. Therefore work is underway to develop a manual control system incorporating a teach/record/playback facility. This will allow utilisation of the skills and knowledge of workers with years of metal-forming experience.

The human factors research difficulties will centre around the control of a number of axes of motion to produce a resultant deformation which is difficult to visualise or to display. Display format may be based upon operators' internal representations of the process; display state will be determined from pre-set variables and controlled machine motions, and modified through an historic knowledge base. Research parallels can be drawn with elements of work on process control, vehicle guidance, robotics and teleoperation.

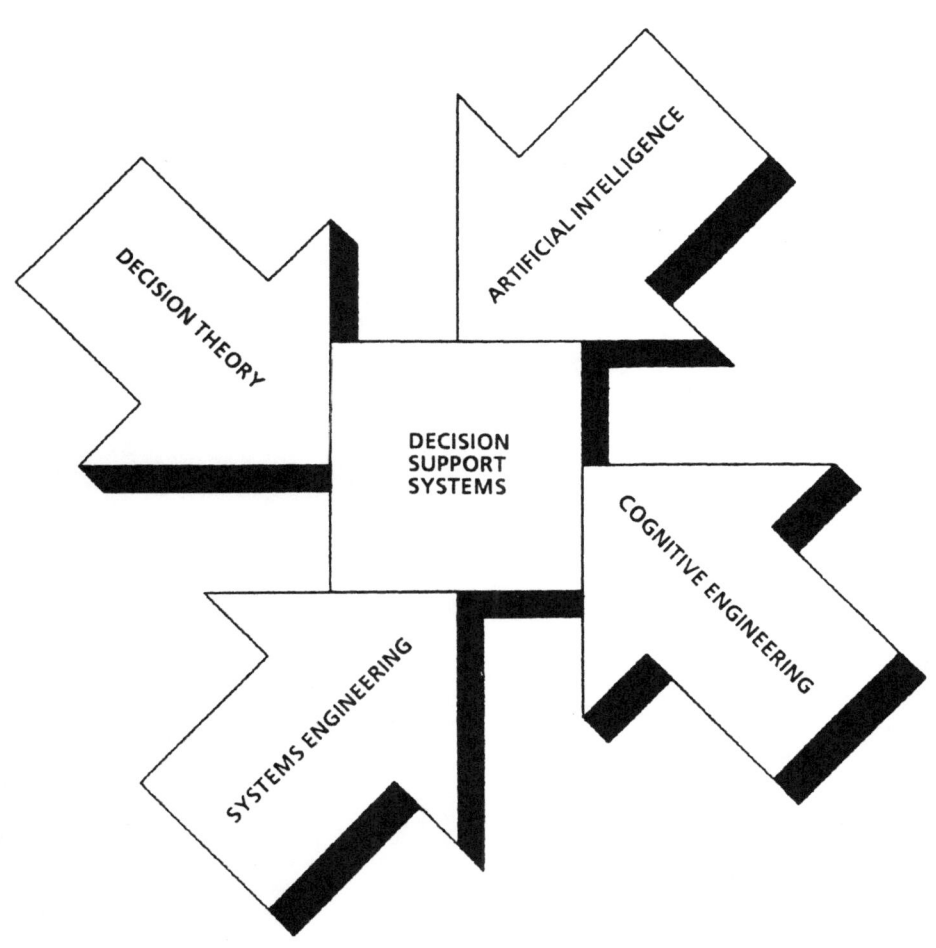

DECISION THEORY

ARTIFICIAL INTELLIGENCE

DECISION
SUPPORT
SYSTEMS

COGNITIVE ENGINEERING

SYSTEMS ENGINEERING

LIST OF PARTICIPANTS

Dr. Aniello Amendola
CEC - JRC Ispra
I-21020 Ispra (Varese), Italy

Dr. George Apostolakis
University of California, 5532 Boelter Hall
Los Angeles, CA 90024, USA

Mr. Klaus Barentsen
University of Aarhus, Institute of Psychology
Asylvej 4
DK-8240 Risskov, Denmark

Professor Nicholas Belkin
Rutgers University
School of Communication, Information and Library Studies
4, Huntington Street
New Brunswick, NJ 08903, USA

Dr. Giancarlo Bello
TEMA (ENI Corporation)
26 Via Medici Del Vascello
I-20128, Milano, Italy

Dr. Kevin Bennet
Westinghouse Research & Development
1310 Beulah Road
Pittsburg, PA 15668, USA

Mr. Ugo Bersini
CRC Euratom
CEC - JRC Ispra
I-21020 Ispra (Varese), Italy

Dr. James Buck
University of Iowa
Department of Industrial and Management Engineering
2353 Cae Drive, Iowa City, IA 52240, USA

Dr. Pietro Cacciabue
CEC - JRC Ispra
I-21020 Ispra (Varese), Italy

Dr. Michael J. Coombs
Department of Psychology and Computing Research Laboratory
New Mexico State University
Las Cruces, New Mexico, NM 88003, USA

Dr. Jose Craveirinha
Departamento de Engenharia Electrotécnica
Universidade de Coimbra
P-3000 Coimbra, Portugal

Dr. V. Grayson Cuqlock
US Army Human Engineering Laboratory
Aberdeen Proving Ground
Aberdeen, MD 21078, USA

Dr. Veronique de Keyser
Universite de Liege, Institut de Psychologie
PART 4000, Liege 1, Belgium

Dr. Anthony Debons
University of Pittsburgh
135 North Bellefield Avenue
Pittsburgh, PA 15260, USA

Dr. Francois Decortis
Universite de Liege, Institut de Psychologie
PART 4000, Liege 1, Belgium

Dr. Didier Dubois
Universite Paul Sabatier, Langages et Systemes Informatique
118 Route de Narbonne
F-31062, Toulouse, France

Dr. Keith Duncan
UWIST, Department of Applied Psychology
Llwyn-y-Grant, Penylan
Cardiff CF3 7UX, Wales

Dr. David Embrey
Human Reliability Associates
1 School House, Higher Lane, Dalton, Parbold
Lancashire WN8 7RP, England

Dr. Halim Ergunalp
Telecam A.S.
P. K. 28, Luleburgaz, Turkey

Dr. Peter Essens
Institute for Perception TNO
P.O. Box 23
NL-3769 Soesterberg ZG, Netherlands

Dr. Baruch Fischhoff
Decision Research
1201 Oak Street
Eugene, OR 97401, USA

Professor Paolo Garbolino
Scuola Normale Superiore
Piazza del Cavalieri
I-56100, Pisa, Italy

Dr. Sergio F. Garriba
CESNEF - Politecnico di Milano
Via Ponzio 34/3
I-20133 Milano, Italy

Mr. William Gaver
UCSD, Institute for Cognitive Science C-015
La Jolla, CA 92093, USA

Dr. David I. Gertman
OECD Halden Reactor Project
P.O. Box 173
N-1751 Halden, Norway

Mr. Len P. Goodstein
Risö National Laboratory
DK-4000 Roskilde, Denmark

Dr. Giovanni Guida
Dipartimento di Eletronica, Politecnico di Milano
Piazza Leonardo da Vinci 32
I-20133, Milano, Italy

Dr. Manfred Härter
CCR Karlsruhe, P.O. Box 2266
D-7500 Karlsruhe, BRD

Dr. Erik Hollnagel
Computer Resources International
Vesterbrogade 1A
DK-1620 Copenhagen V, Denmark

Mr. Conny B.-O. Holmström
OECD Halden Reactor Project
P.O. Box 173
N-1751 Halden, Norway

Mr. Gunnar Hovde
Risö National Laboratory
DK-4000 Roskilde, Denmark

Mr. Michael Jepsen
Computer Resources International
Vesterbrogade 1A
DK-1620 Copenhagen V, Denmark

Professor Gunnar Johannsen
University of Kassel, Laboratory for Man-Machine Systems
Mönchebergstrasse 7
D-3500 Kassel, BRD

Dr. Michael D. Kelly
BDM Corporation
7915 Jones Branch Drive
McLean, VA 22102, USA

Dr. Karl-Friedrich Kraiss
Forschungsinstitut für Anthropotechnik
Neuenahrerstrasse 20
D-5307 Werthoven, Bonn, BRD

Dr. Steen F. Larsen
University of Aarhus, Institute of Psychology
Asylvej 4
DK-8240 Risskov, Denmark

Professor Jacques Leplat
Laboratoire de Psychologie du Travail
41 Rue Gay Lussac
F-75005 Paris, France

Mr. Zoltan Leskowsky
University of Toronto
Department of Industrial Engineering
Toronto, Ontario M5S 1A4, Canada

Professor Morten Lind
University of Aalborg, Institute for Electronic Systems
Strandvejen 19
DK-9000 Aalborg, Denmark

Dr. Giuseppe Mancini
CEC - JRC Ispra
I-21020 Ispra (I-21020 (Varese), Italy

Professor Marino Mazzini
University of Pisa, Dipartimento di Costruzione Meccaniche
Via Diotisalvi 2
I-56100 Pisa, Italy

Mr. Dennis Mitchell
University of Toronto
Department of Industrial Engineering
Toronto, Ontario M5S 1A4, Canada

Professor Neville Moray
University of Toronto
Department of Industrial Engineering
Toronto, Ontario M5S 1A4, Canada

Professor Donald A. Norman
UCSD, Institute for Cognitive Science C-015
La Jolla, CA 92093, USA

Dr. Leena Norros
Technical Research Centre of Finland
Otakaari 7B
SF-02150 Espoo 15, Finland

Dr. Sue Norton
Hughes Aircraft Company
8000 East Maplewood, Suite 226
Englewood, CO 80111, USA

Mr. Fridtjov Öwre
OECD Halden Reactor Project
P.O. Box 173
N-1751 Halden, Norway

Dr. Luciano Olivi
CEC - JRC Ispra
I-21020 Ispra (Varese), Italy

Dr. Philip Oziard
Generale de Service Informatique
25 Boulevard Amiral Bruix
F-75015 Paris, France

Mr. Robert Pashby
Loughborough University, Department of Chemical Engineering
Ashby Road, Loughborough
Leicestershire, LE11 3TU, England

Dr. Daniel Power
University of Maryland, College of Business
College Park, MD 20742, USA

Mr. Jens Rasmussen
Risø National Laboratory
DK-4000 Roskilde, Denmark

Professor James Reason
University of Manchester, Department of Psychology
Manchester, M13 9PL, England

Dr. Antonio Rizzo
Istituto di Psicologia CNR
Via Monti Tiburtini 509
I-00157 Roma, Italy

Mr. Giuseppe Salvato
SNIA Fibre, Sistemi Arce Tecnica
Via Friuli 55
I-20031 Cesano Maderno, Italy

Dr. Penelope M. Sanderson
University of Illinois at Urbana-Champaign
Department of Mechanical and Industrial Engineering
Urbana, IL 61801, USA

Professor Roger W. Schvaneveldt
Department of Psychology and Computing Research Laboratory
New Mexico State University
Las Cruces, New Mexico, NM 88003, USA

Professor Hendrik G. Stassen
Delft University of Technology
Mekelweg 2
Delft, Netherlands

Dr. Erland Svensson
National Defense Research Institute, Human Studies Dept.
P.O. Box 13400
S-58013 Linköping, Sweden

Dr. Maud Thanderz
National Defense Research Institute, Human Studies Dept.
P.O. Box 13400
S-58013 Linköping, Sweden

Dr. Salvatore Tucci
Universita di Pisa, Dipartimento di Informatica
Corso Italia 40
I-56100 Pisa, Italy

Dr. Giuseppe Volta
CEC - JRC Ispra
I-21020 Ispra (Varese), Italy

Dr. Micheal Wachter
Elektronik System Gesellschaft
Vogelweideplatz 9
D-8000 Munich 80, BRD

Professor Willem A. Wagenaar
University of Leiden, Department of Experimental Psychology
P.O. Box 9509
NL-2300 Leiden RA, Netherlands

Dr. Edward Walker
BBN Laboratories
10 Moulton Street
Cambridge, MA 02238, USA

Mr. Michael Waygood
Army Personnel Research Establishment
Operator Performance Division
c/o RAE, Farnborough, Hampshire, England

Dr. Maurice White
UNIT/NTH, Marine Technology Center
N-7000 Trondheim, Norway

Dr. John Wilson
University of Nottingham
Department of Production Engineering
University Park, Nottingham NG7 2RD, England

Dr. David D. Woods
Westinghouse Research & Development
1310 Beulah Road
Pittsburg, PA 15668, USA

NATO ASI Series F